Measuring Gene Expression

Measuring Gene Expression

Matthew B. Avison, PhD

Senior Lecturer in Microbiology
University of Bristol,
Department of Cellular & Molecular Medicine,
School of Medical Sciences,
University Walk,
Bristol, BS8 1TD, UK.

Taylor & Francis
Taylor & Francis Group

Published by:
Taylor & Francis Group
In US: 270 Madison Avenue
           New York, NY10016
In UK: 4 Park Square, Milton Park
           Abingdon, OX14 4RN

ISBN: 0-4153-7472-3

A catalog record for this book is available from the British Library.

Library of Congress Cataloging-in-Publication Data

Avison, Matthew B.
   Measuring gene expression / Matthew B. Avison.
      p. ; cm.
   Includes bibliographical references and index.
   ISBN 0-415-37472-3 (alk. paper)
  1. Gene expression.   I. Title.
   [DNLM: 1. Gene Expression. 2. Genetic Techniques. QU 475 A959m 2007]
   QH450.A95 2007
   572.8'65—dc22

                                      2006013485

Editor: Elizabeth Owen
Editorial Assistant: Kirsty Lyons
Editorial
Production Manager: John Foley
Typeset by: Phoenix Photosetting, Chatham, Kent, UK
Printed by: MPG BOOKS Limited, Bodmin, Cornwall, UK

Printed on acid-free paper

10 9 8 7 6 5 4 3 2 1

Taylor & Francis Group, an informa business

Visit our web site at http://www.garlandscience.com

Contents

Abbreviations

α-CTD	the C-terminal portion of the α subunit of RNA polymerase
A	adenine
AMV	avian myoblastosis virus
ANOVA	analysis of variance
APS	ammonium persulfate
A-site	acceptor site
ATP	adenosine triphosphate
BP	binding protein
BSA	bovine serum albumin
C	cytosine
cAMP	cyclic adenosine monophosphate
CCD	charged coupled device
cDNA	copy DNA
CRP	cAMP responsive protein
2D-DiGe	two-dimensional difference gel electrophoresis
DD RT-PCR	differentiated display RT-PCR
DIG	digoxogenin
DMSO	dimethylsulfoxide
DNA	deoxyribonucleic acid
DSC	differential subtraction chain
ECF	enhanced chemifluorescence
ECL	enhanced chemiluminescence
EGTA	ethylene glycol bis(2-aminoethyl ether)-N,N,N'N'-tetraacetic acid
ELISA	enzyme-linked immunosorbant assay
ESI	electrospray ionization
G	guanine
GAP-DH	glycerol-3-phosphate dehydrogenase
GFP	green fluorescent protein
GTP	guanosine triphosphate
hnRNA	heteronuclear RNA
HPLC	high-performance liquid chromatography
IF	initiation factor
IgG	immunoglobulin

LCR	locus control region
MALDI	matrix-assisted laser desorption ionization
MAR	matrix attachment region
M-MuLV	Maloney murine leukemia virus
mRNA	messenger RNA
NHS	sulfo-*N*-hydroxysuccinimide
nt	nucleotides (when used as a unit)
ONPG	*ortho*-nitrophenyl galactoside
PBS	phosphate buffered saline
PCR	polymerase chain reaction
P-site	peptidyl site
PVDF	polyvinylidine difluoride
qPCR	quantitative PCR
RAP-PCR	RNA arbitrarily primed PCR
RD-PCR	restriction display PCR
RFLP	restriction fragment length polymorphism
RIN	RNA integrity
RNA	ribonucleic acid
RNP	ribonucleoprotein
rRNA	ribosomal RNA
RT-PCR	reverse transcription PCR
SDS-PAGE	sodium dodecyl sulfate polyacrylamide gel electrophoresis
SIP PCR	sequence independent primed PCR
snRNA	small, nuclear RNA
SSC	standard saline citrate buffer
T	thymine
TAE	Tris-acetate-EDTA
TBE	Tris-borate-EDTA
TBP	TATA-box binding protein
TF	transcription factor
tRNA	transfer RNA
U	uracil
UV	ultraviolet

Preface

In the last 10 years, molecular biology research has changed remarkably. Once, experiments could only be targeted to a small number of well-characterized genes. Now, the free availability of huge amounts of genomic sequence data begs the question: 'what do all these genes do?' An increasingly popular way of addressing this question is to find out when certain genes are expressed and how their expression is regulated. So, increasingly, biologists are measuring gene expression, and my primary aim in writing this book was to give them a thorough understanding of the methods involved, how they were developed, and how they can be adapted to ask different questions.

Global approaches for measuring gene expression such as transcriptomics, proteomics, and comparative gene expression analysis have developed apace in the post-genomic era. These have been described in detail, explaining their pitfalls and placing the emphasis on experimental design and interpretation of the data generated. Whilst they are potentially very powerful, these global methodologies are frequently associated with poor resolution, and are very susceptible to experimental error. So we cannot lose sight of techniques for measuring the expression of a single gene, or a small group of genes. These methods are important in their own right, and are essential to confirm gene expression differences flagged up in global experiments. Therefore, this book gives a thorough description of northern hybridization, RT-PCR and reporter gene assays. So, whatever the scale of your current gene expression experiments, and however it may change in the future, this book is designed to help you understand what to do, why you are doing it, how it can all go wrong, how to know when it has, and what to do about it.

I am grateful to a host of people who have helped me write this book. To the staff of Taylor & Francis who have guided me with the minimum of fuss; to my colleagues at the University of Bristol who made allowances for my absence at meetings; to members of my research group for not openly questioning me when I said I was 'working from home'; to my students for asking the questions that informed much of its content; to my wife, Helen, for her encouragement and support during the whole project, and to my parents for pretty much everything else. I hope you enjoy reading it half as much as I enjoy having finished writing it!

Matthew B. Avison,
Christon, Somerset. 14th May 2006

Gene expression and its control

1

1.1 Introduction

What makes a cell act and look a certain way? Why do two cells express different phenotypes? When talking about two different individuals, one might start by highlighting differences between the complements of genes present in each. In bacteria, such differences are very pronounced. The presence of mobile DNA elements, with 'plug-and-play' attributes, seemingly randomly scattered amongst members of a population provide phenotypic variation to the population as a whole (Avison and Bennett, 2005). Smaller genetic variation, caused by point mutations, deletions and duplications are found in populations of bacteria and eukaryotic organisms alike. Mutations within coding sequences can alter, or even destroy the activity of the protein produced, and many examples of phenotypic variation caused by such mutations are known in our own species. Most help give us a wonderfully varied society; some cause terrible genetic disease (Avison and Bennett, 2005; Crow, 2000).

Phenotypic variation is not just about differences in functional gene complement, though. Normally the genotypes of, for example, brain and liver cells from the same individual are identical, and yet their phenotypes are remarkably different. Similarly, bacterial cells can express strikingly diverse phenotypes dependent upon the environmental conditions in which they find themselves. To explain this, one must look away from the physical makeup of a genotype and instead, consider its expression. The study of how a genotype produces a phenotype is widely referred to as functional genomics, and with the current availability of huge amounts of genome sequence information, this is a real growth industry. Functional genomics represents a massive field of study encompassing classical and molecular genetics (knocking genes out and over-expressing them in order to learn more about the functions they encode), structural genomics (systematically over-expressing and purifying recombinant proteins to see what they look like) and studies on gene expression (find out when a gene is expressed and you might learn more about the function it encodes). An important part of this last facet of functional genomics is the methodology associated with measuring gene expression, and this is where this book aims to help.

A quip, uttered by many embittered practitioners of functional genomics is that you only know you've made it in this business if you coin a new 'ome'. An 'ome of your own', if you will. There are many weird and wonderful omes: 'metabolome' (the sum of the metabolic pathways working in a cell at any one time) and 'regulonome' (the sum of all the transcriptional regulatory pathways working in a cell at any one time) for example, but the three omes you need to understand for this book are 'genome' (obvious, the sum total of genes in a cell), 'transcriptome' (the sum total of RNA molecules produced at any one time in a cell) and 'proteome' (the sum total of proteins produced at any one time in a cell). Analysis of genomic complement is the job of comparative genomics experts and that book is for another time. In this book, we will concentrate on measuring the transcriptome and proteome of a cell, or any subdivision of each. Don't forget, though, that since many genes encode proteins whose roles are to regulate the expression of other genes, there is clearly an inter-relation between genotype and gene expression. Changes in gene expression do not solely result as responses to external stimuli; they can occur due to mutation at regulatory loci. For example, loss of function mutations in transcriptional repressor genes will lead to constitutive over-expression of genes upon which the encoded repressor protein would normally act. Indeed, a very significant reason for measuring gene expression is as part of

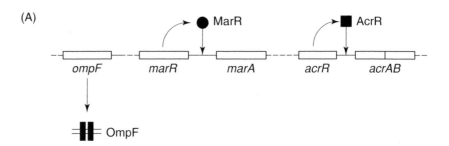

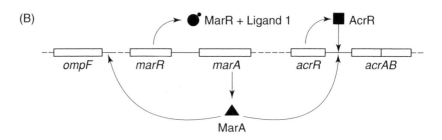

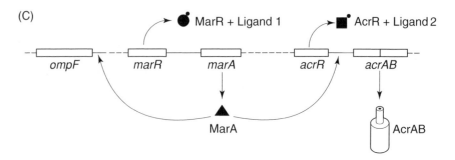

a series of experiments aimed at characterizing a gene expression regulatory function. For example, you might simply be interested in proving that a known regulatory system controls the expression of a test gene or group of genes; this would probably involve introduction of mutations into a cell, which constitutively activate or repress the regulatory mechanism, and measurements of expression levels for each test gene. Alternatively, you might be interested in which genes have their expression controlled by a poorly characterized regulator; again, you would probably mutate the regulator, but in this case, would take a global look at the effects on gene expression. *Figure 1.1* illustrates an example regulatory system in *Escherichia coli*, the Mar regulon (Miller and Sulavik, 1996) and how measurements of gene expression have been used to characterize its function.

The other main reason for measuring gene expression is for pure 'blue-sky' functional genomics. You might be interested in which genes are expressed in a particular growth condition, as a fore-runner to hypothesizing what function a gene might encode. Alternatively, you might want to know which genes are differentially expressed in a diseased or differentiated

Figure 1.1

Complications in functional genomics illustrated by the Mar regulon. The multiple antimicrobial drug resistance (Mar) regulon consists of two transcriptional regulators, MarR and MarA. The gene *ompF* and the operon *acrAB* encode a porin, through which antimicrobial drugs enter the cell, and an efflux pump which exports antimicrobials, respectively. Transcription of *acrAB* is under repression from the local regulator, AcrR and is activated through MarA binding upstream. Transcription of *ompF* is repressed when MarA binds upstream. MarR is a repressor for transcription of *marA*. Thus in (A), MarR binds to the promoter for *marA*, and represses transcription. AcrR binds to the promoter for *acrAB*, and represses transcription. There is no MarA to repress *ompF* transcription. Antimicrobials could flow into the cell through OmpF, and there would be no AcrAB available to pump them out again. In (B) an inducing ligand binds to MarR, reporting the presence of a toxic compound within the cell (e.g. an antimicrobial drug). This causes a conformational change in MarR, and *marA* transcription becomes derepressed. MarA then blocks transcription of *ompF*, but cannot significantly activate transcription of *acrAB*, because AcrR is still bound at the promoter, and its repressive effect is dominant. In this state, further antimicrobial entry would be limited, but the antimicrobial already inside the cell will not be pumped out. In (C) a second regulatory ligand has built up sufficiently to bind to AcrR and de-repress *acrAB* transcription. However, in the absence of MarA, transcription of *acrAB* would be low. In this case, however, MarA is available to activate transcription of *acrAB*, causing active efflux of the antimicrobial present within the cell. This illustrates the idea of multiple signals linking into a regulatory pathway. It also illustrates some of the inherent problems of studying regulation of gene expression. A deletion of *marR* would cause OmpF production to stop, so it might be concluded that MarR is an activator of *ompF* expression. Furthermore, mutations lead to activation of MarR, and so production of MarA may not always lead to production of AcrAB, thus it may be missed that MarA regulates *acrAB* transcription. In this figure, part of the Mar regulon has been isolated. In truth, MarA also has an activatory ligand, and there are a number of other transcriptional regulators that can substitute for MarA functionally, but do so in response to different ligands, and their expression is controlled by multiple different upstream regulators. In eukaryotes, things are likely to be even more complex!

cell, in order to predict which are the key functions in causing disease or differential phenotype.

You can probably think of many other possible experimental scenarios, but it doesn't matter why you want to measure transcriptomic or proteomic complement, the same fundamental problems apply.

It is important that you realize from the start that the transcriptome and (to a lesser extent) the proteome are highly dynamic, and many of the methodological 'hoops of fire' you will have to jump through whilst measuring gene expression are primarily to make sure you catch an accurate picture of the ome as and when you want to take it, and that you minimize the introduction of omic artifacts (particularly those that you are unaware of) simply by doing the experiment a certain way. Even if you are very careful, however, I cannot guarantee that tried and tested methods will work for your particular cell type under your particular growth conditions. Sometimes, the ome does not keep still long enough for you to picture it and no matter how hard you bang your head on the desk, you will not be successful in taking a reliable picture. So, I have included advice concerning experimental design for each type of method, and how to check the statistical significance and reproducibility of the data you are generating, so that you might be better equipped to know when to give up and try something else.

Just because you think you have isolated a regulatory mechanism, and even have manipulated it specifically by mutation, the fact that the expression of a proposed target gene changes in the mutant does not prove that the target gene has its expression controlled directly by the regulatory mechanism. There is always a chance that your pet regulatory mechanism affects the activity of a secondary regulatory mechanism, and that this is the regulator of the target gene (*Figure 1.1*). Keep an open mind.

The ultimate ome when looking at a phenotype is the proteome. Who cares if a protein-coding gene is transcribed if the transcript never becomes translated? There are a number of studies where scientists have determined the transcriptome and proteome in parallel under the same growth conditions; in some they have obtained worryingly inconsistent results (e.g. Anderson and Seilhamer, 1997). Many people forget this, and because transcriptomics is methodologically more straightforward (and perhaps more easily automated) this is the predominant approach for measurement of global gene expression. Beware! For a gene to be truly expressed, a functionally active protein must be produced. Because it is often not possible to measure the amount of a particular protein in a cell, for example if it does not have an easily assayable activity, or if you do not have an antibody suitable for quantification of the protein, it is common to check for differential translation of a transcript flagged up by a transcriptomics experiment by fusing the gene to a reporter gene. These either encode an enzyme that is easily assayable, or a peptide tag whose concentration can be measured using a commercially available antibody. This is generally satisfactory, but be aware that the use of reporter genes is an artificial situation as set out in Chapter 6.

There are many other problems of biological and experimental variability particular to the individual methods for measuring gene expression, and these will be described in the various chapters of this book. The remainder of this chapter is aimed at explaining why the transcriptome and proteome are both so dynamic, by providing a brief overview of the myriad regulatory and mechanistic pathways involved in controlling the expression of a gene. This is a long way from a thorough explanation of gene expression and its control, however, and the reader should look elsewhere for more detailed examples that may be pertinent to their particular field of research.

1.2   An overview of the mechanics of transcription

Protein/DNA interactions

Transcription, and its control, is all about protein/DNA interactions. The conversion of DNA into an RNA transcript is a complex process, which needs to be timed correctly, and completed efficiently in order to correctly process the information content of DNA and ultimately express the phenotypes that it encodes.

DNA can be thought of as a whole lot of information coated in sugar and phosphate. The outer casing of deoxyribose sugar and tough phosphate backbone protects the information contained within the hydrogen-bonding potential of nitrogenous bases, and so secures this information content for generations to come. But all this protection poses the question: How can the cell access its own genetic information? Well the answer is through proteins that interact with the DNA and poke about through the cracks in the backbone to sense the sequence of bases below. The cracks are called the 'major groove' and 'minor groove', and into these grooves most DNA binding proteins fit their sensor regions. But in order to get close in the first place, DNA binding proteins need to be able to interact with the predominantly negatively charged phosphate backbone of DNA, and because of this, most DNA binding proteins have large, curved, positively charged surfaces, which allow the protein to fit snugly around the phosphate backbone. The positioning of the DNA binding protein allows the sensor region to fit into the major and/or minor groove, and interact with a number of bases, often through end face interactions with the ring structures that form a part of them, and sometimes indirectly with the hydrogen bonding network (*Figure 1.2*). Just as with the idea of an enzyme active site, these sensor regions have an amino acid sequence, which positions various critical chemical groups in such a way that if they happen to interact with a particular arrangement of bases, the result is a conformational change in the whole protein, and some output (Choo and Klug 1997; Harrison, 1991). The classical example of this would be a type II restriction enzyme, where the output would be cleavage of the phosphodiester bond at an appropriate position relative to the sequence of bases that the sensor region of the enzyme has evolved to recognize.

How DNA binding proteins find their correct target sites is not clear. Do they simply lock onto the DNA backbone and slide up and down constantly reading the sequence until they happen upon the appropriate target? Or do they work more like enzymes, and constantly collide with their substrate, hopping on and off the DNA, testing the nucleic acid sequence until a productive hit occurs? Or is it a mixture of the two? For the purposes of this book, however, the answer to this fascinating question (Halford and Marko,

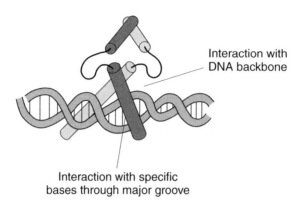

Interaction with
DNA backbone

Interaction with specific
bases through major groove

Figure 1.2

A helix-turn-helix protein binding to its DNA target. The figure shows a homo-dimer of helix-turn-helix transcription factors interacting with their target inverted repeat sequence through the major groove of DNA. The helices are represented as cylinders. Each turn of DNA is approximately 10–11 base pairs dependent upon the super-helical structure; the DNA binding region of each monomer interacts with a five-base-pair sequence. The two sequences are thus separated by five to six bases, putting them both in the same orientation with respect to the DNA. However, the two proteins point in opposite directions with respect to one another (i.e. they point away from each other in the figure) meaning that the sequences to which each binds are arranged in opposite orientations on (and so opposite strands of) the DNA. This would represent a typical inverted repeat. The dimerization domain is usually the binding site for effector ligands, or for some covalent modification such as phosphorylation. Such changes influence DNA binding by actively promoting or discouraging dimerization, or by causing a conformational change in the dimer that affects DNA binding ability.

2004) is not particularly important. What needs to be remembered is that particular proteins have domains that respond to particular nucleotide sequences, associating sequence specific DNA recognition with a variety of possible enzymatic functions. Alternatively, the DNA binding domains form part of proteins that simply provide a core for an enzyme complex to form around.

RNA polymerase enzymes and the mechanics of transcription

The transcription of a DNA molecule into an RNA copy is achieved by a DNA-dependent RNA polymerase enzyme. Essentially, this enzyme catalyzes the joining of a single-stranded chain of ribonucleotides via phosphodiester bonds, with information concerning the order of the ribonucleotides being provided by the sequence of the DNA molecule being transcribed (Gelles and Landick, 1998). RNAs produced include those which encode proteins (coding RNAs: hnRNAs in eukaryotes and mRNAs in prokaryotes) and those that have a structural role in the cell, including those that are involved in translating coding RNAs into proteins (see Section 2.2). Each RNA has a defined sequence and length, and the region of DNA encoding an individual RNA molecule is referred to as a gene. In prokaryotes, some mRNAs carry information for the translation of more than one protein. Such polycistronic mRNAs are encoded by 'operons' of genes, which are transcribed as if they were single genes; that is, there is only one start point and one end point for transcription of each operon. To confuse matters a little, some genes within operons can be transcribed as individual units as well as being transcribed as part of the operon as a whole.

DNA is a double-stranded molecule, where each strand is 'complementary'. This means that the hydrogen bond potentials of the bases on each strand are such that they pair up. Deoxycytosine will only pair with deoxyguanine; deoxyadenine will only pair with deoxythymine and vice versa. The orientation of one DNA strand to the other, with respect to the chemical linkages of their sugar phosphate backbones, is anti-parallel. DNA sequences are generally written in the 5' to 3' direction, with the 5' representing the phosphate end of a deoxynucleotide monomer, and 3' representing the sugar-hydroxyl end, which is linked to the next 5' phosphate in the DNA sequence in the form of a phosphodiester bond. The two anti-parallel DNA strands are referred to as the 'coding' and 'non-coding' strands, and the sequence of a coding DNA molecule, when written, is always that of the coding strand. However, it is the non-coding strand that carries the true coding information of DNA, being the template for RNA production. The term 'coding strand' is therefore something of a misnomer, but it is used because it represents the sequence of the RNA molecule encoded.

Transcription is broken into three phases: initiation, elongation and termination. The initiation phase is concerned with where (and as we shall see below, when) an RNA polymerase enzyme starts transcription on a DNA molecule. If the site of transcriptional initiation were not tightly regulated, then the whole process would become very inefficient, producing lots of useless RNA. Furthermore, since RNA synthesis requires a great deal of energy, this would be a waste. Thus the site of transcription initiation defines the start of a gene (or operon). To define the end of a gene (or operon) transcriptional termination must also occur at a defined position in the DNA molecule. Between the initiation and termination phases of transcription is the elongation phase (Gelles and Landick, 1998).

The mechanics of transcriptional elongation

The elongation phase of transcription is a simple, cyclical process where an RNA molecule is built, one ribonucleotide at a time, with a newly inserted ribonucleotide being joined to the existing RNA chain via a phosphodiester linkage between the 5' phosphate of the new ribonucleotide and 3' sugar hydroxyl at the end of the existing RNA. This means that RNA polymerases produce RNA in a 5' to 3' direction, but work in a 3' to 5' direction with respect to the non-coding strand of the DNA being transcribed, which is their template (Gelles and Landick, 1998).

The approach taken to transcribe a DNA sequence by an RNA polymerase enzyme is to open up the molecule and form a so-called 'open complex' or 'transcription bubble', where the two DNA strands are separated and RNA polymerase gains access to the non-coding strand of the DNA. Reading the sequence information carried on the non-coding strand in this transcription bubble is a simple matter of the enzyme exposing DNA bases sequentially. So-called 'complementary' pairs of bases are attracted because they share a particular geometry of potential hydrogen bonds. Uracil pairs with deoxyadenine, adenine pairs with deoxythymine, guanine pairs with deoxycytosine and cytosine pairs with deoxyguanine. Thus, if a mixture of the four ribonucleotides is washed into the DNA transcription bubble, in the vast majority of cases, only the correct one will hydrogen bond with the exposed DNA base. All the RNA polymerase then has to do is to catalyze joining of the nucleotide onto the nascent RNA molecule by forming a phosphodiester bond, and then push the transcription bubble forward in a 5′ to 3′ direction with respect to the coding DNA strand, exposing the next DNA base, where the whole process repeats itself. The transcription bubble is approximately 30 nucleotides long, and within it is the DNA/RNA hybrid together with the separate coding DNA strand, which needs to be stabilized. As the 5′ end of the transcription bubble moves towards the 3′ end of the coding DNA strand, the DNA/RNA hybrid must be dissociated and the two DNA strands re-associated so that the bubble remains the same size. It is not surprising to find that RNA polymerase enzymes are actually multi-protein complexes given that a number of functions are being carried out at once. Indeed, the size of the transcription bubble is dictated mainly by the size of the enzyme complex that must be accommodated within it. The general mechanism of RNA polymerase mediated transcription is set out in *Figure 1.3*.

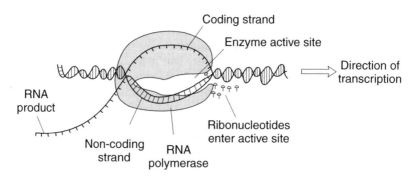

Figure 1.3

A general illustration of transcription. The RNA polymerase is thought of as a clamp, which bears down on the DNA molecule at a promoter and causes open complex formation. The two DNA strands are separated and free ribonucleotides diffuse into the active site, where they are used to synthesize a nascent RNA product.

Transcriptional termination

In bacteria, the termination of transcription of a gene or operon is initiated in one of two main ways. So-called Rho-dependent termination relies on a small protein, Rho, recognizing a sequence in the DNA at the point transcription must be terminated. Rho actively displaces RNA polymerase from the DNA by unwinding the RNA/DNA hybrid in the transcription bubble, causing transcription to cease. More commonly, however, termination of transcription is Rho-independent. Here, an inverted repeat in the DNA sequence followed by a run of five to ten thymines (in the coding strand) means that when transcribed into a single-stranded RNA molecule, the inverted repeat sequences are attracted through their complementarity and cause folding into a so-called 'hairpin' structure through hydrogen-bond formation. The hairpin is formed within the transcription bubble, which destabilizes the interaction of RNA polymerase with the DNA sequence, and the run of uracils on the mRNA actually signals for the polymerase to dissociate from the DNA (Das, 1993). Rho-independent transcriptional termination in prokaryotes is illustrated in *Figure 1.4*.

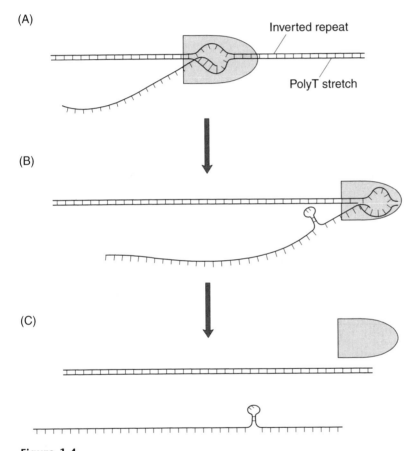

Figure 1.4

Rho-independent transcriptional termination in bacteria. RNA polymerase proceeds until it comes across a termination signal. In (B) a region of DNA containing an inverted repeat has been transcribed, and this has resulted in the formation of a hairpin in the nascent RNA product once it has excited the RNA polymerase enzyme complex. This, in itself, is not sufficient to cause termination of transcription (i.e. it is a fairly frequent event that occurs by chance). However, the presence of this hairpin primes the RNA polymerase to recognize a run of thymines (polyT stretch), and if one occurs within close proximity to the inverted repeat, RNA polymerase stops transcription and dissociates from the DNA (C).

Transcriptional termination in eukaryotes (*Figure 1.5*) is a far less well understood process. Specific sequences within the emerging nascent RNA are recognized by two proteins, CstF and CPSF, which are associated with and targeted to the RNA, through interaction with the transcription initiation complex specific to coding RNAs. These proteins are responsible for

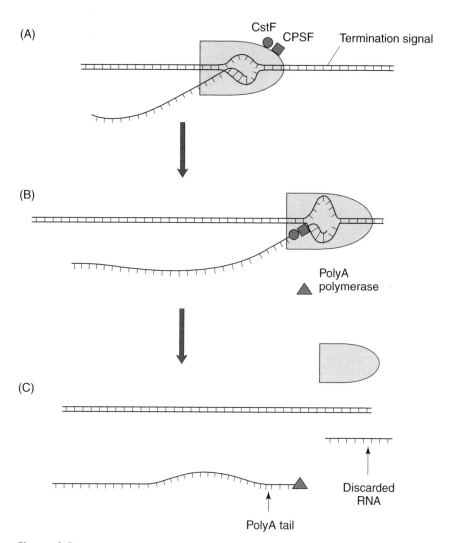

Figure 1.5

Transcriptional termination in eukaryotes. The eukaryotic RNA polymerase carries a number of protein factors around with it. These include termination factors such as CstF and CPSF. When a particular seqence of DNA (the termination signal) has been transcribed, these proteins bind to its copy in the nascent RNA, and cleave the transcript (B). PolyA polymerase is then recruited (this is also likely to be associated with RNA polymerase) and this synthesizes a polyadenine tail onto the RNA. The RNA polymerase becomes destabilized once the RNA transcript has been cleaved, and it limps along for a few hundered bases, producing a 'junk' RNA product as it goes, until it finally falls off the DNA (C).

cleaving the hnRNA product, thus terminating effective transcription. Once the RNA molecule has been cleaved from the RNA polymerase, it becomes less tightly associated with the DNA, so whilst transcription to produce an essentially useless, and ultimately wasted, RNA molecule continues for a couple of hundred nucleotides, the RNA polymerase eventually dissociates from the DNA molecule.

In a process that is coupled with cleavage, CPSF recruits the enzyme polyA polymerase, which adds a long (50–250 residues) polyadenine tract onto the 3' end of the cleaved hnRNA. For an unknown reason, though, some hnRNAs do not become polyadenylated so it cannot be guaranteed that all coding RNAs are polyadenylated (denoted polyA$^+$). It is likely that non-coding RNA molecules are cleaved in order to stimulate RNA polymerase dissociation in a similar way to non-coding RNAs, but they are not subjected to polyadenylation (von Hippel, 1998).

Bacterial RNA polymerases

In its simplest state, the bacterial RNA polymerase is composed of four subunits. Two are identical, the alpha subunits, and two are similar but not identical, the beta and beta prime subunits. This RNA polymerase is often referred to as an $\alpha_2\beta\beta'$ complex, or the core RNA polymerase complex (Finn et al., 2002). This core complex can catalyze the elongation phase of transcription, but is incapable of recognizing the start of a gene. Transcriptional initiation is dependent upon the sigma subunit, which recognizes the 'promoter' sequence, which marks the start of a gene. The RNA polymerase holoenzyme is referred to as an $\alpha_2\beta\beta'\sigma$ complex. The sigma subunit anchors the RNA polymerase to the promoter, holding it in place long enough for the core enzyme subunits to unwind the DNA and form the initial transcription bubble, called 'open complex formation'. Once transcriptional elongation has started, the sigma subunit dissociates (Finn et al., 2002).

Sigma subunits recognize specific promoter sequences, which are generally composed of two elements: a TA-rich region which is close to the transcriptional initiation site, and a promoter specific region which is usually 15–20 nucleotides further upstream. In each bacterium, a number of different sigma factors exist, and each can recognize a different specific promoter sequence. The classic and best studied promoter is the σ^{70} promoter from Escherichia coli. This is defined by the consensus sequences TTCAGA, centered at 35 nucleotides 5' proximal to the transcriptional start site (the −35 box) and TATAAT, centered at 10 nucleotides 5' proximal to the transcriptional start site (the −10 box). Other sigma factors have variations on this theme, or in some cases use sequence elements that are 150 or so nucleotides 5' proximal to the transcriptional start site, and which can only come into sufficient proximity with the −10 box to allow the sigma factor to interact with both, if the DNA molecule is bent and distorted. For this to happen, specific proteins responsible for DNA bending must be able to bind to recognition sites within the −10 and −150 boxes. The 'strength' of a promoter (i.e. how efficient transcriptional initiation is at a promoter) is defined primarily by how close the sequences of its sigma factor binding sites are to the consensus sequences for those sites. This dictates the strength of the interaction of the sigma factor (and so the RNA polymerase complex) with the promoter, and so the likelihood that transcription will be initiated (Finn *et al.*, 2002). The binding of the sigma subunit to promoters can be affected by other factors designed to control the rate of transcriptional initiation, and these will be discussed in Section 1.5.

The alpha subunit of the bacterial RNA polymerase is not catalytic, but is essential for formation of the core RNA polymerase complex. Furthermore, its presence can help to improve the efficiency of transcriptional initiation at some promoters, because the C-terminal portion of the subunit (α-CTD) favors interaction with specific sequences of DNA; so-called 'up-elements'. If one of these is present around 50 nucleotides 5′ proximal to the transcriptional start site, RNA polymerase will interact more strongly with the promoter, and initiation of transcription will be more likely to occur. The alpha subunit also plays an important role in transcriptional control through interaction with proteins whose role is to affect the rate of transcriptional initiation, as will be discussed in Section 1.5.

The catalytic power of bacterial RNA polymerase is located in the beta and beta prime subunits, which are some of the largest bacterial proteins. Another, loosely associated subunit, the omega (ω) subunit is believed to fold around the beta and beta prime subunits and stabilize them, but has no direct role in transcription (Finn *et al.*, 2002). The structure of bacterial RNA polymerase, and its interaction with promoter sequences is illustrated in *Figure 1.6*.

Eukaryotic RNA polymerases

In eukaryotes, three distinct RNA polymerase complexes are responsible for transcription. RNA polymerase I (Paule and White, 2000) transcribes a single structural rRNA precursor that represents about 80% of all primary

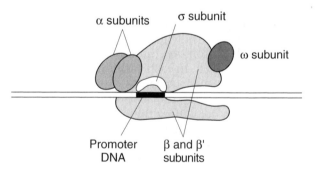

Figure 1.6

The *Escherichia coli* RNA polymerase complex. The complex consists of the core RNA polymerase that includes two catalytic subunits, β and β' (which together form the clamp that closes around the DNA and forms the open complex) and two ancillary (α) subunits, which have a role in promoter recognition and transcriptional control. Two additional subunits are involved. The σ subunit recognizes the promoter sequence and dissociates shortly after the start of transcription. Different promoters are recognized by different σ subunits. The role of the ω subunit is not entirely understood, but it is believed to stabilize the catalytic subunits.

transcript in the cell. This precursor rRNA is modified and processed to create three rRNA molecules: the 18S, 28S and 5.8S rRNAs. RNA polymerase II (Lee and Young, 2000) produces coding RNAs, and some small, stable RNAs such as snRNA. RNA polymerase III (Geiduschek and Kassavetis, 2001) transcribes the production of the 5S rRNA and tRNA molecules and a few small structured RNAs. More details about these various RNA species are given in Section 2.2.

The three RNA polymerases have much in common. There are five subunits shared by all three and five more that are very similar in all three; these include the two large subunits that do the main job of transcription, which are structurally related to the beta and beta prime subunits of bacterial RNA polymerases (Ebright, 2000). In addition, there are polymerase-specific subunits, which number four for RNA polymerase I, two for RNA polymerase II and seven for RNA polymerase III. These subunits are really only responsible for the recognition of promoter sequences, which are different for the different classes of genes to be transcribed by the distinct RNA polymerase types. However, unlike the sigma subunits of bacterial RNA polymerases, these initiating subunits do not contact promoters directly, but they interact with a cognate pre-initiation complex, formed between a group of transcription factors (TFs) and a promoter DNA sequence. At the heart of each of these complexes, whatever the RNA polymerase or the specifics of the promoter, is the TATA-box binding protein (TBP), since TA rich sequences are commonly found in eukaryotic promoters close to the transcriptional start site, as seen with the -10 boxes of prokaryotic promoters. Other sequence elements linked to this TATA box, recognized by promoter-specific TFs, dictate which pre-initiation complex will form at a promoter, and so define the different promoters to be transcribed by different RNA polymerases (Lee and Young, 1998, 2000).

The RNA polymerase I pre-initiation complex consists of two multi-subunit TFs, each binding to one of a pair of distinct sequence elements that together make up the promoter. The polymerase III pre-initiation complex also requires two multi-subunit TFs, though transcription of 5S rRNA by this RNA polymerase requires a third (Paule and White, 2000). The RNA polymerase II pre-initiation complex is far more complex, however (*Figure 1.7*). It requires a basic component of six general TFs: TFIIA, B, D, E, F and H, and each is made up of one to fourteen polypeptides. The actual promoter recognition is performed by TFIID, which is composed of the TATA-box binding protein and 13 additional factors. The other TFs fold around TFIID to create the pre-initiation complex (Green, MR, 2000). RNA polymerase II can bind to this complex, but transcription cannot always begin because of the influence of a large variety of other factors. These include chromatin structural elements adjacent to the promoter, the presence of enhancers or regulatory protein factors which will all be discussed in Section 1.6. The C-terminal domain of RNA polymerase II is important for interacting with these regulatory moieties and with RNA processing factors (see Section 1.3).

Once a competent pre-initiation complex has been formed and an appropriate RNA polymerase has attached to it in a functionally active state, transcription is initiated by separating the DNA strands to form an open complex. Eventually, transcription elongation begins, but does not always

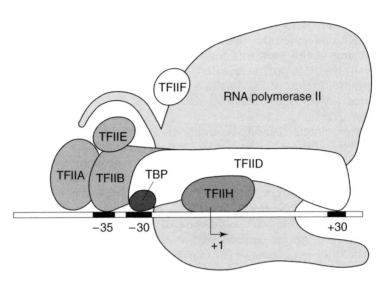

Figure 1.7

The eukaryotic general transcription factor/RNA polymerase II complex. Eukaryotic promoters are recognized by transcription factors. In the diagram, the promoter is recognized by TFIID, which includes the TATA-box binding protein. Once TFIID has bound to the promoter, other TFs associate around it, some of which recognize additional promoter sequence elements, thereby stabilizing the TF complex. Finally, RNA polymerase II associates and, if the structure of the DNA allows (i.e. if the DNA is not condensed into chromatin and repressors are not bound), transcript will commence through open complex formation at the point marked '+1'.

proceed at a constant rate. The polymerase can move quickly, slowly, or can even pause, and other protein factors binding to sequence-specific elements are required to regulate these events (Conaway *et al.*, 2000).

1.3 Post-transcriptional modification, processing and nuclear export of coding RNA

Post-transcriptional modification of coding hnRNA molecules to produce mRNAs that are then exported from the nucleus is a feature exclusive to eukaryotic cells. Polyadenylation of the 3′ ends of hnRNAs, as discussed when dealing with transcriptional termination (above , and see *Figure 1.5*) is one example (Minvielle-Sebastia and Keller, 1999), and is the last event to occur. The first processing event is capping of the 5′ end of an RNA with an inverted 7-methyl-guanosine moiety. Capping is a process that is undertaken by a triumvirate of enzymes, and occurs only for coding RNAs after around 25 nucleotides have been synthesized by RNA polymeraseII. There is a very small number of hnRNAs that are not capped, though the reasons for this are not clear. Capping is kept exclusively to coding RNAs because the capping enzymes can only interact with RNA polymerase II, which has an extended C-terminus to which many RNA processing factors bind. The cap has two main roles. First, it is involved in the export of mRNAs from the nucleus and their transportation to the cytoplasm, where they are translated. The second role of the cap is to form an anchoring point for the assembly of the translation initiation machinery. The cap may also protect an mRNA molecule from attack by exonuclease enzymes (Shatkin and Manley, 2000).

Splicing is an essential processing event in the conversion of hnRNA into mRNA. It is not clear why, but most protein coding genes in eukaryotes contain introns, which are regions of DNA that do not participate in encoding the protein product of the gene. It is possible that introns are involved in controlling transcriptional elongation rate, or that they are involved in the packaging of DNA into cells. Either way, there needs to be a mechanism by which the introns can be discounted when the transcript derived from the gene is translated into protein. Since the RNA polymerase cannot simply jump over introns, producing an intron-free transcript, the introns must be removed from the hnRNA primary transcript before translation can occur. The way this is forced to happen is that translation takes place outside of the nucleus and removal of the introns, or splicing, takes place within the nucleus (Lewis and Tollervey, 2000).

Splicing is a complex process, and can be achieved in a number of different ways. None of the known splicing mechanisms need be discussed here in any great detail. However, the basic principle of splicing (*Figure 1.8*) is that specific, short, RNA sequences flank intron sequences within hnRNAs, and these sequences define the limits of the intron to be removed. The sequences are 'read' by splicing factors; proteins that interact with the C-terminal extension of RNA polymerase II. The splicing factors are the focus for the formation of ribonucleoprotein complexes (RNPs), which involve small nuclear RNAs and a number of different proteins. A different RNP marks each end of the intron. Splicing involves bringing together the RNPs to form a splicosome, so that the ends of the intron come together to form a loop of RNA. Then, essentially, the splicosome catalyzes a pair of nucleophilic attacks from one end of the intron to the other in such a way that the phosphodiester backbone of the RNA molecule is broken at each end of the intron and then rejoined excluding the intron, with the intron being removed as a circle. Different splicing mechanisms are catalyzed by different splicosomes, and are targeted to different introns with different flanking sequences.

Once an hnRNA molecule has been processed into an mRNA molecule it is exported from the nucleus into the cytoplasm (Daneholt, 1997). This occurs following the binding of the cap-binding complex of proteins to the 7-methylguanine cap at the 5' end, and the polyA binding protein to the polyadenine tail at the 3' end. mRNAs cannot be exported if still associated with RNP complexes, since this indicates that splicing has not occurred, and the mRNA is not mature. Export of mRNAs is difficult because of their size. It requires the involvement of a large number of proteins called RNA export factors, which effectively coat the mRNA and pull it through the nuclear pore complex. As the mRNA is exported, these export factors are sloughed off and remain in the nucleus. Once in the cytoplasm, the cap-binding complex is replaced by the translation initiation factor eIF4E, which can also bind to 7-methylguanine. Once the entire mRNA is through the nuclear pore, the polyA binding proteins associated with the 3' end of the mRNA (which enters the cytoplasm last) interacts tightly with eIF4E,

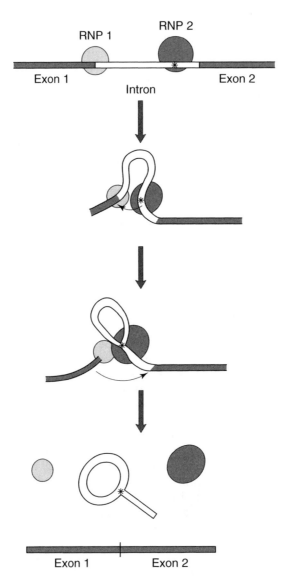

Figure 1.8

A simplified illustration of the process of splicing. Two different small nuclear RNA/protein (RNP) protein complexes form either side of the intron, attracted to specific sequences in the RNA. The two RNPs come together, pulling a catalytic adenine residue (marked with a star), whose geometry is distorted such that it can donate electrons to the RNA backbone at the opposite end of the intron. This causes cleavage of the RNA, and formation of a loop structure. The free RNA end donates electrons to the RNA backbone at the other side of the intron, effectively splicing out the intron. The whole process is catalyzed by the RNP complexes themselves.

folding up the mRNA, and signaling that the entire mRNA has exited from the nucleus.

1.4 An overview of the mechanism of translation

Like transcription, translation of an mRNA into protein is a cyclical process that involves the expenditure of a large amount of energy in the form of ATP and GTP hydrolysis. The mechanics of translation is very similar in prokaryotes and eukaryotes, but there is one fundamental difference. Transcription and translation are couples in prokaryotes, with the transcription machine, the ribosome, associating with the nascent mRNA and beginning to translate it, whilst the mRNA is still being produced by RNA polymerase. Because of this coupling, it is not surprising to find that the over-riding point of control of gene expression in prokaryotes is at the level of transcription, and that the rate of translation is only limited by the availability of ribosomes and the amino acid building blocks used to make proteins. In contrast, transcription in eukaryotes occurs in the nucleus, and translation occurs in the cytoplasm. Thus, the process of translation requires more effort to initiate in eukaryotes than prokaryotes. This is because when transcription and translation are linked, the mRNA to be translated is held in place by RNA polymerase so that the ribosome can associate; when the mRNA is free in the cytoplasm, however, the ribosome needs to make more effort in order to find it. This is a potential reason why eukaryotic mRNAs have evolved to be more stable than prokaryotic ones – they need to hang around long enough to be translated – and it almost certainly explains how translational control of gene expression in eukaryotes is possible, and indeed, at least for short-term responses, perhaps is more important than the control of transcription.

The initiation of translation

In DNA, specific 'promoter' sequences signal the point at which transcription should start by forming a structure to which the RNA polymerase complex can bind. In the same way, sequences within RNA provide structures to which ribosomes associate, and so initiate translation. These sequences come in two parts: the ribosome binding sequence and the initiation codon. The ribosome binding sequence positions the ribosome so that translation can be initiated if there is an initiation codon on the mRNA at an appropriate position relative to where the ribosome binds. Only when these two elements come together can translation be initiated.

Initiation of translation in bacteria

In *E. coli*, the ribosome binding sequence is called the Shine–Delgarno sequence, which is 'GGAGGU'. That said, however, translation often initiates successfully even if the Shine–Delgarno sequence is not a perfect match with this consensus. As is the case for nonperfect promoter sequences, however, the strength of the ribosome/mRNA interaction will be lower if the Shine–Delgarno sequence is not perfect, reducing the rate of translational initiation, and so probably the amount of protein produced in a given period of time. Initiation of translation in bacteria involves the interaction of the small ribosomal subunit, (the so-called 30S ribosomal subunit) with the Shine–Delgarno sequence. The 30S ribosomal subunit is a ribozyme, which consists of 21 proteins, together with a molecule of 16S rRNA (see Section 2.2). It is a sequence within this 16S rRNA that causes binding of the 30S ribosomal subunit to the Shine–Delgarno sequence, because it is complementary (i.e. has the opposite base sequence of 'ACCUCC') to the Shine–Delgarno sequence, meaning that the two RNA sequences hydrogen bond together. This interaction is stabilized by proteins known as 'initiation factors' (IFs), IF1 and IF3. These proteins do not form part of the ribosome *per se*, but are loosely associated and dissociate once translation has been initiated. Next, a third initiation factor, IF2 encourages the association of the initiator amino acid to the 30S ribosomal subunit/mRNA complex. Amino acids to be used as building blocks for translation are provided joined to tRNA molecules, which are said to be charged with the amino acid. There are many different tRNA molecules in a cell; each folds into a distinctive clover-leaf structure due to high levels of internal sequence complementarity, since all tRNAs share a high degree of sequence identity. Each different tRNA has a specific amino acid with which it can be charged and has a hypervariable region, which includes a so-called 'anticodon' triplet of RNA nucleotides that is not part of a complementary region, and so is free to hydrogen bond with a complementary nucleotide sequence in the mRNA to be translated. It is the specificity of tRNA/mRNA anticodon/codon interactions that means the correct amino acids are added sequentially, so facilitating synthesis of the correct protein. The initiator amino acid in bacteria is a modified version of methionine called *N*-formylmethionine. The association of the 30S subunit with the Shine–Delgarno sequence, puts the portion of the 30S subunit to which the tRNA charged with *N*-formylmethionine binds, immediately adjacent to the first (initiator) mRNA triplet codon, which is usually AUG, but can be GUG. At this point, a molecule of GTP binds to IF2, then follows association of the 50S ribosomal subunit, which consists of 31 proteins, the 23S rRNA and the 5S rRNA. Finally, the GTP molecule is hydrolyzed to GDP and the initiation factors dissociate (Green and Noller, 1997). The initiation of translation in bacteria is illustrated in *Figure 1.9*.

The initiation of translation in eukaryotes

Eukaryotic mRNAs do not have true ribosome binding sequences in the same way that prokaryotic messages do. Instead, interaction of the small ribosomal subunit with the mRNA upstream of the initiation codon is dependent upon the cap-binding complex of translation initiation factors, which form around the 7-methylguanine cap structure at the 5' end of the mRNA to be translated. So, in terms of translation initiation in eukaryotes (*Figure 1.10*), the cap itself might be thought of as the ribosome binding sequence. The cap-binding protein is called eIF4E; the other member of the cap-binding complex is eIF4G. Another two proteins, eIF4A and eIF4B associate with the cap-binding complex by binding to part of eIF4G, and together form the translation initiation complex known as eIF4F. Once the small ribosomal subunit (the 40S ribosome, consisting of 33 proteins and

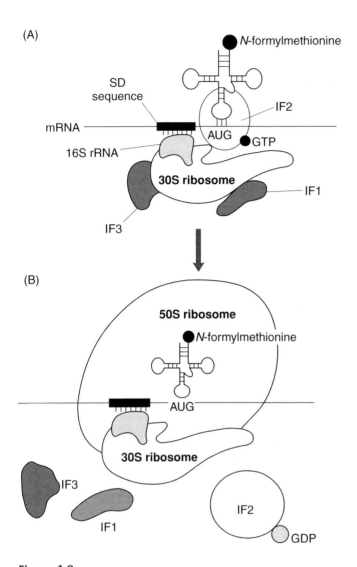

Figure 1.9

Translational initiation in prokaryotes. In (A) the 30S ribosome locates the ribosome binding (Shine–Delgarno) sequence through complementary base pairing with the 16S rRNA component of this ribosomal subunit. This interaction between ribosome and mRNA is strengthened by initiation factors (IFs) 1, 2 (which carries an N-formylmethionine-charged tRNA complementary to the AUG, and rarely GUG, start codon) and 3. Once this has occurred, the formation of the entire ribosome occurs through joining of the 50S ribosomal subunit, and transfer of the f-met-tRNA into the P site (see Figure 1.11) of the ribosome. This last process is catalyzed by IF2, and involves the hydrolysis of bound GTP to GDP.

the 18S rRNA) has bound to the eIF4F complex via another translation initiation factor, eIF3, the ribosome, together with eIF4A and eIF4B, breaks away from the cap-binding complex, and scans along the mRNA, with eIF4A

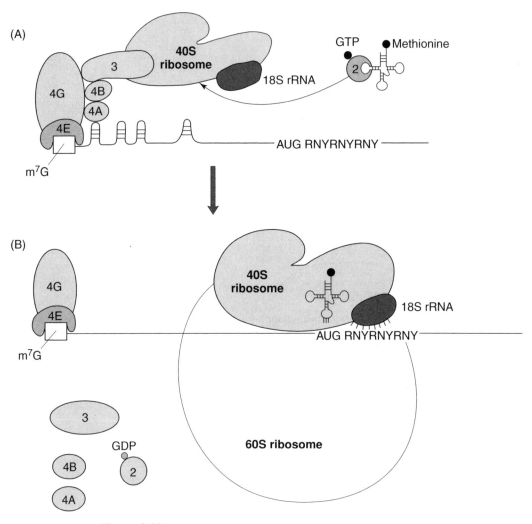

Figure 1.10

Translational initiation in eukaryotes. There is no ribosome binding site in eukaryotic messages. The ribosome is targeted to mRNAs through a complex of translation initiation factors (eIFs). The cap binding complex forms the core of this complex, because it associates with the 7-methyl-guanine cap at the 5' end of most mRNAs. The gap between the cap-binding complex and the 40S ribosomal subunit is bridged by eIF3. The initiator methionine tRNA is loaded onto the 40S ribosomal subunit before initiation of translation via eIF2, in a process catalyzed by the hydrolysis of GTP into GDP. The eIF–ribosome complex breaks away from the cap-binding complex and moves along the mRNA in a process driven by the RNA helicase activity of eIF4A, which also irons out any secondary structure in the 5' untranslated region of the mRNA (which can slow down the rate of transcriptional initiation). Once the correct initiation AUG, in the appropriate downstream consensus is reached (B), the large ribosomal subunit associates, eIFs dissociate and translation begins. The correct initiation site is sensed through interactions between the initiator tRNA, and the 18S rRNA components of the small ribosomal subunit. 'R' represents any purine, 'Y' represents any pyrimidine, 'N' represents any base.

removing secondary structure within the untranslated region of the message, until the first AUG initiator codon is reached. Initiation of translation can be very weak at some AUG sequences if the downstream context is not correct, however. Ribosomes favor a regular pattern of triplet codons that have a 'purine, any base, pyrimidine' arrangement. It is thought that this helps keep the reading frame correct, and it is apparent that this signal is particularly strong in the first few codons of a coding sequence. The 18S rRNA portion of the small ribosomal subunit probably interacts with this regularity, thus strengthening the association between ribosome and mRNA, and making translation more likely to initiate. In this way, if the first AUG in an mRNA is not in front of a strong consensus, translational initiation can be abortive, and the ribosome can carry on searching further downstream. Initiation of translation occurs at the AUG initiation codon by attraction of a methionine-charged tRNA, which recognizes the AUG by virtue of the fact that it has the correct anticodon sequence. This met-tRNA has already been put into position within the 40S ribosomal subunit prior to interaction with the eIF4F complex due to the action of another initiation factor, eIF2, which is dependent upon GTP hydrolysis to work. Once the initiator tRNA/ribosome complex is fixed onto the initiation codon, this stimulates association of the large, 60S ribosomal subunit, which consists of 49 proteins, together with the 5S rRNA, the 5.8S rRNA and the 28S rRNA molecules, and dissociation of all initiation factors (Sachs and Varani, 2000).

The elongation and termination of translation

Once translation has been initiated, both in prokaryotes and eukaryotes translation proceeds in an essentially identical manner (Wilson and Noller, 1998) (*Figure 1.11*). The initiator tRNA-amino acid is present in part of the ribosome called the peptidyl site (P-site). Immediately upstream of the P-site is the acceptor site (A-site) of the ribosome, which surrounds the next codon on the mRNA. The appropriate charged tRNA approaches the A-site and its anticodon hydrogen bonds with the mRNA. Next, the peptidyl transferase center of the ribosome catalyzes peptide bond formation between the two adjacent amino acids freeing the initiator-tRNA and associating the nascent peptide with the second tRNA, and the whole ribosome moves down the mRNA three nucleotides, moving the second tRNA/peptide complex into the P-site, and the initiator tRNA into the exit site (E-site) of the ribosome. The third charged tRNA now enters the A-site, and a peptide bond is formed between the C-terminal end of the peptide in the P-site and the new amino acid in the A-site, transferring the peptide onto the tRNA in the A-site. The ribosome moves down the mRNA once more, the initiator tRNA dissociating from the E-site to be replaced by the second tRNA, and the third tRNA, joined to the peptide moving into the P-site. The process then continues in a cyclical manner, synthesizing the polypeptide, until a terminator codon is reached on the RNA (Frank, 2000). In bacteria, here, release factors enter the A-site and stimulate dissociation of the ribosome from the mRNA, and release of the protein product (Green and Noller, 1997). Similar release factors are also used to terminate translation in eukaryotes, and here, the termination codon is located immediately

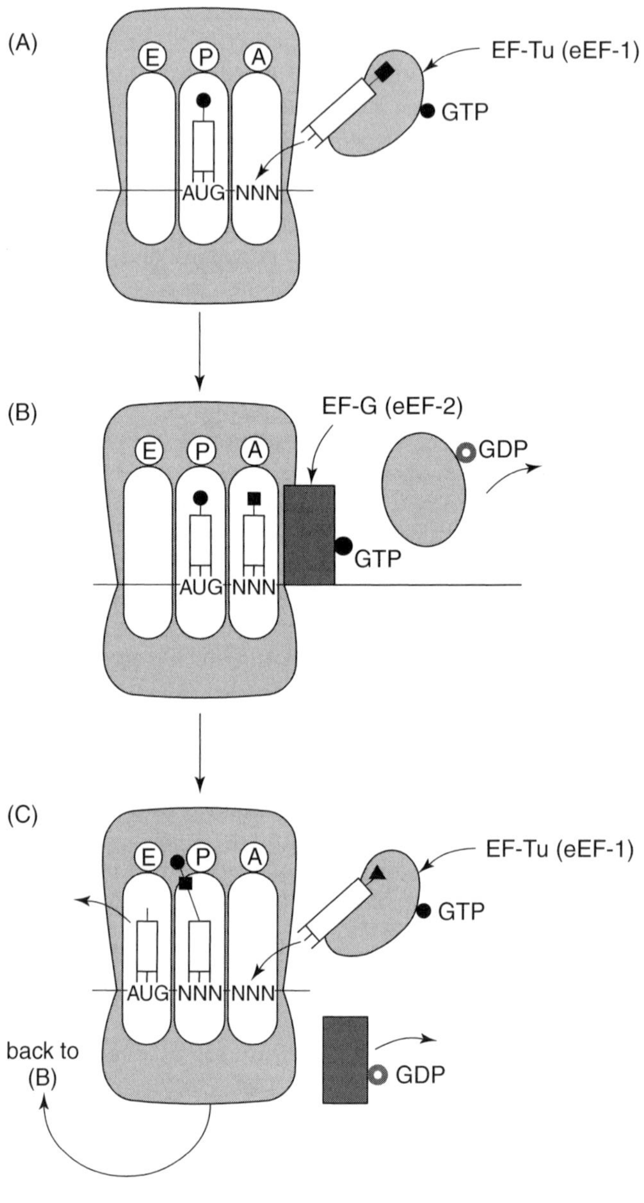

Figure 1.11

The mechanics of translation in prokaryotes and eukaryotes. The figure illustrates the mechanism of translational elongation set out in the text. The two elongation factors drive the process forward through delivering the next charged tRNA to be incorporated into the growing polypeptide, and providing the force that moves the ribosome along the mRNA. In (A) the initiator tRNA is bound in the P-site (i.e. immediately after initiation of translation (Figures 1.9 and 1.10). The delivery of tRNAs costs energy in the form of GTP hydrolysis. In (B) EF-G (eEF2) associates with the ribosome and in (C) drags the ribosome along the mRNA one codon, again with the energy being provided by GTP hydrolysis. At this point, the tRNA in the E-site dissociates and a new charged tRNA is delivered. The cycle of translation from this point is from (B) to (C) and back to (B); until a terminator codon is reached where an uncharged, termination tRNA is delivered to the A-site. The peptidyl transferase center of the ribosome cannot join the polypeptide in the P-site onto the terminator tRNA, and the polypeptide is released.

upstream of the polyadenine stretch found at the ends of most eukaryotic mRNAs. This polyadenylated region is actually bound to the cap at the 5′ end of the mRNA through interaction of the polyA⁺ binding protein with the eIF4G component of the cap binding complex. Thus the mRNA molecule is looped around, which sterically hinders progression of the ribosome beyond the termination codon, and may actively encourage re-initiation of transcription back at the beginning of the message (Frank, 2000).

The process of translational elongation is made more rapid and efficient due to the interaction of elongation factors with the ribosome at each turn of the protein synthesis machine. The process is driven by the hydrolysis of GTP, carried by these elongation factors. This also helps to improve the accuracy of translation, because the force generated by GTP hydrolysis actively favors dissociation of incorrectly hydrogen-bonded charged tRNAs in the A-site before peptide bond formation can occur (*Figure 1.11*). The two elongation factors used in bacteria are EF-Tu and EF-G, whose close relatives in eukaryotes are called eEF1 and eEF2. Essentially, EF-Tu (eEF1) controls insertion of the new charged tRNA and EF-G (eEF2) controls translocation of the ribosome following peptide bond formation (Green, R, 2000).

1.5   Control of transcription in prokaryotes

Without the influence of outside effector proteins, the rate of transcriptional initiation in prokaryotes, which is the prime rate-determining step of transcription as a whole, is regulated by the closeness of the match between the promoter sequence upstream of a transcriptional unit and the consensus promoter sequence for a particular RNA polymerase sigma subunit. Further enhancement of the rate of transcriptional initiation can be obtained if an 'up-element' is present upstream of the promoter, to which the C-terminal portion of the alpha subunit of RNA polymerase binds (Finn *et al.*, 2002). Additional proteins, often referred to as transcription factors, can affect, either positively or negatively, the binding of RNA polymerase to a promoter, and so can affect gene expression rates. The various methods for transcription factors to regulate transcriptional initiation in prokaryotes are illustrated in *Figure 1.12*.

Transcriptional repressors

A whole host of transcription factors control the rate of transcriptional initiation, but the most simple of these are the transcriptional repressors. These proteins bind to specific recognition sequences that are located in close proximity to the −10 box of the promoter. They bind very tightly, and in so doing physically block interaction of RNA polymerase with the promoter, preventing transcriptional initiation.

There are many different types of transcriptional repressors, but most have two domains. One which binds to the DNA at a particular recognition sequence, and another that receives some control input. The most common form of control input would be the interaction of a small ligand with the control domain, causing a conformational change within the DNA binding domain and either preventing or promoting DNA binding, and

(A)

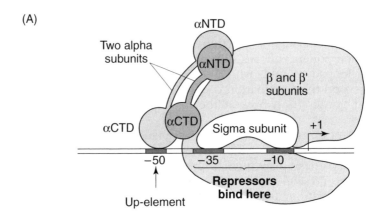

(B)

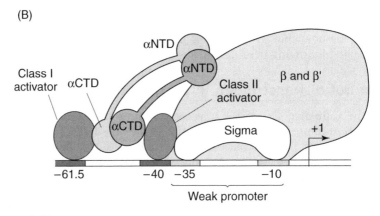

Figure 1.12

An illustration of control points in prokaryotic transcriptional initiation. In (A) a strong bacterial promoter is illustrated. There is a good match to the consensus sequences to which the sigma subunit of RNA polymerase binds. Furthermore, there is an up-element (and A–T rich sequence) to which the C-terminal domain of the alpha subunit (α-CTD) can bind. This strong interaction between promoter and RNA polymerase means an open complex is likely to form, and transcription can begin. The rate of transcription at strong promoters is usually regulated by the actions of transcriptional repressors, which bind to sequences that overlap the promoter more strongly than the sigma subunit of RNA polymerase binds to the promoter, thereby preventing significant binding of RNA polymerase, open complex formation, and so transcriptional initiation occurs. In (B) a weak promoter has a poor consensus to the sigma factor binding site and no up-element. RNA polymerase is held down onto the promoter, thereby stimulating transcriptional initiation, through the interaction of transcriptional activator proteins with various subunits of RNA polymerase. Class I activators bind upstream of the promoter and interact with α-CTD. The position marked at −61.5 is that of the famous activator cAMP receptor protein (CRP), though some activators bind further upstream. Class II activators bind very close to the promoter and interact either with the sigma subunit or with α-CTD. Some promoters are linked to binding sites for both class I and class II activators and repressors.

transcriptional repression. The classic example of this is the lactose operon repressor, LacI, which, in the absence of the lactose derivative allolactose, binds to its recognition sequence and prevents RNA polymerase from transcribing the operon of genes that encode proteins essential for lactose uptake and utilization. In the presence of lactose, and so allolactose, however, the latter binds to the control domain of LacI, and induces a conformational change within LacI, preventing it from binding to its target DNA, and therefore allowing RNA polymerase to bind to the *lac* operon promoter and initiate transcription. LacI illustrates two of the extra complexities common to many repressors. First, the primary unit of DNA binding is as a dimer, with the individual LacI proteins forming in a tail to tail arrangement. This means that a tandem array of LacI binding sites is required, present as an inverted repeat of the single binding sequence. Second, LacI dimer binding sites exist in addition to that present overlapping the −10 sequence. Binding of LacI dimer at these sites, coupled with interaction between the dimers bound at the different sites to form a tetramer, bends the intervening DNA, providing further steric hindrance of RNA polymerase/promoter interactions (Kercher *et al.*, 1997).

Other factors that can influence the ability of repressors to bind to DNA are phosphorylation of the control region, interaction of another protein with the control region, or proteolytic degradation of the entire repressor. Proteolysis is generally carried out by an additional domain within the same protein. For example, in the SOS response of *E. coli*, the presence of DNA damage induces the auto-proteolytic cleavage of the transcriptional repressor LexA, which therefore means de-repression of the genes whose expression LexA regulates, and a stimulation of the production of a number of proteins whose role is to overcome the DNA damage (Little, 1991).

A final type of control event is the cascade type event, which I refer to as 'transcriptional networking'. If the expression of a gene encoding a transcriptional repressor is controlled by another transcriptional regulator, then any situation where transcription of the repressor is off will mean de-repression of transcription of the gene whose expression is regulated. A good example of this is the mar system in *E. coli*. The transcriptional repressor MarR regulates the expression of the *marA* gene, encoding another transcriptional regulator. Thus if MarR is active as a DNA binding protein, it will repress production of MarA, which will not be able to regulate production of target genes in a ligand-dependent manner. However, in the presence of the appropriate ligand, MarR will no longer be a repressor of *marA* transcription producing downstream effects on gene expression (see *Figure 1.1*). This sort of transcriptional networking event means that multiple layers of environmental control are integrated into a particular transcriptional response (Miller and Sulavik, 1996).

Transcriptional activators

Many weak promoters exist that cannot support sufficient RNA polymerase binding to give efficient levels of transcriptional initiation. The most common problem is a −35 box sequence bearing very little, if any, resemblance to the consensus sequence of the sigma subunit of the RNA polymerase holoenzyme. What is required here is a transcriptional activator protein whose role is to literally hold down the RNA polymerase complex onto the promoter, so that transcription can be initiated. Such a protein must be able to interact with DNA in a regulated manner, and also to interact with the RNA polymerase enzyme.

As with transcriptional repressors, there are many types of transcriptional activator. The classic activator is cAMP-responsive protein (CRP) which interacts with the lactose operon in *E. coli*. This protein binds to a recognition sequence which is centered around 62 nucleotides upstream of the transcriptional start site associated with the lactose operon promoter. The protein interacts with the C-terminal portion of the alpha subunit of RNA polymerase (α-CTD), thus strengthening RNA polymerase/promoter interactions and facilitating more efficient transcriptional initiation. Bacterial promoters that benefit from the interaction of a transcriptional activator positioned at −60 to −80 relative to a transcriptional start site with α-CTD are known as class I promoters. Association of CRP to the lactose operon promoter is controlled by the presence of the ligand, cAMP, which binds to the control domain of CRP. No cAMP, no protein/DNA interaction, and so no enhancement of promoter activity (Ebright, 1993).

Other activators may bind to recognition sequences that are positioned about 40 nucleotides upstream of the transcriptional start site. Promoters that work in this way are known as class II promoters. In these cases, the possibility exists that the activator will interact with either α-CTD alone, α-CTD and the N-terminus of the alpha subunit, α-CTD and the sigma subunit, or with the sigma subunit alone. Some activators can work at both class I and class II promoters, but only those that are able to interact with α-CTD alone or in combination with another part of the RNA polymerase holoenzyme (Browning *et al.*, 2002). It should be remembered that proteins which are predominantly transcriptional activators can also work as repressors at some promoters where the sequences that they are programmed to recognize and bind to happen to overlap with the −10 box. Indeed, there are examples of transcriptional regulators (e.g. the LysR class of regulators) that work as repressors in the absence of a particular control ligand and activators in the presence of the control ligand (or vice versa). This happens because in one situation (e.g. when the ligand is absent), the regulator assumes a conformation causing it to occlude RNA polymerase from binding to the promoter, repressing transcription. However, in the other situation (e.g. when the ligand is present and binds to the control region) the regulator assumes a different conformation, and now actively recruits RNA polymerase to bind to the promoter, thus switching on gene expression (Jacobs *et al.*, 1997).

As with repressors, transcriptional activators generally work as dimers, formed head to head, and so bind to inverted repeats of their recognition sequence of five to seven nucleotides, though some activators, particularly those from the two-component system response regulator class, work as head to tail dimers, and so bind to direct repeat sequences.

Non-protein-mediated regulation of transcription

There are a number of regulatory mechanisms that impinge on the rate of transcriptional elongation in bacteria. For example, the stringent response that, put simply, provides a readout of the physiological state of the cell in the form of the concentration of the molecule ppGpp (guanine tetraphosphate). If the levels of this compound rise, a general stress response occurs, which has many effects, not least to generally reduce the rate of transcription. It is thought that ppGpp interacts with the β′ subunit of RNA polymerase, possibly via the ω subunit where it has its desired effects (Finn *et al*., 2002). ppGpp also stimulates RNA polymerase to bind alternative sigma factors (see below).

Another method of regulating transcription indirectly is through changes in local supercoiling of the DNA being transcribed (Dorman, 2002). Supercoiling is a property of circular DNA molecules whereby the DNA strands are over wound (i.e. negatively supercoiled) or under wound (i.e. relaxed). A simple circle of DNA is the most relaxed form, but takes up a lot of space, so the introduction of negative supercoils makes the structure as a whole more dense, and so more able to fit into a bacterial cell. The introduction of negative supercoils is performed by a group of DNA topoisomerase enzymes, and the process requires the expenditure of energy. The DNA becomes increasingly wound up, like a spring and more and more negative supercoils are added, causing stresses and strains within the molecule. These stresses can have dramatic effects on transcription, because it is very difficult to form open complexes (i.e. melt DNA into a transcription bubble) if the molecule is very tightly wound up. On the other hand, a very stressed region can melt spontaneously, which will potentially increase transcriptional initiation rates. Hence the influence of supercoiling on transcription is undoubtedly complex, and is due to the presence of specific DNA sequences adjacent to the supercoiled-controlled promoters.

The introduction of negative supercoils into DNA is dependent upon ATP hydrolysis, so as ATP levels fall in a cell, its DNA becomes more relaxed. Other factors that affect supercoiling are osmolarity and oxygen potential. Osmotic stress and anaerobiosis induces negative supercoil formation. It is not certain how these external factors affect supercoiling levels, but it is very likely through control of the expression of topoisomerase genes, so that the ratio of enzymes that put in and remove negative supercoils is altered under some circumstances compared with others. Transcription itself causes an introduction of negative supercoils upstream from the transcription bubble opened up in a DNA molecule. This means that the rate of initiation of transcription of one gene can be affected by whether or not adjacent genes, and particularly those that are oppositely oriented on the genome, are being transcribed (Dorman, 2002).

Alternative sigma factors

As discussed, several different sigma factors can be produced by each bacterial cell, and each recognizes a different promoter sequence, facilitating the transcription of a different subset of genes whose promoters they recognize. Therefore, controlling the levels of the different sigma subunits is an excellent way of controlling the rates of transcriptional initiation for different groups of genes. The vast majority of genes are associated with promoters that are recognized by one, so-called housekeeping, sigma factor. The level of this sigma factor is kept fairly constant during most growth conditions.

However, under some stress conditions, the transcription of a gene encoding a so-called alternative sigma factor can be induced, which means that in addition to the genes whose promoters are recognized by the housekeeping sigma factor, genes whose promoters are recognized by the alternative sigma factor can be transcribed. Of course, just the fact that these genes can be expressed does not mean that they will be expressed, since expression may be controlled by a host of transcriptional regulators, as described above.

Sometimes, as well as causing the potential expression of a group of genes, invoking the production of an alternative sigma factor must be coupled with prevention of expression of genes whose promoters are recognized by other sigma factors. The most efficient way of doing this is to either sequester or degrade other sigma factors in the cell. This is the role of the anti-sigma factor, a protein encoded by a gene whose expression is dependent upon a particular alternative sigma factor. In this way stresses such as heat shock can be overcome, or the production of proteins can be achieved in a temporally controlled manner. An excellent example of the latter is control of the synthesis of the flagellum in various bacteria (Chilcott and Hughes, 2000). This extra-cellular appendage has a number of key components, each made up of a group of proteins, and it is not possible to make the second component until the first has been made, and so on. The key is alternative sigma factors. Upon the first signal, the production of the first alternative sigma factor is induced, and expression of genes encoding proteins that make up the first component of the flagellum begins. In addition, production of a second alternative sigma factor starts, since its gene has a promoter recognized by the first alternative sigma factor. However, production of the second sigma factor is slow, through other brakes on its transcription, and its concentration does not get above the required threshold until the first component of the flagellum has been made. At this point, the second alternative sigma factor stimulates transcription of genes encoding the second component of the flagellum, plus an anti-sigma factor that degrades the first alternative sigma factor, immediately stopping transcription of genes whose products make up the first flagellar component. Now, therefore, the cell concentrates its resources on producing the second flagellar component, plus a third alternative sigma factor that will eventually allow production of the third flagellar component, and an anti-sigma factor that will degrade the second alternative sigma factor and so prevent further production of the second flagellar component. The process repeats itself until the entire flagellum has been built.

Attenuation

Attenuation is a system of regulation specific to the transcription of operons encoding amino acid biosynthesis enzymes. The paradigm of this type of regulation is control of expression of the tryptophan biosynthesis operon (Yanofsky, 2000). Transcription of this operon is controlled in multiple ways. First, negatively by a transcriptional repressor, TrpR, which works as described above for LacI, but will only bind to DNA, and so repress transcription, if tryptophan is bound. If tryptophan is abundant, there is no need to make more tryptophan, meaning that expression of the tryptophan biosynthesis operon can be suppressed.

Attenuation represents a more subtle control of transcription of the *trp* operon, and requires the ability to sense the concentration of tRNAs charged with tryptophan, and so indirectly the concentration of tryptophan in the cell. Control is brought about by the ribosome. Immediately upstream of the first gene in the *trp* operon, *trpE*, but downstream of the promoter is the attenuator region. This comprises a gene encoding a very short protein, a so-called 'leader peptide' (which is rich in tryptophan amino acids) and a short non-coding sequence. The attenuator region contains several inverted repeat sequences meaning that when transcribed from the *trp* operon promoter into a single-stranded mRNA, the molecule has internal complementarity and will fold up due to hydrogen bond formation to produce secondary structure. One of the inverted repeats is a Rho-independent terminator region. In the presence of tryptophan, the existence of plenty of tryptophan tRNAs means that the leader peptide-encoding mRNA region of the attenuator can be translated efficiently. Movement of the ribosome along this region does not affect the secondary structure or the function of the Rho-independent transcriptional terminator region. So, in the presence of tryptophan, RNA polymerase will fall off the *trp* operon DNA sequence before it reaches *trpE*, blocking transcription of the operon. However, in the absence of tryptophan, there will be no free tryptophan tRNA, meaning that the ribosome will not be able to translate the leader peptide mRNA and the ribosome will stall, sitting on the mRNA. The presence of a stalled ribosome destabilizes the secondary structures formed in the attenuator mRNA, meaning that the Rho-independent terminator structure is lost. Thus, in the absence of tryptophan, a stalled ribosome on the attenuator mRNA allows transcription of the entire *trp* operon to proceed and tryptophan biosynthesis is initiated.

1.6 Control of transcription in eukaryotes

As discussed above, the whole process of transcriptional initiation in eukaryotes, and particularly for the transcription of protein-coding genes, is far more complex than it is in prokaryotes. However, as in prokaryotes, transcription of the vast majority of protein-coding genes is controlled by sequence-specific DNA-binding proteins. The added complication is that eukaryotic DNA is very tightly packed into chromatin, meaning that general TFs essential for the RNA polymerase holoenzyme assembly simply cannot access promoters, and so initiation of transcription is blocked. This is not the case to the same extent in bacteria, though it is becoming increasingly apparent that chromatin does affect gene expression even in bacterial cells.

Genes that are not expressed (most often because they are not transcribed) are said to be silent, and the majority of genes in a eukaryotic cell at any given time are silent. This means that there is a subtle difference in emphasis between prokaryotes and eukaryotes, from control of transcription predominantly by DNA-binding proteins that are repressors, which is very common in bacteria, to control of gene expression by DNA binding proteins that are transcriptional activators, which predominates in eukaryotes. Transcriptional activation (and to a certain extent, repression) in eukaryotes can be facilitated in many different ways, and is a hugely complex field of study. Indeed, virtually all genes are subjected to multiple layers of transcriptional control. Here, we only have space to refer to the general principles that apply, without being able to list detailed examples of each.

Transcriptional regulation due to control of chromatin structure

As discussed above, silencing eukaryotic coding genes normally occurs because the genes, together with their promoters, are tightly packed with histone proteins into chromatin, preventing interaction of general TF complexes with promoters, and so the recruitment of RNA polymerase II to initiate transcription. Many transcriptional activators are general activators of the expression of a group of genes clustered together in one region of a chromosome. Such activators often exert their effects by binding to a specific DNA sequence and recruiting one of two histone modifying enzyme complexes (or occasionally both). The first is the histone remodeling complex, which uses ATP hydrolysis to essentially push apart histone proteins; the second is the histone acetyl-transferase, which transfers acetyl groups onto the histone proteins, making them change shape. The result of both of these processes is that a region of chromatin becomes less condensed. Indeed, the region can be large, since once the histone-modifying enzyme complexes have been recruited, they can fan out from the transcriptional activator binding site, modifying histones as they go. It is likely that this process of opening up the chromatin complex increases the ability of general TF complexes to interact with adjacent promoters, and so to form a focus for RNA polymerase binding and transcriptional initiation (*Figure 1.13*).

Locus control regions (Fraser and Grosveld, 1998) are DNA sequences associated with general activation of gene expression in a particular region of a chromosome. Many chromosomal regions are associated with multiple locus control regions, and the expression level of a gene is generally directly associated with the number of locus control regions linked to it. The method by which locus control regions work is well understood, but is believed to be through binding of a protein to the locus control region, which recruits histone acetyl transferase, thereby reducing chromatin density, and generally activating transcription in the vicinity. Control of this type of activatory mechanism must come though controlling the expression or activity of the locus control region binding protein. Since such regions are commonly used to switch on groups of genes in a tissue-specific manner, it is thought that the former is more likely, which highlights the multi-layered 'transcriptional networking' seen in eukaryotic cells, as well as to a lesser extent in prokaryotes.

Since local changes in chromatin structure can exert their transcriptional activatory effects over large areas of DNA, it is important to have mechanisms which specifically block the effects of locus control regions, and so the transcription of some genes that happen to be linked to them, but that do not need to always benefit from enhanced transcription. One such negative regulator is the so-called matrix attachment region (Hart and

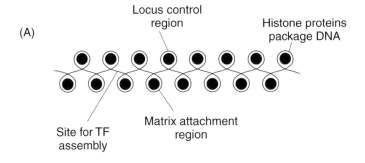

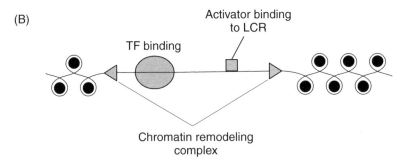

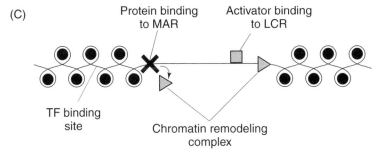

Figure 1.13

Control of chromatin structure in eukaryotes. In (A) the packaging of DNA into chromatin is illustrated, occluding a promoter, and preventing transcription factor binding. In (B) a protein has bound to a locus control region (LCR), and recruited histone remodeling enzymes (to form a chromatin remodeling complex), which has fanned out and freed up the TF binding site, thus allowing transcription to be initiated. In (C) a protein has bound to the matrix attachment region (MAR), and has blocked passage of the chromatin remodeling complex, thereby isolating the particular promoter illustrated from the effects of this form of transcriptional activation.

Laemmli, 1998). These represent DNA sequences to which a protein binds. Binding is believed to block the progression of histone acetylation from, for example, a locus control region, putting up a barrier against enhancement of transcription from the point of the matrix attachment region and beyond. The way to produce an 'island' gene that is silent in a 'sea' of modified chromatin is to have one distant locus control region and one adjacent matrix attachment region either side of the gene to remain silent (*Figure 1.13*). Clearly the barrier formed at a matrix attachment region is dependent upon the expression and activity of protein/s that bind to it, and so can be tightly regulated.

Transcriptional activators that bind a long way from promoters

Enhancers are DNA sequences found often at sites remarkably distant (>1000 bp) from the core RNA polymerase II promoter, which was discussed in Section 1.2. Each promoter can be affected by a whole host of enhancers, thus providing an extra level of complexity of control not possible in prokaryotes. It is not certain how they work to increase the rate of transcriptional initiation, and so to stimulate gene expression. Indeed, it is likely that different enhancers exert their effects in different ways (Bulger and Groudine, 1999). One possibility is that they provide a binding site for RNA polymerase II, which can then move along the DNA and so more efficiently find its promoter, to which the large transcription factor complex centered around TFIID is positioned. This seems unlikely, however, since it is equally probable that RNA polymerase will stumble across the promoter/TF complex directly, as that it will encounter the enhancer sequence. Another more likely possibility is that enhancers are binding sites for protons that recruit histone remodeling complex, thereby reducing the density of chromatin in the region between the promoter and the enhancer, opening up the DNA sequence, and making promoter/TF interaction, and so RNA polymerase/TF interaction more likely. In this way they may work like mini-locus control regions. Another mechanism for enhancer-mediated activation of transcriptional initiation, is that the enhancer provides a binding site for proteins that bend DNA, plus binding sites for transcriptional activators, such that the DNA sequence between the enhancer and the promoter loops over, and the transcriptional activators bound at the enhancer come into close proximity with the TF/RNA polymerase complex at the promoter, thus stimulating assembly of the TF complex at the promoter and/or recruitment of RNA polymerase II to the TF/promoter complex. This is similar in many ways to the actions of a rare group of transcriptional activators of prokaryotic genes, which also have distant upstream transcriptional activator binding sites and require DNA bending to bring the transcriptional activator and the RNA polymerase into close proximity.

Because enhancers can work over large distances, each has the potential to activate the transcription of many genes. Thus for general transcriptional activation due to histone acetylation, as for locus control regions, above, there needs to be the possibility of an 'opt-out' for some genes that do not need to be transcriptionally enhanced along with their neighbors. The classic mechanism for this is the presence of an insulator DNA sequence between the enhancer and the gene to remain silent (Geyer and Clark, 2002). The binding of a protein to the specific insulator can block histone remodeling complex from passing by, thus limiting the effects of the enhancer on genes distal to the insulator. Alternatively, if the enhancer works by DNA bending and protein/protein interactions, then the particular insulator might attract a protein that prevents bending of the intervening DNA, and interaction of the protein bound to the enhancer with the RNA polymerase/TF/promoter complex, sometimes through a second protein called a 'mediator'. The possible effects of enhancers on transcription initiation rates through DNA bending and protein/protein interactions are set out in *Figure 1.14*.

Local control of transcriptional initiation

Since RNA polymerase II cannot locate its promoter without a large complex of general TFs getting there first, the assembly of this complex at

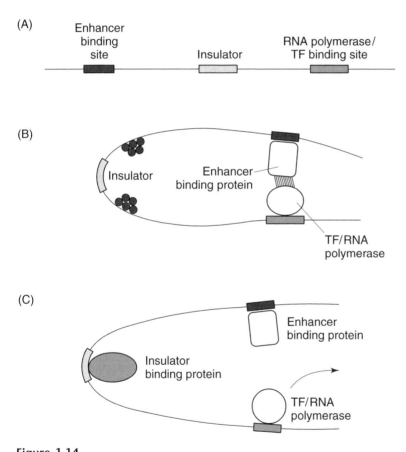

Figure 1.14

Control of transcriptional initiation at a distance. In (B) an enhancer binding protein has bound to its target, and, probably aided by other proteins that bind and bend the DNA (represented as circles), has become associated with the RNA polymerase/transcription factor complex (often through a mediator protein, not illustrated). This stabilizes the RNA polymerase/TF/promoter complex, and makes transcriptional initiation more likely. In (C) binding of a protein to an insulator has prevented the DNA bending sufficiently so that the enhancer binding protein and the RNA polymerase can interact. In this case, transcriptional initiation is less likely to occur than is the case in (B).

the promoter is a possible point of transcriptional control. In exactly the same way that some enhancer-binding proteins can activate transcription, locally bound transcriptional activator proteins can interact with the RNA polymerase II through a so-called mediator protein, which binds to the large C-terminal domain of one of the RNA polymerase II subunits. This stimulates the rate of transcription by putting the RNA polymerase II enzyme in close proximity with the promoter, and so its associated general TF complex, making TF/RNA polymerase/promoter complex formation more likely to occur, and increasing the rate of transcriptional initiation. In many ways, this mechanism resembles transcription activation in prokaryotic cells. Similarly, repressor binding proteins can associate with specific DNA sequences overlapping the promoter, and can therefore prevent TF/promoter interaction, and so transcriptional initiation. Furthermore, transcriptional repressors can bind away from the promoter, but can interfere with transcriptional initiation through binding to the TF complex, inhibiting it from binding to RNA polymerase. Alternatively, repressors can work by blocking the binding of transcriptional activators either to the activator DNA binding site (i.e. by binding site occlusion) or to the TF/RNA polymerase complex through blocking the TF/RNA polymerase binding domain on the activator. Finally, transcriptional repressors can work by binding to sites close to promoters and recruiting enzymes that cause chromatin condensation. These would include histone deacetylase (Brivanlou and Darnell, 2002).

In addition to repression of transcription being caused by repressor DNA-binding proteins, transcriptional initiation can be inhibited through blocking the binding of TF/RNA polymerase complex, and/or transcriptional activators upstream of a gene due to methylation of GC-rich regions known as CpG islands (Bird, 2002). The means by which control of CpG methylation is exerted is not understood, but it is typically a mechanism that switches off large regions of chromosomes, even whole chromosomes, rather than regulating small numbers of genes. It is likely that methylation of DNA interferes with TF/promoter interactions, and/or open complex formation by RNA polymerase. CpG methylation is very stable, and the template DNA strand can remain methylated following replication and cell division, hence this type of transcriptional control can be 'inherited' by a daughter cell, and is often known as 'imprinting', or more commonly nowadays, 'epigenetic control of gene expression' (Bird, 2002).

Control of transcriptional elongation

Control of transcription in eukaryotes is primarily at the level of the initiation phase, but at least some control of transcriptional elongation rate is exerted over large numbers of, if not all, genes. The processivity of the RNA polymerase enzyme along a DNA template is dependent upon contact of the enzyme with the DNA at three sites. The RNA polymerase keeps strong contact with the DNA immediately 5' proximal to the transcription bubble, relatively weak contact with the RNA/DNA hybrid immediately 5' proximal to the point at which RNA synthesis is being catalyzed, and an intermediate strength of contact with the nascent RNA strand in a more 5' proximal position than the site of RNA/DNA contact. These contacts allow enough stability so that the RNA polymerase does not simply fall off the template, but not so much stability that it cannot process down the template and synthesize RNA, because the interactions must be broken and remade further downstream.

The secondary structures formed in the transcribed attenuator region of the tryptophan biosynthesis operon (above) affect the interaction of RNA polymerase with the template and nascent RNA molecules and lead to RNA polymerase stalling, and ultimately to the enzyme falling off the template. Pausing of RNA polymerase in eukaryotes is thought to be induced by very strong interactions between the polymerase and the template 5′ proximally to the transcription bubble and so-called 'pause sites', which have evolved to become specific DNA sequences to which RNA polymerase unusually binds strongly. The end result of pausing is at least a reduction in the rate of elongation of transcription of a gene, and at worst, premature transcriptional termination. To control this pausing event, and so the rate of transcriptional elongation, proteins exist that, when activated to bind to RNA polymerase, reduce the strength of the polymerase/pause site interaction, thus preventing pausing and speeding up the rate of transcriptional elongation. Furthermore, the rate of RNA synthesis catalyzed by RNA polymerase II can be increased by phosphorylation of the extended C-terminal arm of the polymerase in response to general external signals (Uptain *et al.*, 1997).

1.7   Post-transcriptional control of gene expression

The steady-state level of an mRNA in a bacterium depends upon the rate of its synthesis by RNA polymerase and the rate of its degradation by RNase enzymes. In eukaryotes, the level of each mRNA in the cytoplasm depends on the net rate of hnRNA synthesis, hnRNA to mRNA processing and nuclear export and on the rate of mRNA degradation by RNases. Control of the rate of hnRNA (mRNA in prokaryotes) synthesis is dealt with under control of transcriptional initiation and elongation (Section 1.6); control of hnRNA processing in eukaryotes and export of mRNA from the nucleus is dealt with below. It is clear, in addition, that different mRNA molecules have different rates of digestion by nucleases once synthesized. This is due primarily to the sequences of the RNA molecules, and the resultant structural implications of those sequences, causing differential occlusion of potential nuclease recognition and cleavage sites. Furthermore, and particularly in eukaryotic cells, proteins exist that can alter RNA stability in a regulatable manner, including those that actively shorten the polyadenine tail on eukaryotic mRNAs, thus reducing stability (see below). RNAs are targeted for removal of polyadenine tails because of specific sequence elements found within the mRNA coding sequence, and the sorts of mRNAs affected in this way are those encoding proteins that are only needed for very brief periods during the cell cycle (Jacobson and Peltz, 1996).

Control of hnRNA processing and the nuclear export of mRNA

Given the complexity of the splicing and nuclear export processes, it is not surprising to find that they represent targets for control mechanisms that regulate gene expression. However, control of splicing is not really designed to control gene expression rates (i.e. production of the hnRNA transcript of the gene) but to affect the nature of the mRNA(s) being produced. Since different splicing forms of the same hnRNA can produce related, but different, mRNAs, encoding related but different protein products, the control of this process will clearly affect the transcriptome and, more obviously, proteome of the cell. For example, the human genome contains around 30 000 genes, but these encode around 100 000 proteins. The amplification is achieved by alternative splicing (Graveley, 2001). The control of alternative splicing is likely to be very complex, since there is often tissue dependency of the splice form of mRNA produced. It will almost certainly involve proteins that recognize specific sequences associated with different splice sites, and either promote, or repress, splicosome formation at some of them.

Control of other forms of RNA processing, such as capping and polyadenylation is generally dependent upon the rate of transcription of a gene, since the processes are linked. There is, however, the potential to alter the activities of the enzymes involved in these processes, both generally, affecting the overall processing levels of the transcriptome, or more often specifically affecting the level of polyadenylation that occurs following the termination of transcription of some messages but not others. Furthermore, it is known that enzymes, which remove adenine moieties from polyadenine tails on mRNAs can be controlled to specifically target only some mRNAs. This leads to particularly variable levels of polyadenylation of some mRNAs, dependent upon particular external signals. A reduction in polyadenine tail length can reduce the level of mRNA stability because the polyadenine tail protects an mRNA from exonuclease attack. In addition, because the polyadenine tail physically links the sites of translational termination and initiation, the length of the polyadenine tail can affect the frequency of translational re-initiation following termination, ultimately affecting the amount of protein produced in a given period of time.

Once processed, mRNAs need to be exported from the nucleus. This process involves a large number of different proteins, and because of this it is not surprising to find that their expression levels, and typically their levels of phosphorylation, can affect the rate at which they export mRNAs, and so ultimately the rate of translation.

Control of translational initiation

As described in Section 1.4, there are two main critical points in the initiation of protein synthesis in eukaryotic cells. The first is the formation of the cap-binding complex, thus facilitating formation of the eIF4F translational initiation complex, which is essential for tracking the small ribosomal subunit along the untranslated region of an mRNA until it encounters the AUG translational initiation codon. The second critical process is the loading of methionine-charged tRNA onto the small ribosomal subunit, which occurs before the ribosome associates with the eIF4F translational initiation complex. Both these processes are under tight regulatory control.

Control of cap-binding complex formation, and so eIF4F complex formation is facilitated through small inhibitory binding proteins that interact with the cap-binding protein, eIF4E, and block its association with eIF4G, and so formation of the cap-binding complex. Whether or not these eIF4E binding proteins (BPs) bind to eIF4E, and so inhibit translational initiation depends upon their phosphorylation state, which in turn depends upon the activation of cell signaling cascades in response to certain environmental and hormonal signals. Once phosphorylated, eIF4E BPs dissociate from eIF4E, and eIF4G can associate in its place, triggering the first stage in translational initiation (Dever, 1999).

Methionine-tRNA loading of the ribosome is dependent upon the protein eIF2 (see Section 1.4). The catalysis of loading is driven by hydrolysis of GTP bound to eIF2 to form GDP, which remains bound. Thus, once eIF2 has catalyzed one round of methionine-tRNA loading it becomes inactive and, if all molecules of eIF2 in a cell are present in the GDP bound form, translational initiation will stop. Exchange of GDP for GTP on eIF2 is catalyzed by another protein, eIF2B, the guanine nucleotide exchange factor. Thus control of the methionine loading rate, and so translational initiation rate, can be exerted through controlling the activity of eIF2B. As with eIF4E BPs, control of eIF2B activity is by differential phosphorylation of certain amino acids within the protein; some phosphorylation events inactivate the protein, some activate it. The pathways responsible for activating eIF2B by phosphorylation overlap with those that result in the phosphorylation of eIF4E BPs, providing a concerted signaling cascade that activates translational initiation at multiple sites (Dever, 1999). These key points of translational initiation control are illustrated in *Figure 1.15*.

Control of translational elongation

As with control of transcriptional elongation, control of translational elongation is facilitated by activating or inhibiting the catalytic activity of the machinery involved. However, as described in Section 1.4, the bulk of catalysis in protein synthesis is actually the formation of peptide bonds, which is mediated by a reaction center buried deep within the ribosome, and is thus fairly inaccessible to control by external factors, e.g. through phosphorylation. So, control of elongation rate is exerted through control of eEF2 activity, which is essential for movement of the ribosome along the mRNA. Since the kinetics of peptide-bond formation are far from rate limiting, the rate of translation is dependent upon the rate at which the ribosome moves along the mRNA. Control is exerted primarily through phosphorylation of eEF2 by a dedicated kinase in response to various external stimuli, which makes the protein more active and so speeds up the ribosomal translocation rate. There is some overlap between the signaling events that lead to phosphorylation of eEF2 and activation of eIF2B, but there are differences, which allows some external signals to predominantly affect initiation rate, and others to predominantly affect elongation rate (Browne and Proud, 2002).

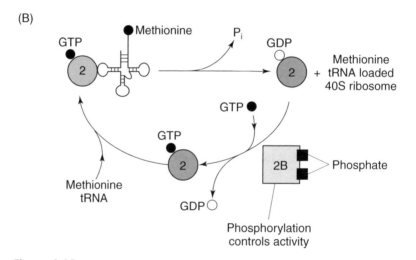

(A)

4E BP

4E

m⁷G

No assembly of initiation complex

4G

4E

m⁷G

Assembly of initiation complex

Phosphorylated 4E BP

(B)

Methionine

GTP

2

P_i

GDP

2

Methionine
+ tRNA loaded
40S ribosome

GTP

GTP

2

2B

Phosphate

Methionine
tRNA

GDP

Phosphorylation
controls activity

Figure 1.15

Control of translational initiation in eukaryotes. Two main areas of control are illustrated. In (A) the interaction of eIF4E with eIF4G, which is essential for formation of a translational initiation complex, is blocked by the stronger association of eIF4E with a small binding protein, a 4EBP. Different groups of tissues produce different 4EBPs. Phosphorylation of 4EBPs, through a variety of signal transduction pathways, results in its dissociation from eIF4E, and a translational initiation complex can form. In (B) control is exerted through recycling of eIF2, which delivers charged initiator tRNAs to the ribosome. Recycling involved removal of GTP, and replacement with GTP, and this process requires eIF2B, whose catalytic activity is controlled by phosphorylation, both positively and negatively dependent upon the site of phosphorylation, and so the particular signal transduction pathways that are activated in the cell.

Further reading

Alberts, B, Johnson, A, Lewis, J, Raff, M, Roberts, K and Walter, P (2002) *Molecular Biology of the Cell*, 4th Edn. Garland Science, New York, NY.
Lewin, B (2000) *Genes VII*. Oxford University Press, Oxford.
Prescott, LM, Harley, JP and Klein, DA (2002) *Microbiology*, 5th Edn. McGraw-Hill, New York, NY.

References

Anderson, J and Seilhamer, J (1997) A comparison of selected mRNA and protein abundances in human liver. *Electrophoresis* 18: 533–537.
Avison, MB and Bennett, PM (2005) Bacterial Genetics. In: *Topley & Wilson's Microbiology and Microbial Infection – Bacteriology*, 10th Edn (eds S.P. Borrellio, P.R. Murray and G. Funke). Hodder Arnold, London, pp. 80–135.
Bird, A (2002). DNA methylation patterns and epigenetic memory. *Genes Dev* 16: 6–21.
Brivanlou, AH and Darnell, JE (2002) Signal transduction and control of gene expression. *Science* 295: 813–818.
Browne, GJ and Proud, CG (2002) Regulation of peptide chain elongation in mammalian cells. *Eur J Biochem* 269: 2360–2368.
Browning, D, Lee, D, Green, J and Busby, S (2002) Secrets of bacterial transcription initiation taught by the *Escherichia coli* FNR protein. In: *Signals, Switches, Regulons and Cascades: Control of Bacterial Gene Expression* (eds D.A. Hodgeson and C.M. Thomas). Cambridge University Press, Cambridge, pp. 127–142.
Bulger, M and Groudine, M (1999) Looping versus linking: toward a model for long-distance gene activation. *Genes Dev* 13: 2465–2477.
Chilcott, GS and Hughes, KT (2000) Coupling of flagellar gene expression to flagellar assembly in *Salmonella enterica* serovar typhimurium and *Escherichia coli*. *Microbiol Mol Biol Rev* 64: 694–708.
Choo, Y and Klug, A (1997) Physical basis of a protein–DNA recognition code. *Curr Opin Struct Biol* 7: 117–125.
Conaway, JW, Shilatifard, A, Dvir, A and Conaway, RC (2000) Control of elongation by RNA polymerase II. *Trends Biochem Sci* 25: 375–380.
Crow, JF (2000) The origins, patterns and implications of human spontaneous mutation. *Nat Rev Genet* 1: 40–47.
Daneholt, B (1997) A look at messenger RNP moving through the nuclear pore. *Cell* 88: 585–588.
Das, A (1993) The control of transcriptional termination by RNA binding protiens. *Annu Rev Biochem* 62: 893–930.
Dever, TE (1999) Translation initiation: adept at adapting. *Trends Biochem Sci* 24: 398–403.
Dorman, CJ (2002) DNA topology and regulation of bacterial gene expression. In: *Signals, Switches, Regulons and Cascades: Control of Bacterial Gene Expression* (eds D.A. Hodgeson and C.M. Thomas). Cambridge University Press, Cambridge, pp. 41–56.
Ebright, RH (1993) Transcriptional activation at class I CAP-dependent promoters. *Mol Microbiol* 8: 797–802.
Ebright, RH (2000) RNA polymerase: structural similarities between bacterial RNA polymerase and eukaryotic RNA polymerase II. *J Mol Biol* 304: 687–698.
Finn, RD, Orlova, EV, van Heel, M and Buck, M (2002) Structures of multisubunit DNA-dependent RNA polymerases. In: *Signals, Switches, Regulons and Cascades: Control of Bacterial Gene Expression* (eds D.A. Hodgeson and C.M. Thomas). Cambridge University Press, Cambridge, pp. 73-103.

Frank, J (2000) The ribosome – a macromolecular machine par excellence. *Chem Biol* 7: 133–141.

Fraser, P and Grosveld, F (1998) Locus control regions, chromatin activation and transcription. *Curr Opin Cell Biol* 10: 361–365.

Geiduschek, EP and Kassavetis, GA (2001) The RNA polymerase III transcription apparatus. *J Mol Biol* 310: 1–26.

Gelles, J and Landick, R (1998) RNA polymerase as a molecular motor. *Cell* 93: 13–16.

Geyer, PK and Clark, I (2002) Protecting against promiscuity: the regulatory role of insulators. *Cell Mol Life Sci.* 59: 2112–27.

Graveley, BR (2001) Alternative splicing: increasing diversity in the proteomic world. *Trends Genet* 17: 100–107.

Green, MR (2000) TBP-associated factors (TAFIIs): multiple, selective transcriptional mediators in common complexes. *Trends Biochem Sci* 25: 59–63.

Green, R (2000) Ribosomal translocation: EF-G turns the crank. *Curr Biol* 10: 369–373.

Green, R and Noller, HF (1997) Ribosomes and translation. *Ann Rev Biochem* 66: 679–716.

Halford, SE and Marko, JF (2004) How do site-specific DNA binding proteins find their targets? *Nucleic Acids Res* 32: 3040–3052.

Harrison, SC (1991) A structural taxonomy of DNA-binding domains. *Nature* 353: 715–719.

Hart, CM and Laemmli, UK (1998) Facilitation of chromatin dynamics by SARs. *Curr Opin Genet Dev* 8: 519–525.

Jacobs, C, Frere, JM and Normark, S (1997) Cytosolic intermediates for cell wall biosynthesis and degradation control inducible beta-lactam resistance in Gram-negative bacteria. *Cell* 88: 823–832.

Jacobson, A and Peltz, SW (1996) Interrelationship between the pathways of mRNA decay and translation in eukaryotic cells. *Ann Rev Biochem* 65: 693–739.

Kercher, MA, Lu, P and Lewis, M (1997) Lac repressor–operator complex. *Curr Opin Struct Biol* 7: 76–85.

Lee, TI and Young, RA (1998) Regulation of gene expression by TBP-associated proteins. *Genes Dev* 12: 1398–1408.

Lee, TI and Young, RA (2000) Transcription of eukaryotic protein-coding genes. *Annu Rev Genet* 34: 77–137.

Lewis, JD and Tollervey, D (2000) Like attracts like: getting RNA processing together in the nucleus. *Science* 288: 1385–1389.

Little JW (1991) Mechanism of specific LexA cleavage: autodigestion and the role of the RecA co-protease. *Biochimie* 73: 411–421.

Miller, PF and Sulavik, MC (1996) Overlaps and parallels in the regulation of intrinsic multiple-antibiotic resistance in *Escherichia coli*. *Mol Microbiol* 21: 441–448.

Minvielle-Sebastia, L and Keller, W (1999) mRNA polyadenylation and its coupling to other RNA processing reactions and to transcription. *Curr Opin Cell Biol* 11: 352–357.

Paule, MR and White, RJ (2000) Survey and summary: transcription by RNA polymerases I and III. *Nucleic Acids Res* 28: 1283–1298.

Sachs, AB and Varani, G (2000) Eukaryotic translation initiation: there are at least two sides to every story. *Nat Struct Biol* 7: 356–361.

Shatkin, AJ and Manley, JL (2000) The ends of the affair: capping and polyadenylation. *Nat Struct Biol* 7: 838–842.

Uptain, SM, Kane, CM and Chamberlin, MJ (1997) Basic mechanism of transcript elongation and its regulation. *Ann Rev Biochem* 66: 117–172.

von Hippel, PH (1998) An integrated model of the transcription complex in elongation, termination and editing. *Science* 281: 660–665.

Wilson, KS and Noller, HF (1998) Molecular movement inside the translational engine. *Cell* 92: 337–349.

Yanofsky, C (2000) Transcription attenuation: once viewed as a novel regulatory strategy. *J Bacteriol* 182: 1–8.

Isolation and analysis of RNA

2

2.1 Introduction

If you want to study the control of gene expression, at some stage you will have to measure transcript levels. Much can happen to a transcript before it is translated into an easily assayable reporter enzyme (Chapter 6), and so to get a true picture of the dynamics of transcription, RNA levels are the place to look. The problem is, RNA molecules are not designed to hang around long enough for scientists to sit and count them. They facilitate rapid responses to external stimuli; they are made and degraded rapidly, and in many cases this property confounds our attempts to catch a fleeting glimpse. The only way to work with transcripts is to put them into a sort of suspended animation. Using various chemicals, one can purify and stabilize RNA molecules long enough to quantify them. However, it should be said from the start that this is an artificial situation, and the fact that different RNA molecules respond to stabilization in different ways is the primary reason why measuring transcript levels can, in some experiments, be a matter of general trends rather than absolute numerical accuracy.

The first aim of this chapter is to introduce the reader to the properties and functions of the different types of RNA molecules found in prokaryotic and eukaryotic cells. Following this, a significant proportion of the chapter is devoted to outlining methods designed for purifying RNA from these cells, since the successful completion of this process is the cornerstone of all the transcript quantification techniques described in later chapters. The final part of this chapter sets out ways of testing the quality of your purified RNA. To get RNA purification right will go a long way to making your gene expression studies reproducible, meaning less worry and more productivity for you. If you are only able to invest significant amounts time, effort and money into perfecting one area of your gene expression analysis, make it this area; this is where the true added value lies.

2.2 The properties of different types of RNA

It is likely that RNA represented one of the first self-replicating molecules, paving the way for life as we know it today (Bartel and Unrau, 1999). RNA is a single-stranded string of the ribonucleotides adenine, cytosine, guanine and uracil (A, C, G and U) connected by phosphodiester bonds along a backbone, and with bases protruding into solution. It is easy to imagine how an RNA molecule sitting in solution could attract single nucleotides to hydrogen bond opposite its bases using the chemically programmed C≡G and A=U pattern of building blocks. However, to make this replicative (i.e. to copy the original template strand) phosphodiester bonds must be formed between the newly attached nucleotides. Such bonds can occur spontaneously, but such is the energetic barrier that it would have been implausible for the complex arrangement of RNA strand and hydrogen-bonded ribonucleotides to sit around long enough for chemistry to happen by chance at each round of replication. What was needed was a catalyst.

It is now apparent that RNA molecules can themselves be catalysts, as are protein molecules. In order to get specificity in terms of electron donation and receipt (i.e. to interact with specific molecules having specific shapes) and to perform specific chemical reactions, a stable structure is needed. Proteins fold into their structures due to various complex interactions between side chains of different amino acids drawing together distant sites within the protein. To a lesser extent, this is something that can occur with RNA molecules, since intra-molecular hydrogen bonding between bases as C≡G and A=U (and sometimes G=U) pairs can pull distant sites within the molecule together to form a secondary structure. Many of these secondary structure conformations are highly stable, and a plethora of possible structures is available dependent upon the specific sequence of ribonucleotides in the molecule. Many such structures have enzyme activity. So-called 'ribozymes' are known that have a variety of functions, including those that can ligate phosphodiester backbones together. If such a ribozyme 'evolved' due to the random deposition of bases, it would be able to catalyze the 'replication' of other RNA molecules.

To a certain extent we are still living in the RNA world. Many structural RNAs exist, whose roles are to bring together other molecules. Their roles could easily be performed by proteins, and yet they are good enough at their jobs for them not to have been replaced during evolution.

Ribosomal RNAs

It has long been known that structured RNA molecules make up part of the protein synthesis machinery, though only in recent years has their role been shown to be any more than just architectural. It is now believed that as well as forming cores for ribosomal protein assembly, ribosomal RNA (rRNA) molecules interact with mRNAs that are to be translated, and tRNAs that carry amino acids. Clearly, whilst proteins could evolve to interact with the bases of RNA, nucleic acids have ready built hydrogen-bond potential and orientation, and existed long before proteins, so there was no need for proteins to perform this function. This is also the case with the involvement of tRNAs in mRNA decoding during protein synthesis.

Ribosomal RNAs are highly structured, comprise more than 80% of the total RNA in a cell, and are the only RNAs that form discrete bands following electrophoresis and staining of a total preparation of cellular RNA (see Section 2.10). In prokaryotic cells there are three rRNAs, each encoded by a discrete gene (often these genes are in multiple copies): 5S (approximately 120 nucleotides), 16S (approximately 1540 nucleotides) and 23S (approximately 2900 nucleotides). The 16S rRNA forms part of the 30S ribosomal subunit, and the other two molecules form part of the 50S ribosomal subunit. In eukaryotic cells, there are four rRNAs: 5S (approximately 120 nucleotides), 5.8S (approximately 160 nucleotides) and 28S (approximately 4700 nucleotides) forming part of the large, 60S ribosomal subunit and the 18S rRNA (approximately 1700 nucleotides) forming part of the small, 40S ribosomal subunit.

Coding RNAs

Messenger RNA (mRNA) accounts for around 5% of all RNA in a cell. Molecules vary in size from a few hundred to several thousand nucleotides. mRNA molecules are very unstable relative to structured RNAs, because exposed, single-stranded sections are more efficient targets for nuclease attack than are double-stranded, folded RNA sequences. It would be wrong to think, however, that mRNA has no structure. Due to the same forces that cause structural RNAs to fold, mRNAs have the potential to fold, if the appropriate sequence complementarities between different parts of the molecule exist. Unlike structural RNAs, however, the sequences of mRNA encoding genes have not evolved to maximize this complementarity. Hence some mRNAs are more stable than others, simply because of the levels of secondary structure formed.

In prokaryotes, one mRNA molecule can encode more than one protein, and mRNAs are translated almost as soon as they are produced by RNA polymerase. This is possible because prokaryotes do not have a physical separation between the transcription and translation machinery. In eukaryotes, transcription of protein encoding genes takes place in the nucleus, and translation happens in the cytoplasm (see Sections 1.3 and 1.4 for details of these processes). Export of mRNA from the nucleus involves association of specific proteins with nascent mRNAs to form so-called ribonucleoprotein (RNP) complexes. The presence of mRNAs within RNP complexes makes the mRNAs more stable, and is one reason why eukaryotic mRNAs are more stable than those from prokaryotes (which have half-lives of <5 min). The main reason for this differential stability, however, is the modification of coding RNA molecules in eukaryotic cells. The 5′ end is modified by the addition of an inverted 7-methylguanine residue, and the 3′ end is modified by the addition of a polyadenine tail. As well as assisting in mRNA export from the nucleus and in the initiation of translation, both help to fortify mRNA molecules against exonuclease attack, and neither are found associated with prokaryotic mRNAs (see Section 1.3).

In eukaryotes, coding RNAs are made as a precursor transcript, which includes both intronic and exonic sequences. This precursor, known as heteronuclear RNA (hnRNA) is processed to mature mRNA via the splicing process (see Section 1.3). The processing of hnRNA into mRNA in the nucleus, and the export of the mature mRNA into the cytoplasm are both important points for the regulation of gene expression in eukaryotes. Therefore, to experimentally isolate transcriptional control from other mechanisms of controlling mRNA production, it is essential to measure nuclear hnRNA levels, since these are controlled solely at the level of transcription. This means separating nuclei from the cytoplasmic contents of a cell, leaving the nuclei intact and their hnRNA constituents in place. The cytoplasmic RNA pool is then purified from the cytoplasmic contents, and the hnRNA pool is purified from the purified nuclei which are first lysed.

Small, structured RNAs

There are a large number of small RNA molecules having different sequences in eukaryotic cells, and to a lesser extent in prokaryotes. These are all encoded in DNA, and transcribed by RNA polymerase variants in the same way that protein coding genes are encoded. The most well known are the various tRNA molecules, which collectively form a pivot between mRNA and protein synthesis. Whilst the general structure of all tRNAs, the famous clover leaf, is the same, subtle sequence differences allow different molecules to be charged with different amino acids, and cause them to interact with different mRNA codons using complementary base pairing. tRNAs account for about 10% of total cellular RNA.

In eukaryotic cells, small nuclear RNAs (snRNAs) are produced, and account for < 1% of cellular RNA. They range in size, but are all < 100 nucleotides and have numerous roles, though almost exclusively they bring together RNA molecules, or parts thereof, using hydrogen bonding in a complementary base-pairing manner. The best studied of these processes is in hnRNA splicing, where they form a pivotal part of the splicosome, and bring together sequences that flank introns within hnRNA, so that they can be joined, and the intronic sequence removed (see Section 1.3).

2.3 Purification of RNA: an introduction

A flow diagram, illustrating the basic steps of RNA isolation and purification is set out in *Figure 2.1*. When working with nucleic acids, the two main enemies of the embattled biologist are mechanical shearing and nuclease enzymes. Both turn your nucleic acid preparations into a pool of nucleotide mush. When working with RNA, shearing is much less of a problem, and nucleases are much more of a problem than when working with genomic DNA. Shearing forces act in a length-dependent manner; RNA molecules are generally shorter than genomic DNA molecules. It is the RNase, therefore, that will ruin your experiments and send you home cursing. They are everywhere: in the cell of course (that's what gives RNA molecules such a short half-life), but also all over you. In an ancient first line of defense against microbial attack, nucleases are secreted in your sweat. RNases are very stable, and are one of the last enzyme activities to degrade when a cell dies, thus since dust is mainly made up of dead cells, it is a very rich source of RNases. From these sources, they get into reagents, onto equipment, into plastic-ware, and ultimately, into your RNA preparation where they wreak havoc.

Before you start

The first thing to think about before working with RNA is 'what are the possible sources of RNase contamination', and then you need to decide how to minimize the risks. Of course, it's not possible to cover all potential

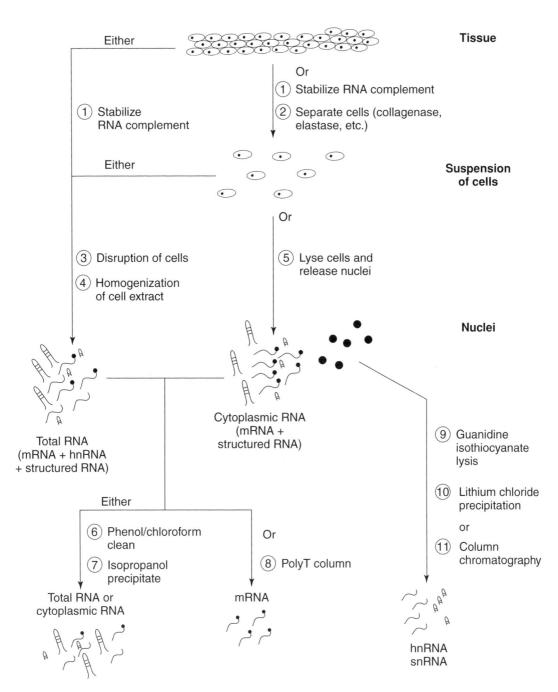

Figure 2.1

A schematic to show the various possible approaches to isolating and purifying RNA from various sources. Tissues (including plant and fungal tissues) and tissue culture cells can be treated in the same way: either disrupt them as a unit or separate out the cells. A suspension of cells applies to tissue culture cells growing in suspension as well as cells separated from a tissue or monolayer through the actions of enzymes. These cells can be disrupted or lysed to release nuclei if you want to separate cytoplasmic and nuclear RNA. Bacterial and yeast cells can be treated in the same way as nuclei.

experimental approaches here, but some general problem areas, and possible solutions, are listed below.

Gloves

Wear latex or vinyl gloves all the time when preparing for and performing an RNA purification protocol, or any experiment involving RNA. Do not use gloves from an old packet that has been left open to gather dust. You must change your gloves regularly, and particularly when they become contaminated with cell lysate, sweat (i.e. you forget and scratch the end of your nose) or if you touch anything else that you cannot guarantee is a minimal RNase risk (e.g. door handles).

Surfaces

Since most RNases come from cells that are disrupted (i.e. the cells are lysed, and all membranes, including the membranes of organelles are degraded) it is not necessary to worry about RNases in growth chambers (culture flasks, Petri dishes etc.). However, from the point of disruption, any glassware, plastic-ware or metal surfaces that come into contact with the sample should be considered RNase-contaminated and treated accordingly before proceeding with the RNA preparation protocol chosen.

Glassware

All glassware that will come into direct contact with samples containing RNA (including that used for making up and storing buffers used in RNA preparation and analysis) together with the metal parts of homogenizers and blenders that come into contact with RNA-containing samples should be thoroughly cleaned in the usual way and then rinsed with 0.1% v/v diethyl pyrocarbonate (DPEC) (Berger, 1975) in molecular biology grade water and autoclaved as near to the time of their use as possible. DPEC is an excellent nuclease inhibitor, working by alkylating proteins, but it is potentially toxic, so make sure you are aware of the risks before splashing it around. Autoclaving is essential to degrade residual DPEC, which can inhibit downstream enzymatic reactions if it is allowed to contaminate your RNA preparation.

Nuclease-free consumables

The easiest way to reduce the problems of RNases on plastic-ware is to purchase certified nuclease-free polypropylene consumables. These will be more expensive than standard consumables, but at least you can blame someone else if the experiment doesn't work. Of course, once you open a pack, make sure it is sealed again and does not become contaminated with dust. If financial constraints mean nuclease-free consumables are not an option, or if you need a specialist type of plastic-ware that is not available nuclease free, then soak the plastic-ware with 0.1 M NaOH containing 1 mM EDTA for a few minutes, then rinse in 0.1% v/v DPEC in water before autoclaving. Be careful because acrylic (e.g. used to make most electrophoresis equipment) can become damaged with DPEC.

Pipettes

You should be aware that automatic pipettes are a common source of nuclease contamination. Seriously consider using filter pipette tips. Whilst they are more expensive than nonfilter tips, they prevent the spread of nucleases from sample to sample via the barrels of automatic pipettes. In addition, most pipettes can be stripped down and the plastic barrels removed and treated as for plastic-ware, above, and autoclaved. Check the manufacturers' instructions that come with your automatic pipette before doing this. If possible, use a separate set of pipettes for RNA work, or perhaps consider purchasing extra barrels, so that one can be used whilst the other is being autoclaved prior to RNA work.

RNase-free water

To make RNase-free water, add 0.1% v/v DPEC to molecular biology grade water and leave for 12 h at room temperature. Then autoclave to degrade the DPEC. Use only RNase-free water when making up buffers and solutions for RNA purification and analysis protocols.

RNase inhibitors

A number of RNase inhibitors are available (*Table 2.1*). You should consider adding them to reagents into which cells will be disrupted or lysed (see Section 2.5) unless guanidine salts are being used for simultaneous disruption or lysis and isolation of RNA, since guanidine salts very rapidly degrade all proteins, including RNases (see Section 2.6). Other reagents that can be used during cell lysis or disruption, whose presence will reduce RNase activity as a function of their general action on proteins are EDTA, sodium dodecyl sulfate, polyvinyl sulfate and proteinase K (requires a period of incubation), though none of these agents will remove all types of RNase activity. Combinations will be required. Furthermore, none are instantaneous in their action.

Table 2.1 Some common RNase inhibitors for use in disruption/lysis buffers

Agent	Active concentration	Comments
Aurintricarboxylic acid	10 μM	Broad spectrum activity
Diethyl pyrocarbonate	0.1% v/v	Toxic. Can modify RNA bases. Reacts with Tris-containing buffers
8-Hydroxyquinoline	0.1% w/v	Toxic
Human placental RNase inhibitor	1 U μl^{-1}	Narrow spectrum, sensitive to oxidative damage. Many recombinant mutant variants, with improved properties, are available from different manufacturers

It is not necessary to treat bench surfaces with RNase inhibitor solutions. Indeed, given the toxic nature of many of these compounds, this is a risk to health not worth taking. However, it is good practice to make a larger than required area of bench top available for RNA experiments, to free that area from all laboratory equipment, supplies and documents, and to wipe down that area of bench top with ethanol immediately prior to use.

2.4 Stabilizing the RNA complement prior to harvesting cells

As soon as you start to centrifuge or otherwise roughly treat cells whilst harvesting them prior to disruption or lysis, their normal gene expression patterns can become altered. Centrifugation is a rather unusual thing for a cell, and is likely to invoke a stress response. Stress responses involve dramatic changes in gene expression and, commonly, a mass shut-off of transcription. If the half-life of a transcript were less than 5 min, as is typically the case in bacteria, and its production were shut off during the harvesting of cells, each cell could have dramatically fewer molecules of that transcript immediately prior to disruption than was the case during growth in the experimental conditions. If all transcripts behaved identically in this regard, there would not be a great problem, but all transcripts are not equal. The expression of some genes is under tighter control than that of others during stress responses, and transcript half-lives are sequence specific, because of length and secondary structure constraints, and in some cases, due to the actions of sequence specific RNases. Therefore, the profile of transcripts in a harvested cell may not be representative of the relative abundances of transcripts in that same cell during growth in the conditions being tested in the experiment. This means that the differential stability of RNA molecules derived from different genes may become a key issue for your experiments much sooner than you might first imagine.

Snap freezing to stabilize RNA content

The easiest way to deal with this problem is to stabilize the RNA content of cells whilst still in their experimental growth conditions and before harvesting takes place. Until recently this was not an option, except when dealing with organs or pieces of tissue, rapidly dissected out, wrapped in aluminum foil and dropped into liquid nitrogen to be snap frozen. This very quickly stabilizes the RNA complement, but the problems do not entirely go away. Freezing breaks open organelles within eukaryotic cells, releasing RNases into the cytoplasm, and this makes the RNA complement less stable upon defrosting than if the sample had not been frozen at all. Thus, it is imperative that the sample is not allowed to defrost prior to cell disruption into a buffer containing RNase inhibitors (see Section 2.5 and *Table 2.1*). As far as snap freezing other cells when they have been harvested and pelleted by centrifugation goes, it is only really useful if harvesting and disruption need to be undertaken on different days. This is because it is the harvesting process during which most stresses are put on the cells.

Chemical stabilization of RNA content in human and animal cells

In recent years, true RNA stabilization has become a reality in some cases thanks to the introduction of a number of reagents that permeate cells and block both RNA synthesis and degradation prior to cell harvesting. In the case of tissue pieces, they are dissected out and immersed in the stabilizing reagent. 'RNA*later*' is the regent marketed by Qiagen for this purpose, and it can be purchased ready aliquotted into sample collection tubes or in larger volumes for aliquotting as needed. Tissue samples must be less than 0.5 cm in at least one dimension to allow permeation of the reagent, though in theory, the other two dimensions can be as large as required. Ten micro-liters of RNA*later* reagent are needed per 1 mg of tissue, and the entire sample must remain submerged at all times until the time of disruption, which can be up to a week later if the samples are stored at room tempera-ture. Furthermore, samples can be stored for 2–4 weeks at $4°C$, and for many months at $-80°C$, and can be freeze–thawed multiple times without loss of RNA stability. In this latter case, it is advisable to store the tissue overnight at $4°C$ and then remove the tissue from the RNA*later* reagent, transfer it to a clean tube and snap freeze it before archiving.

For tissue culture cells, the RNA*later* regent can also be used to stabilize RNA content. However, in this case, the cells must be harvested, pelleted by centrifugation and resuspended in phosphate buffered saline (PBS) to a density of 10^6 cells per μl before RNA*later* can be added (5 volumes of RNA*later* to 1 volume of cells in PBS). Hence, apart from stopping the need for cell disruption and RNA purification to follow immediately after cell harvest, the stabilization reagent here has limited value. No doubt, however, products are in development to fill this gap in the market. Keep an eye out for them being launched.

Chemical stabilization of RNA content in bacteria

'RNA*protect*' is the reagent devised for stabilizing RNA content within bacterial cells grown in liquid culture. It is particularly important to stabi-lize RNA in bacteria, because it is in these cells that RNA turnover is fastest, and control of transcription is under the tightest control and so the transcriptome is most likely to be materially affected by stress responses. The reagent is provided as a concentrate, which is added to culture medium immediately prior to cell harvest at a quantity of two parts reagent to one part culture. It is more economical to remove a volume of bacterial culture sufficient to produce the required amount of RNA and add the RNA*protect* reagent to the removed culture rather than adding the reagent to the culture as a whole. How much culture you will require in order to give an appropriate yield of RNA may be a matter of trial and error if you are working with bacteria that do not fall into the Enterobacteriaceae or *Bacillus*/*Staphylococcus* groups, though some suggestions as to numbers of cells are given below in Section 2.5 and *Table 2.2*. According to the manufacturer's literature, RNA*protect* is suitable for use with many different species of bacteria, including Gram positive bacteria, mycobacteria, and Gram negatives. The reagent works most efficiently when added to minimal medium grown cells (since no doubt it is affected by the complex

Table 2.2 Typical yields of RNA from various sources (from the Qiagen RNEasy RNA isolation handbook)

Cell cultures (10^6 cells)	Rat tissues (10 mg)	Bacteria (10^9 cells)	Yeast (10^7 cells)	Plant (100 mg leaves)
NIH/3T3 (10 µg)	Embryo (13 day) (25 µg)	E. coli (55 µg)	S. cerevisiae (25 µg)	Arabidopsis (35 µg)
HeLa (15 µg)	Brain (10 µg)	B. subtilis (30 µg)		Maize (25 µg)
COS-7 (35 µg)	Heart (10 µg)			Tomato (65 µg)
LMH (15 µg)	Kidney (40 µg)			Tobacco (65 µg)
Huh (15 µg)	Liver (40 µg)			
	Spleen (40 µg)			
	Thymus (40 µg)			
	Lung (10 µg)			

constituents of rich broth media), but the reagent works in a rich medium as well. If a minimal medium is not available for the specific species of cells you work with, it might be sensible to choose a more refined and defined broth, such as Müller–Hinton or Isosensitest broth rather than a more general yeast extract type broth such as LB. These will have fewer complex polymer components, meaning that the RNA*protect* will have more chance to permeate the cells. Furthermore, since there is much batch specificity in complex yeast extract based broths, dependent upon the exact mix of each extract used, results of different experiments would be more reproducible if a refined and defined broth were chosen; this is particularly important for transcriptome analysis.

Stabilization of RNA in plant and fungal cells

Currently, there is no reagent on the market for use in chemical stabilization of the RNA content in plant or fungal cells, even following their harvest. If RNA stabilization is required when working with these cells, drop the polypropylene centrifuge tube containing pelleted cells into liquid nitrogen to snap freeze and store the cells at –80°C. If you use frozen pellets to isolate RNA, seriously consider using a rapid mechanical cell disruption method (yeast cells), or a pestle and mortar (plant cells and tissues and filamentous fungi), so that the pellet does not thaw out prior disruption of the cells (see Section 2.5).

2.5   Harvesting and lysing or disrupting cells

It's a bit of a pedantic point, but whilst the terms 'cell lysis' and 'cell disruption' are often used interchangeably, they do not mean exactly the same thing. Cell lysis involves breaching the outer envelope of a cell. If the cell were eukaryotic, lysis would leave intracellular organelles intact, thus retaining their specific RNA species if they have them, allowing them to be collected separately. If a eukaryotic cell were disrupted, all membranes would be breached, including those of its organelles, releasing total genomic RNA for simultaneous purification. I will refer to the products of prokaryotic cell lysis and eukaryotic cell disruption collectively as 'cell extracts'. It is also important to note that lysis of prokaryotic cells, and disruption of eukaryotic cells leads to the release of genomic DNA. This, plus the presence of large amounts of various other polymers in certain cells, can make cell extracts very viscous. In this case, it is difficult for protein denaturants and other RNase inhibitors (*Table 2.1*) to access their targets, meaning that the RNA isolation techniques described below (Section 2.6) can result in very low yields. So in these cases, homogenization of the cell extract is required to reduce viscosity.

It is not possible in this section to cover all potential harvesting, lysis/disruption and homogenization methods specific to every cell type you could choose. However, if stabilization reagents are not being used, it is advisable that whatever harvesting technique is chosen, the time from the end of cell growth to the point of lysis/disruption is as short as possible. Do not harvest cells, resuspend them in lysis/disruption reagent, and then leave them on the bench for any significant period of time prior to lysis/disruption. The use of an ice bucket may do more harm than good, since RNases retain significant levels of activity at 0°C, but such a low temperature may invoke a cold-shock response in the cells (which are still alive, remember) inevitably altering the transcriptome. Accordingly, get all your reagents prepared and your equipment organized before you begin and get on with the job of harvesting and lysis/disruption as efficiently as possible. An ice bucket, and ice cold lysis/disruption reagents (unless enzymes are being used) are good ideas, since once the cells are lysed/disrupted, cold-shock is not a concern, and any reduction in RNase activity due to the low temperature is welcome.

Disrupting cells within human and animal tissues

Pieces of tissue need shearing forces to break them up into individual cells and to disrupt the cells. Furthermore, if this is done correctly, the extract is homogenized at the same time. Two technologies are particularly appropriate for simultaneous disruption and homogenization. In both cases, the tissue must be weighed first. The best way to do this is to weigh the tube you will use to homogenize the tissue before and after adding the tissue and calculate the difference. Add 20 µl of disruption reagent per milligram of tissue to the homogenization tube and ensure the tissue is submerged. The disruption reagent must contain some means of inhibiting RNases (see *Table 2.1*), since these are released upon disruption of the cells.

The most common tool for mechanical disruption and homogenization of tissues is the rotor–stator homogenizer, which employs a very rapidly rotating probe to generate shearing forces that rip open the cells and homogenize the resulting mixture. For tissue samples that weigh less than 50 mg, the best homogenization tube would be a 1.5 ml polypropylene tube (commonly referred to as an Eppendorf tube). In this case the rotor–stator homogenizer probe size should be 5 mm, since larger probes do not work efficiently in 1.5 ml microcentrifuge tubes. Fifty milligrams of tissue is perfectly sufficient for most small-scale RNA isolation and purification methods, and would yield around 50–200 µg of total RNA (*Table 2.2*). This would provide enough material for a useful number of RT-PCR reactions, for array hybridization, but not really for northern blotting. If larger preparations are required, and so larger tissue samples are needed, larger homogenization containers will be needed, together with proportionally larger rotor–stator probes. Follow the manufacturer's instructions that come with your rotor–stator homogenizer concerning the maximum depth-to-surface diameter of a solution to be homogenized, which will influence your choice of homogenization containers. Whatever the sample size, two bursts of homogenization, each for 15 s is sufficient for most tissue types. Some tissues, particularly spleen, brain and lung may well require longer and/or more frequent bursts. You should refer to the instructions that come with your rotor–stator homogenizer, but the end result should be a lump-free solution which is easy to pipette through an aperture diameter equal to that of a 250 µl capacity pipette tip. If the solution plugs the tip, or does not move into and out of the tip with the required ease, more homogenization and/or dilution of the sample with more disruption reagent will be required.

Another popular method for simultaneous disruption and homogenization of cells within animal tissues involves the use of bead grinding technology. Basically, the tissue is placed in a tube with disruption buffer (20 µl per mg of tissue) and either a single stainless steel (5 mm diameter) bead, or an equal amount of acid-washed glass beads (0.5 mm diameter) to tissue, is added. The tube is capped and shaken vigorously by a machine, grinding the tissue and its constituent cells against the side of the tube and causing them to be disrupted and the resulting solution to be homogenized. A number of different manufacturers make these machines. They are particularly useful for RNA isolation because they cause heating of the sample during the grinding process, which assists in denaturing proteins. The shaking speed and the length of shaking time needed to completely disrupt and homogenize cells in a tissue will vary from tissue to tissue. The same end result concerning the consistency of the resulting solution is desired following this method, than is the case when using a rotor–stator homogenizer and so in the absence of any other information, it is the apparent viscosity of the solution that should be your guide when you consider whether or not more shaking speed or time is required.

Disruption of tissue culture cells

Tissue culture cells grown in suspension should be pelleted by centrifugation in the manner appropriate for your particular choice of cell type and the pellet should be re-suspended in disruption reagent to a density of

1×10^4 cells per µl. Simply pipetting up and down using a 1 ml capacity pipette tip is sufficient to re-suspend the cells. If using a guanidine isothiocyanate containing reagent for simultaneous cell disruption and RNA isolation (see Section 2.6) simply vortexing the cells will be enough to disrupt them. Further homogenization may be necessary if the disrupted cell suspension appears viscous, however, and this will be obvious if it is difficult to pipette the solution. Homogenization is most easily achieved by pouring the sample into an appropriately-sized syringe (make sure you have removed RNase contamination by using DPEC treated water and autoclaving prior to use) and passing the solution back and forth through a 20 gauge (9 mm diameter) needle, five to ten times.

You may choose not to simultaneously disrupt tissue culture cells and isolate their RNA; for example, if you choose an acid phenol containing reagent to isolate RNA from the cell extract (see Section 2.6). In this case, you must choose a disruption reagent containing an RNase inhibitor (*Table 2.1*). Here, the cell pellet should be re-suspended in the disruption reagent to the same density described above. Once re-suspended, the cells can be disrupted and homogenized using either rotor–stator homogenization or bead grinding as described for the disruption of cells within tissue samples.

Many tissue culture cells are grown in monolayer culture. If you use a culture dish to grow cells, it is easiest to simply aspirate the medium (make sure there is not a large volume of medium left) and add disruption reagent (1 µl for every 1×10^4 cells to be disrupted) direct to the monolayer and use a 'rubber policeman' (a piece of rubber bung skewered on a needle) to shear the cells from the surface of the dish and simultaneously disrupt and homogenize them. Remember, the disruption reagent must contain an RNase inhibitor (*Table 2.1*). Following the use of a rubber policeman, it may be necessary to undertake further homogenization using a syringe and needle, as set out above, particularly if the suspension appears viscous or contains lumps. A white/cream precipitate is common when using guanidine isothiocyanate containing disruption reagents (see Section 2.6) due to precipitated protein, so do not worry if such a precipitate is seen, even following homogenization.

If you are using tissue culture cells grown in flasks, then it will not be possible for you to disrupt the cells *in situ*. The monolayer should be treated with trypsin to release individual cells, which should then be treated as if they were cells growing in suspension for the purposes of disruption and homogenization, as set out above. Different cells require different trypsinization protocols, and it is beyond the scope of this book to delve into all the possibilities. Take advice if you are unsure.

Gentle lysis of animal cells and separation of nuclei from cytoplasmic contents

As described in Section 2.2, the transcription of protein encoding genes in eukaryotic cells produces hnRNA and occurs in the nucleus. Processing of this hnRNA into mRNA is an important point of control of gene expression and also occurs in the nucleus. As they are produced, mature mRNA molecules are exported from the nucleus into the cytoplasm. Therefore, the only way to separate these different coding RNA species, and so to look at these different facets of control of gene expression, is to separate nuclei from the rest of the cell, and then use the purified nuclei to isolate hnRNA and the separated cytoplasm to purify total RNA, which will include mRNA. The process of separation requires gentle lysis of the cells using isotonic solublization of their plasma membranes with a non-ionic detergent into a buffer containing RNase inhibitors (Favaloro *et al.*, 1980). The lysate is subjected to differential centrifugation in order to pellet nuclei, which of all organelles, pellet most readily because they are the largest. The supernatant can then be used as a source of total cytoplasmic RNA, and the nuclei can be lysed separately and used to extract hnRNA using the same method described for disruption of entire animal cells. Clearly the potential for DNA contamination of the RNA product in nuclear extracts is far higher than when isolating total genomic RNA, so particular care should be taken to check the DNA content of the nuclear hnRNA sample, and to take appropriate action (see Section 2.9).

Harvesting and lysis of bacterial cells

Harvesting bacterial cells grown in batch culture is a simple matter of centrifuging the culture and then aspirating the supernatant. Do not be tempted to use a chilled centrifuge for this unless you have stabilized the RNA content first (see Section 2.4), it will only make the effects of stress responses on transcript levels more pronounced. Some people are keen to wash the cells in an equal volume of re-suspension buffer as supernatant is removed and re-centrifuge the cells prior to lysis. I have never personally seen the benefit of this, and whilst it may be worthwhile if a solely lysozyme-mediated lysis method is being used, when a chemical lysis method is being used it is almost certainly not necessary, and indeed, if RNA has not been stabilized, it may cause more problems due to stress-induced transcriptome changes.

The best way to lyse bacterial cells is to use guanidine isothiocyanate (see Section 2.6) which isolates cellular RNA at the same time. However, some pre-digestion of the cell wall helps to improve the efficiency of cell lysis, and this is almost essential for good RNA yields from Gram positive bacteria. For cell wall pre-digestion, add lysozyme (0.4 mg ml^{-1} final for Gram negative bacteria or 3 mg ml^{-1} final for Gram positive bacteria) to your chosen re-suspension buffer. Use 2 ml of this lysozyme containing buffer to resuspend the cells from 10 ml of bacterial culture having an OD$_{600}$ of 0.5. You should adjust the amount of re-suspension buffer used in a linearly proportional manner depending on the OD$_{600}$ of the culture you are working with. This value is based on the known *Escherichia coli* cell density of 10^9 cells per ml at an OD$_{600}$ of 1.0 (i.e. 1 ml of buffer is used to re-suspend 5×10^9 cells). For bacteria other than *E. coli*, you may need to calculate how many cells per milliliter are present in a 1.0 OD$_{600}$ suspension using a simple serial dilution and viable count experiment. Following this, you can calculate the volumes of culture needed for each RNA preparation based on the number of cells to be lysed. In the cases of bacteria other than *E. coli*, however, it is probably safest to reduce the cell numbers used for each RNA preparation to around 10^8 cells per ml of re-suspension buffer, since some bacteria have more total RNA per cell than *E. coli* and it is not sensible to try and purify too high a concentration of RNA, since loss will be experiment specific, making end results variable. The best method for re-suspension of bacterial cell pellets is to pipette up and down rapidly using a 250 µl capacity pipette tip until cell clumps are no longer visible. The cell suspension should then be left on the bench for 5 min (Gram negative bacteria) or 15 min (Gram positive bacteria) to allow partial cell wall digestion by the lysozyme enzyme. It is important not to allow significant lysis to occur unless RNA stabilization reagents have been employed prior to cell harvest, or the re-suspension buffer has been supplemented with RNase inhibitors (*Table 2.1*). In the absence of RNase inhibitors, lysed cells means lost RNA. Lysis is best induced by adding guanidine isothiocyanate, which both isolates and stabilizes cellular RNA (i.e. it denatures RNases) as described in Section 2.6. Bacterial cell extracts benefit from homogenization to improve the efficiency of RNA isolation, and so the yield of purified RNA. If the cell lysate appears very viscous, pour the sample into an appropriately sized syringe (make sure you have removed RNase contamination using DPEC treated water and autoclaving prior to use) and pass the solution back and forth through a 20 gauge (9 mm diameter) needle five to ten times.

Disruption of yeast

Yeast cells can be treated very much like bacterial cells in terms of cell disruption and RNA isolation. In this case, enzymatic pre-digestion of the cell wall uses the enzyme zymolase. Pelleted cells are re-suspended to a density of 1×10^5 cells per µl in re-suspension buffer containing zymolase (500 units of enzyme per ml) and the cell suspension is incubated for 15 min at room temperature to create spheroplasts. These are then lysed by the addition of guanidine isothiocyanate solution. Alternatively, yeast pellets can be re-suspended in disruption reagent without zymolase (but which must have some component that inhibits RNases–*Table 2.1*) to a concentration of 1×10^6 cells per µl and disrupted and homogenized using bead grinding as for animal tissues. For most yeasts, viable count calculations using serial dilutions and growth on solid agar should be performed to determine how many cells are present in a liquid culture having a particular OD_{600}, thus enabling the amount of cells being disrupted to be calculated. For *Saccharomyces cerevisiae* an OD_{600} value of 1.0 means a suspension of around 10^7 cells per ml.

Disruption of plant and filamentous fungal cells

The most common way that plant cells are used is as a chunk of leaf or other plant tissue. These cells are very difficult to disrupt using rotor–stator homogenizers, and whilst bead grinding techniques do work, perhaps the best way to disrupt plant and filamentous fungal cells is to use a pestle and mortar and simply grind them up. To assist in this (as well as to stabilize the RNA content whilst you do it), the material should be weighed and snap frozen in liquid nitrogen. This can be done beforehand (see Section 2.4) or can be done by placing the material into the mortar and covering it with liquid nitrogen. If a sample has been pre-frozen, get it out of the liquid nitrogen Dewer, or –80°C freezer, and place it in the mortar, covering it in liquid nitrogen, which keeps the material frozen. In each case, the frozen material is hard, which allows it to be ground to a fine, thin paste (i.e. with the liquid nitrogen providing for the suspension) with the pestle. Add more liquid nitrogen if required, and then, when the grinding is complete pour the paste into a suitable homogenization tube, which has been sitting on dry ice for a few minutes. Keep the sample on dry ice but allow the liquid nitrogen to evaporate. For best results, the cells should be lysed by the addition of guanidine salts (see Section 2.6), but in this case, guanidine hydrochloride is best, because the stronger denaturant, guanidine isothiocyanate can cause cell extracts from some plant tissues to solidify. Add 10 µl for every 1 mg of material (i.e. before it was frozen) to be lysed. The disruption of plant cells will result in a very viscous solution. It is therefore advisable to homogenize the material using a rotor–stator homogenizer or a bead grinding system, as for animal tissues, though no additional buffer will be needed. In this case, a needle and syringe will not be sufficient. Indeed, the force required to push the glutinous sample through the needle, will make for an excellent ballistic missile!

2.6 Methods for the isolation and purification of RNA

All RNA isolation/purification methods have three major aims: (1) inactivation of nucleases; (2) dissociation of RNA from intimately associated RNA-binding proteins; and (3) separation of RNA from other molecules such as proteins, DNA, polysaccharides and low molecular weight salts. The presence of salts can significantly affect downstream enzymatic reactions. Some isolation procedures start with a cell extract, produced as described in Section 2.5; others involve disrupting/lysing the cells in RNA isolation reagent.

Acid phenol extraction coupled with alcohol precipitation

The first method to be routinely used for RNA isolation was acid phenol extraction (Kirby, 1968). Organic compounds such as phenol denature proteins, whilst leaving nucleic acids intact. This is an excellent way of rapidly inactivating nucleases. In acid conditions (pH 4.5) the phosphate backbones of DNA molecules become fully protonated, uncharged and will consequently only dissolve in organic solvents. Since RNA has exposed bases as well as a phosphate backbone (i.e. is both positive and negatively charged) it remains charged at pH 4.5 and therefore retains water solubility. Accordingly, mixing a solution of DNA, RNA and proteins with organic solvents at pH 4.5 is an excellent method for physically separating RNA from the other components *Figure 2.2*.

To perform RNA isolation by the acid phenol method (*Figure 2.3*), the cells are lysed/disrupted to produce a cell extract, which is homogenized (as described in Section 2.5). The cell extract is then mixed with an equal volume of buffer saturated phenol, chloroform and isoamyl alcohol at a ratio of 50%, 48% and 2% (often referred to as 25:24:1, but this could make you think that 50% of the solution was something else – it isn't), pH 4.5, and the emulsion is shaken vigorously. Following centrifugation, the organic bottom layer contains proteins and DNA, and the upper aqueous

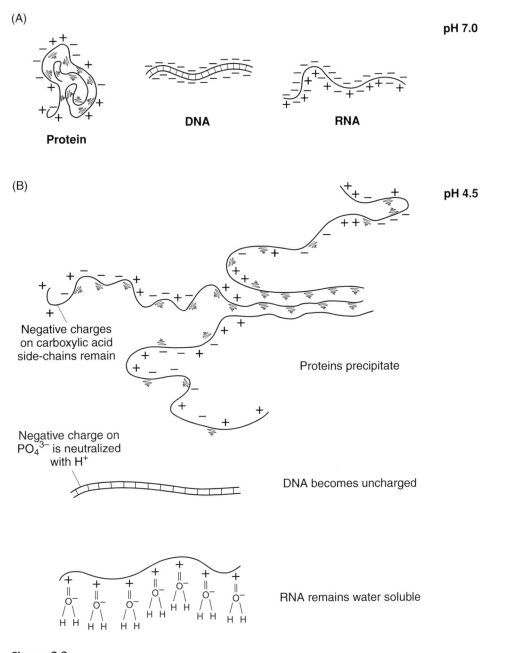

(A)

pH 7.0

Protein

DNA

RNA

(B)

pH 4.5

Negative charges
on carboxylic acid
side-chains remain

Proteins precipitate

Negative charge on
PO_4^{3-} is neutralized
with H^+

DNA becomes uncharged

RNA remains water soluble

Figure 2.2

Why does acid phenol specifically leave RNA in solution? Proteins have surface charge, but are usually held together by hydrophobic interactions. In the presence of phenol, the hydrophobic cores interact with solvent, and this causes proteins to precipitate (become denatured) or, at the very least, dissolve in the organic phase. Charges on proteins remain under acid phenol conditions because the pI values of carboxylic acid side chains are generally <4.5 and remain deprotonated. DNA has a lot of surface negative charge, but the pI of phosphate is >4.5 meaning that in the presence of acid, the charge is wiped out due to protonation. In this case, DNA becomes dissolved in the organic phase. RNA has positive charge in exposed nitrogenous bases, which can interact with water, keeping RNA in the aqueous phase.

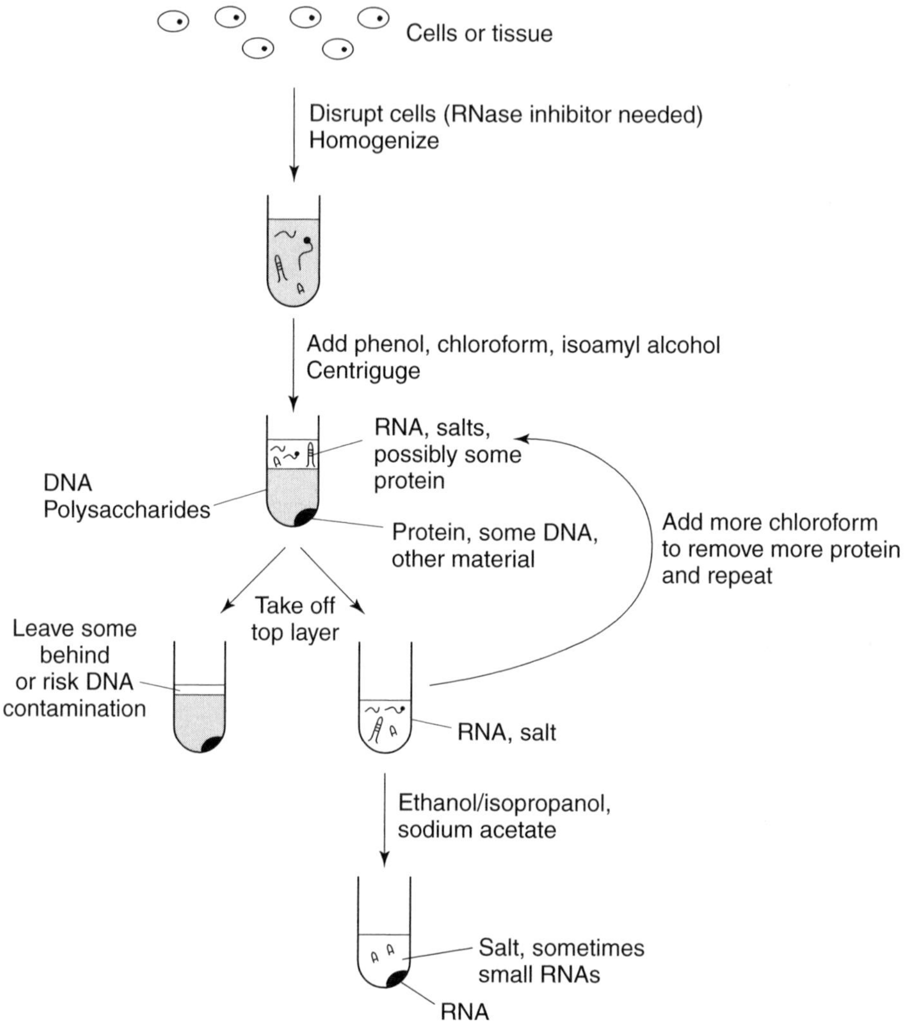

Figure 2.3

The procedure of acid–phenol extraction with chloroform cleaning. The various forms of RNA stay in the aqueous phase and can be removed from the organic phase, which contains DNA, proteins and other contaminants. The three main concerns are contamination of the aqueous phase with DNA, incomplete removal of salt from the RNA pellet and poor precipitation of, particularly, small RNAs.

phase containing RNA is removed. More protein is removed from this aqueous phase using chloroform. RNA is precipitated from the resulting aqueous phase by the addition of sodium acetate (pH 5.2) to a concentration of 0.3 M, and then 2 volumes of absolute ethanol or 1 volume of isopropanol. Precipitation is facilitated by cooling to –20°C for 2 h (ethanol) or 0°C for 15 min (isopropanol) and RNA is pelleted by centrifugation. Finally, the RNA pellet is desalted with ice-cold 70% v/v ethanol. Apart from the long-windedness of the procedure, the use of phenol is potentially hazardous, and the amount of DNA contamination in the final RNA preparation can be high, making the use of this RNA for reverse transcriptase polymerase chain reaction (RT-PCR) problematic. Furthermore, residual phenol is often a problem in the final RNA solution, causing overestimation of the RNA concentration, and inhibiting downstream enzyme reactions. The final problem is that whilst phenol inactivates RNases, this is not instantaneous, and the time taken for inactivation can vary from extraction to extraction, depending upon subtle differences between the compositions of different samples. Accordingly, RNA degradation can be considerable in some cases, and, perhaps more problematically, variable from experiment to experiment. Performing the phenol extraction at 60°C reduces DNA contamination and improves the speed of RNase inactivation, but does nothing for the levels of hazard associated with the procedure! Accordingly, whilst this method is very cheap, it is certainly not quick, and can be considered quite dirty. Better, safer and more reliable methods for RNA isolation are available. The only time I would seriously consider phenol extraction of RNA from a cell extract is when working with tissues that have a very high fat content. In this case, the use of organic solvents cleans away fats, whereas they can cause problems of emulsion formation and fat clotting when using chaotropic agents and inorganic salts.

Guanidine isothiocyanate extraction coupled with lithium chloride precipitation or cesium chloride ultra-centrifugation

The method of RNA isolation (*Figure 2.4*) that has superseded acid phenol extraction in the vast majority of cases, because it is safer and more efficient, involves the use of guanidine isothiocyanate (Chomczynski and Sacchi, 1987) which is a very strong chaotropic denaturant (i.e. it carries strong positive and negative charges). Such molecules are able to neutralize all the charges on the surfaces of proteins, causing them to precipitate and/or become soluble only in organic solvents. Commonly, reducing agents such as β-mercaptoethanol are added in addition to guanidine salts. These break disulfide bonds within and between proteins, assisting denaturation. Nucleic acids have a relative over-abundance of surface charge compared with proteins, which is not entirely neutralized given the concentration of guanidine isothiocyanate used for RNA isolation, meaning that they remain soluble in water. Hence a simple separation of aqueous and organic phases physically separates nucleic acids from proteins. This basic principle is used in many methods for purification of DNA as well as RNA.

The big advantage of using guanidine salts in RNA isolation is that they can in many cases be used to disrupt/lyse cells, extract cellular RNA and denature proteins at the same time. Because of this, their denaturant action, including that imperative inactivation of RNases, is far more rapid than when using acid phenol extraction, where RNA must first be released into a standard lysis/disruption reagent before phenol extraction of the RNA takes place. Even if RNase inhibitors are present in the disruption reagent, they are never going to be quite as good at stabilizing RNA as denaturing all proteins at the point of cell lysis/disruption, which is what disruption in the presence of guanidine salts can achieve. Unlike the acid phenol method,

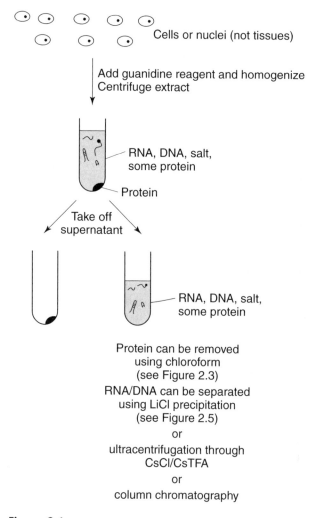

Figure 2.4

The procedure of guanidine isolation of RNA. Guanidine salts rapidly denature proteins, including nucleases. Once this is achieved, there are many different methods to choose for purifying the RNA. Remember, this process removes most proteins, but not all, and does not remove DNA.

however, the use of guanidine salts does not physically separate RNA from proteins and DNA in a single step, so protein contamination must be removed by chloroform treatment and some other method of differentiating between RNA and DNA must be found.

Many commercially available RNA purification kits use a mixture of acid phenol and guanidine isothiocyanate extraction technologies. A mixture of both chemicals is provided in a solution at acid pH, and is used during cell disruption/lysis, allowing for isolation of RNA, with the guanidine salts helping to rapidly inactivate RNases, whilst the acid phenol, upon the addition of chloroform, allows rapid separation of RNA from DNA and protein contaminants. RNA is then precipitated using isopropanol and the pellet desalted with 70% v/v ethanol. Such reagents (e.g. Sigma 'TriReagent') can be used for rapid purification of RNA, and can be safer than using phenol/chloroform extraction, since the presence of the guanidine salt means less phenol is needed, and chloroform, which is very much more volatile than phenol, is added following and not during the often rough process of cell lysis/disruption. They are particularly useful for RNA isolation from very fatty tissues, which can be problematic when using solely guanidine isothiocyanate-containing reagents. The fundamental problem of DNA contaminating the RNA product, seen with all acid phenol-mediated separations of RNA and DNA still exists with this method, meaning that without an additional step the RNA is not suitable for RT-PCR. Such RNA is fine for northern blotting experiments, though, and in this regard such kits for RNA purification represent cheap ways of rapidly purifying reasonable quantities of RNA.

For more reproducible purification of very clean RNA, a more specific method of separating RNA from DNA is needed. One method is to use lithium chloride precipitation (Cathala *et al.*, 1983). RNA can then be selectively precipitated from an aqueous solution containing both DNA and RNA, with the DNA remaining in solution. The basis of this separation is that RNA molecules are smaller in size than most DNA molecules, and so have smaller relative amounts of surface charge. Precipitation with lithium chloride (as with any salt) is caused because the added ions neutralize surface charge, 'salting out' the ability of water to interact with the molecule (*Figure 2.5*). To precipitate RNA selectively, lithium chloride is added to reach a final concentration of 4 M and the solution is incubated at 4°C overnight, with the RNA pellet being recovered by centrifugation. The pellet must be thoroughly washed with 70% v/v ethanol to remove residual lithium chloride, which can adversely affect downstream reactions. This represents an often overlooked method for recovering RNA from a mixture of RNA and DNA and you should consider it before reaching for the vial of DNase I, as is suggested by many commercial RNA purification kit manufacturers (see also Section 2.9). The reason for this is that DNase I lingers in samples, even following heat treatment, and may cause havoc to downstream applications such as cDNA production. So, the use of DNase I means a further chloroform extraction and precipitation of the RNA is required, reducing the yield and increasing the time taken.

Another method of efficiently separating RNA and DNA is the use of cesium chloride density ultra-centrifugation (Glisin *et al.*, 1974). The procedure involves a cell extract (usually produced by guanidine isothiocyanate extraction) being layered on to the surface of a solution of 5.7 M cesium chloride (the volume of cesium chloride solution should be half the volume of cell extract). The extract is then pushed against the cesium chloride solution using centrifugal forces of $250\,000 \times \textbf{\textit{g}}$ for 3 h at 22°C. Because of its relatively high buoyant density, RNA will move through the solution and form a pellet at the bottom of the tube, but DNA, proteins, polysaccharides and other contaminants will not, and will stay above or within the cesium chloride layer. The cesium chloride is then drawn off, exposing the RNA pellet, which needs to be thoroughly desalted with 70% v/v ethanol to remove cesium chloride contaminants. This method is time consuming and

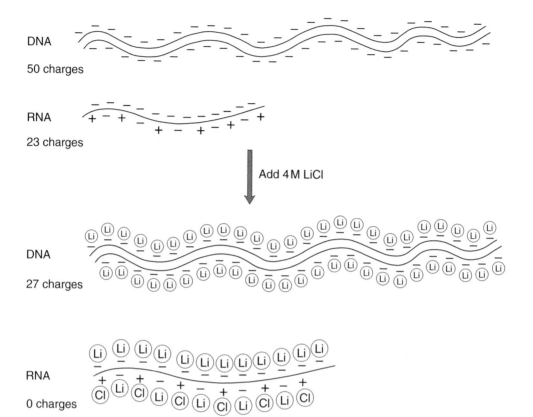

Figure 2.5

Lithium chloride precipitation of RNA. Simply because DNA has more charge per molecule than RNA, it takes fewer lithium chloride ions to neutralize the charge on a typical RNA molecule than on a typical DNA molecule. When the charge is neutralized, the molecule ceases to be soluble in water and will precipitate in an aqueous environment.

scares people because of the massive **g** forces involved. Fewer laboratories have access to an ultra-centrifuge these days, and, whilst this method works and yields high quality RNA, it should be attempted only by those who have access to expert assistance in setting up and running an ultra-centrifuge, since these machines can be lethal if not used correctly. A quicker, safer and apparently very effective method of isopycnic density centrifugation to purify RNA would be to add lithium chloride directly to a guanidine isothiocyanate treated cell extract and then to add cesium trifluoroacetate to the sample followed by room temperature centrifugation in a bench-top microfuge. Under these conditions, the RNA forms a pellet underneath the cesium trifluoroacetate cushion, and all contaminants, DNA, protein etc. sit above the cushion and can be removed. An added bonus with this approach is that cesium trifluoracetate is an excellent chaotropic protein denaturant, and so will inhibit any nucleases not already inactivated by guanidine isothiocyanate treatment.

Matrix-based methods for separating RNA, DNA and proteins

Many commercial RNA purification kits use column based matrix technologies to rapidly separate RNA from proteins, DNA, polysaccharides and other contaminants in cell extracts (usually produced using guanidine isothiocyanate extraction) without the need for organic extraction and time consuming differential precipitation or centrifugation protocols. These are conveniently divided into anion exchange and silica matrix technologies. Manufacturers have adapted both these matrix types for use in micro spin columns, which allow small volume 'minipreps' to be performed, and where short-burst centrifugation steps provide the flow of extracts and reagents through the matrices. Different manufacturers use different buffer components for improved separation of RNA from contaminants, and it is beyond the scope of this book to give a detailed account of each specific buffer solution. Indeed, manufacturers do not like giving out their recipes, since their insider knowledge is what you pay for when buying one of their kits. Hopefully, a general discussion of the chemistries involved in each type of separation method will help you to understand what is going on in the kit you choose.

Anion exchange matrices (*Figure 2.6*) are made up of positively charged (usually DEAE-linked) Sepharose or silica particles to which any molecules with a net negative charge bind. At pH 7.0 proteins and polysaccharide contaminants are weakly negatively charged (if at all) compared with RNA, which itself is weakly negatively charged compared with genomic DNA. This is simply because RNA molecules are smaller than DNA molecules and therefore have shorter phosphate backbones, which carry negative charge. Therefore, when a cell extract is applied to an anion exchange matrix at pH 7.0, proteins and polysaccharides either do not bind, or can be eluted easily using a mid ionic strength buffer (e.g. one containing 0.4 M sodium chloride). The elution of RNA requires the use of 1 M sodium chloride, and DNA remains bound up to a sodium chloride concentration of nearly 2 M at pH 7.0, or can be eluted by using 1 M sodium chloride at pH 4.5 (since acid pH means more protons, which protonate the DNA phosphate backbone and reduce its net negative charge). Anion exchange technologies can thus be used to purify genomic DNA and RNA in parallel from the same tissue, which is very useful when working with precious tissue samples. Of course, nucleic acids in the various eluants need to be concentrated using isopropanol or ethanol precipitation, and it is very important that the pellets are desalted with 70% v/v ethanol before use in downstream applications.

Silica matrices are another highly efficient way of purifying nucleic acids (*Figure 2.7*). In the presence of high concentrations of polyanions (e.g. guanidine) the anions form a chemical bridge between the silica matrix and the phosphate backbone of DNA, making DNA specifically adsorb onto the matrix surface. All other contaminants (including RNA which has a much lower net negative charge) fail to bind. Following washing of the matrix with a high ionic strength buffer to remove any contaminants, the purified DNA can be eluted in water. This is the basis of many commercially available rapid DNA purification kits. However, in the presence of >40% v/v ethanol, silica matrices behave in a very

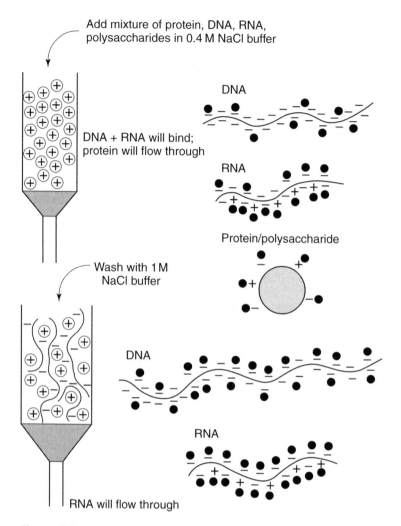

Figure 2.6

Anion exchange chromatography to purify RNA and DNA. In the presence of 0.4 mM NaCl (ions illustrated as black circles), all the surface charge on almost all proteins and polysaccharides will be removed, but RNA and DNA (to a greater extent because DNA molecules are generally larger than RNA molecules) will retain charge, most of it negative, and will therefore bind to anionic (positively charged) beads on a column. However, if the salt concentration is increased to 1 M, this removes all surface charge on RNA, causing it to elute from the column, while the average DNA molecule retains some surface charge, and so sticks to the column. These can be eluted with 2 M NaCl. Problems include small RNA molecules that do not bind to the column initially, and small DNA molecules that can elute with the RNA pool. The beads in the column can be provided as beads in an Eppendorf tube, and here the liquid can be separated from the beads by centrifugation.

(A)

Add DNA, RNA, protein, polysaccharide in polyanionic buffer

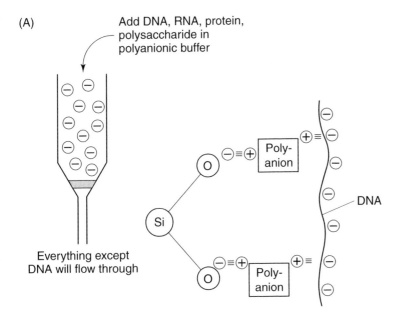

Everything except DNA will flow through

DNA

(B)

Add DNA, RNA, protein, polysaccharide in ethanol

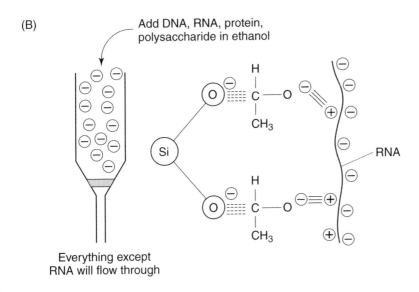

Everything except RNA will flow through

RNA

Figure 2.7

Silica gel chromatography to purify RNA and DNA. In the presence of guanidine polyanions (A), a bridge is formed between the negative charges of the silica beads and the phosphate backbone of nucleic acids. In this case, DNA and RNA stick to the column, though the adhesive strength of a typical RNA molecule is less, and it can be washed away. In the presence of >40% v/v ethanol (B), however, even if polyanions are present, a bridge occurs through the ethanol to join the exposed bases on RNA with the silica beads. Thus RNA becomes adsorbed onto the beads, but DNA and everything else flows through. The interaction of RNA and silica can be broken by adding water. These silica beads can also be provided as a slurry in an Eppendorf, and not just as packing in a column.

different manner. Under these conditions, RNA binds to the silica matrix and DNA remains in solution. The reason for this fundamental difference in behavior is not clear, but is probably explained by the presence of positive charges on the exposed RNA bases, which are not present in double-stranded DNA. These positive charges are likely to associate with a negatively charged chemical bridge between ethanol and the silica matrix. Thus, following lysis/disruption of cells in the presence of guanidine isothiocyanate, an equal volume of absolute ethanol is added to the cell extract and the mixture applied to a silica matrix. The matrix is then washed with 70% ethanol to remove any residual DNA and protein contamination, and the RNA can be eluted in water. It is important to note that very small RNA species (<200 nt) do not bind to silica matrices under these conditions, and anion exchange is the only real matrix technology suitable for purification of these types of RNA molecules. Even in this case, it is probable that small RNAs will elute with significant amounts of negatively charged protein contaminants, so a chloroform clean and isopropanol precipitation step may well be required.

Specific purification of polyA⁺ mRNA molecules

For the vast majority of experiments in which gene expression is being measured, the only type of RNA that needs to be purified is coding RNA. As described above, hnRNA can be purified specifically, since this is the only type of RNA found in nuclei. Furthermore, it is possible to purify mRNA in eukaryotic cells because the vast majority of mRNAs carry polyadenine tracts at their 3' ends. Purification of these so-called polyA⁺ RNA molecules can be achieved by the use of beads attached to polythymine tracts (Aviv and Leder, 1972). When polyA⁺ RNA is mixed with these beads at room temperature, there is complementarity between the polyadenine and polythymine sections, and hydrogen bonds form, causing the polyA⁺ RNA to become associated with the beads (*Figure 2.8*). The beads can then be pelleted, either by centrifugation, or more likely nowadays using a magnetic field (i.e. if the beads are magnetic they will be attracted to the bottom of the tube, when it is placed above an opposite magnetic field). The non-polyadenylated RNA (and other contaminants) remain in the supernatant, which can be removed by aspiration, and the beads washed in low ionic strength buffer (i.e. so that hydrogen bonds are not disrupted) to remove any lingering non-coding RNA contaminants, before the beads are suspended in high ionic strength buffer, which disrupts the hydrogen bonds, releasing polyA⁺ RNA into the supernatant. The salt can then be removed from the polyA⁺ RNA using ethanol precipitation. You should remember when you do this that not every mRNA in a eukaryotic cell is polyA⁺. Also, don't fall into the trap of trying to purify polyA⁺ RNA from prokaryotes, since you will not find any. It is possible to add polyadenine tails to bacterial mRNAs *in vitro* and then purify the polyadenylated RNA using the methods outlined above (Amara and Vijaya, 1997). But given the plethora of methods for purifying RNA from bacteria, and the inherent problems that might be created by the enzyme treatments required to polyadenylate RNA *in vitro*, it is probably best to avoid this sort of approach.

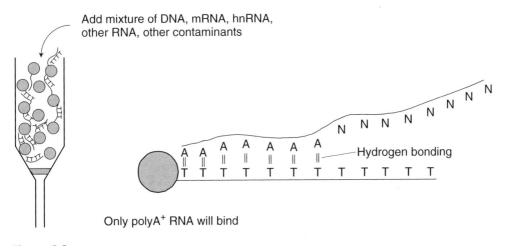

Add mixture of DNA, mRNA, hnRNA, other RNA, other contaminants

Hydrogen bonding

Only polyA$^+$ RNA will bind

Figure 2.8

Polythymine gel chromatography to purify polyA$^+$ mRNA. Sepharose or cellulose beads are chemically linked to polythymine oligonucleotides. The polyadenine tails of polyA$^+$ mRNA from eukaryotic cells (i.e. the vast majority, but not all, of mRNA in a cell) bind to the oligonucleotides via complementary hydrogen bonding; other molecules pass through the column. The mRNA/bead interaction is broken using salt.

Commercial kits for the isolation and purification of RNA

It is likely that you will turn to a commercial kit for RNA purification, rather than following the do-it-yourself protocols described in this book. Indeed, there is a whole host of RNA isolation and purification kits available commercially. They represent a variety of different patented and generic technologies, and because of this, the cost of such kits is highly variable. *Table 2.3* describes some of these kits, and the methodologies that they follow. It is not intended to be a full list of all manufacturer's kits, and readers will need to do some research into what is available, and at what price. It is also not my aim to give a review of which kits work and which kits don't. However, it should be noted that the majority of microarray facilities have their own preferred method for isolating RNA for transcriptomics experiments. In my experience, the Qiagen RNA easy kit is very widely recommended, and since it is coupled with chemical RNA stabilization reagents (Section 2.4) it should be considered (even though it is relatively expensive compared to some other kits) if the downstream application is a 'junk in, junk out' technique like transcriptomics. For more general applications including RT-PCR, all of the kits will be suitable, provided you check for and remove contaminants in the RNA preparation (see Section 2.9).

It is important that whatever kit you choose, you follow the instructions provided to the letter. It is particularly important not to overload the methodology, either by using too much tissue or by trying to purify too much RNA. The end result will be contamination. Cell lysis is the point at which it is most likely to all go wrong. Incomplete lysis and poor homogenization (usually both happen in tandem) result in a very low RNA yield. If

Table 2.3 Commercial total RNA isolation kits

Company	Contact details	RNA isolation kit	Technology
ABgene	www.abgene.com	Total RNA isolation reagent	Phenol/guanidine salt disruption/isolation reagent. Chloroform clean yourself, then isopropanol precipitate
Ambion	www.ambion.com	MagMax – 96	Phenol/guanidine salt disruption/isolation reagent (TriReagent). Use MagMax silica magnetic beads in the presence of ethanol to purify RNA. DNase treatment of the beads is recommended
		RiboPure	As above, but requires a single chloroform (or bromochloropropane) clean, then uses a silica disc filter column in the presence of ethanol instead of magnetic beads
		Different versions have different ways of lysing cells e.g. RiboPure bacteria comes with tubes filled with glass beads for mechanical disruption	
		RNAqueous	As above, but uses guanidine salts alone and no requirement for organic solvents. Plant RNA Isolation Aid also available to help remove polysaccharide contamination
		ToTALLY RNA kit	Acid phenol extraction with guanidine isothiocyanate. Chloroform clean then isopropanol precipitation. LiCl provided for selective precipitation of RNA
		MICROBEnrich	This is a mixture of oligonucleotides bound to beads. They represent capture oligos for mammalian 18S rRNA, 28S rRNA and mRNA (i.e. a polyT oligo). Thus >90% of mammalian RNA can be removed from mixed populations, leaving bacterial RNA behind. Such mixed populations would be obtained during infection experiments, to look for changes in gene expression in bacteria when they come into contact with mammalian cells
		MELT reagent	This reagent includes a cocktail of protease enzymes, which digest cells, and the RNAses within. There is no need for mechanical disruption, and extracts can be stored. Homogenization of the sample will be required. RNA isolation is through one of the kits described above

Company	Website	Kit	Description
Amersham Biosciences (GE Healthcare)	www.amershambiosciences.com	QuickPrep	Guanidine isothiocyanate disruption/isolation reagent. Add LiCl to extract and precipitate RNA. Add cesium trifluoracetate and centrifuge sample. The RNA pellets at the bottom, but nothing else gets past the CsTFA barrier. Centrifugation in bench top centrifuge
		RNA Extraction Kit	As above, but without the LiCl precipitation, so requires isopycnic ultracentrifugation. Scary stuff
Clontech	www.clontech.com	Nucelobond RNA/DNA	Guanidine isothiocyanate disruption. Anion exchange chromatography to separate RNA and DNA
		Nucleospin RNA	As above, but silica matrix column in the presence of ethanol. Comes with DNase I to remove DNA contamination
Invitrogen	www.invitrogen .com	ChargeSwitch	Cocktail of enzymes disrupts cells, degrades protein, including RNases, and also degrades DNA. The resultant RNA is bound to an anionic column; contaminants are washed away. The pH is altered, removing the charge–charge interactions, and the RNA is eluted
		Micro to midi kit	Guanidine isothiocyanate disruption/isolation reagent. Silica spin column for selectively purifying RNA in the presence of ethanol
Promega	www.promega.com	RNAgents	Guanidine isothiocyanate disruption/isolation reagent. Phenol chloroform is then added to selectively purify RNA, which is precipitated with isopropanol
		SV Total RNA isolation kit	Guanidine isothiocyanate disruption/isolation reagent. Silica matrix used to purify RNA in the presence of ethanol. Solutions passed through matrix using centrifugation or vacuum
Qiagen	www.qiagen.com	RNeasy	Guanidine isothiocyanate disruption/isolation reagent. Silica matrix used to purify RNA in the presence of ethanol. Solutions passed through matrix using centrifugation or vacuum. On-filter DNase I treatment may be required to eliminate contaminating DNA

Table 2.3 *continued*

Company	Contact details	RNA isolation kit	Technology
		RNeasy Plus	As above, but before the extract is put onto a silica filter it is passed through an anion exchange filter to mop up genomic DNA (under these conditions, RNA and proteins are not negatively charged and so will not bind). The RNA is then purified using a silica matrix in the presence of ethanol
		RNA/DNA	Guanidine isothiocyanate disruption/isolation reagent. Anion exchange column to separate RNA and DNA; both can be purified from the same sample
Roche	www.biochem.roche.com	High Pure RNA isolation kit	Guanidine isothiocyanate disruption/isolation reagent. Silica matrix used to purify RNA in the presence of ethanol. DNAse I treatment of column to remove contaminating DNA recommended
		TriPure reagent	Phenol/guanidine salt disruption/isolation reagent. Chloroform clean yourself, then isopropanol precipitate
Sigma-aldrich	www.sigmaaldrich.com	GenElute RNA purification	Guanidine isothiocyanate disruption/isolation reagent. Silica matrix used to purify RNA in the presence of ethanol. DNAse I treatment of column to remove contaminating DNA recommended
		TriReagent	Phenol/guanidine salt disruption/isolation reagent. Chloroform clean yourself, then isopropanol precipitate
Stratagene	www.stratagene.com	AbsolutelyRNA kit	Guanidine isothiocyanate disruption/isolation reagent. Silica matrix used to purify RNA in the presence of ethanol. DNAse I treatment of column to remove contaminating DNA recommended
		RNA isolation kit	Phenol/guanidine salt disruption/isolation reagent. Chloroform clean yourself, then isopropanol precipitate

Not all manufacturers are included; you should shop around. Very specialist kits are not included. Most manufacturers produce mRNA isolation kits as well as those described in the table.

you are having problems, consider using mechanical disruption/homogenization of the cells, even if the kit suggests that guanidine isothiocyanate treatment and vortexing should be sufficient. To do this, you will probably need a bead grinding machine, since these are most suitable for the small amounts of cells used in the majority of commercial 'miniprep' type kit reactions.

Whilst most kits are in miniprep formats, an increasing number provide larger scale preparation options. Each miniprep will provide 20–50 µg of RNA, and this is suitable for most downstream applications except northern blotting. It is highly likely to be more cost effective and to cause fewer problems with contamination, therefore, if you run a number of miniprep reactions rather than a single large scale preparation (so long as each miniprep uses starting material from the same sample). Accordingly, the value of midi or maxi prep kits in purification of RNA for experiments aimed at measuring gene expression is limited to situations where sample amounts are not limiting, and northern blotting is the aim.

2.7   Re-solubilization and storage of RNA

Since it has a fairly high isoelectric point, the phosphate backbone of DNA can begin to become protonated if the pH falls much below neutral. This results in insolubility, meaning that DNA is best solubilized in a buffer with a pH of 8.0 (e.g. TE buffer–10 mM Tris-HCl, containing 1 mM EDTA) rather than water, which has no buffering potential and a pH that can easily fall below 7.0. In contrast, RNA is single stranded, so it has both positive and negative charges, making it readily soluble in water. Since the use of water will result in the least number of problems with buffer choice in downstream applications, I strongly recommend you use RNase-free water to dissolve RNA pellets. However, even mRNA molecules are not totally single stranded. Intra- and inter-molecular base complementarity can result in hydrogen-bond formation and secondary structure within and between RNA molecules. This problem gets worse as the concentration of RNA increases, and is maximal when the RNA is pelleted. The result is a reduction of the net positive charge at pH 7.0 of RNA molecules within a pellet, sometimes to the extent that they become insoluble in water. To avoid the need for alkaline buffers to solubilize pelleted RNA, therefore, secondary structure formation must be reduced during the solubilization process. The best way to do this is to heat the RNA during re-solubilization. The heat will break the hydrogen bonds, causing a dramatic reduction in secondary structure, and a consequent increase in solubility. Once in solution at approximately 1 µg µl^{-1}, the RNA will be too dilute to re-precipitate when the temperature is lowered. You will be able to calculate approximately how much water to use to re-solubilize an RNA pellet by using *Table 2.2*, which shows the approximate amounts of RNA present in different amounts of various types of tissues and cells. Once the appropriate amount of water has been added to the tube containing an RNA pellet to give an approximate concentration of 1–2 µg µl^{-1} (usually 50–100 µl) the tube is placed in a heating block set at 70°C for 5 min. Every minute or so, the contents should be mixed by vigorous pipetting. If you have many samples, it is easiest to assign a pipette tip to each tube, and eject the tip back into the tube after each round of mixing, returning to it for the next. Whatever you do, don't use the same tip for more than one sample!

Many manuals give details about the storage of RNA samples. My advice is only store RNA for as long as is absolutely necessary. Even if you use all the RNase inhibitors in the world, a small amount of RNase seems always to remain, resulting in you finding nucleotide mush when you go back to the sample. The way to minimize this is to reduce the temperature. Never store RNA above (i.e. warmer than) −70°C. Theoretically, it should be stable for several months, but I wouldn't use any RNA that was more than a month old. Also, do not freeze–thaw more than once, since every time the sample is thawed out, RNases can strike. It is also imperative, and often forgotten, that you re-determine the concentration of the RNA sample following storage and defrosting, even if storage has been for a short period of time. Thus, in summary, isolate and purify the RNA as close to the time of its use as possible, put it into the freezer as quickly as possible and defrost it and quantify the RNA concentration (see Section 2.8) immediately before you use it.

2.8 Quantification of RNA concentration using a spectrophotometer

Spectrophotometric analysis is the most common method used for determining RNA concentration. RNA and DNA strongly absorb UV light with a wavelength of 260 nm. The approximate extinction coefficient for RNA at this wavelength is 40 mg l^{-1} per absorbance unit (A_{260}) when using a quartz cuvette with a path length of 1 cm. The exact extinction coefficient of RNA is dependent upon the sequence of each molecule. RNA from very GC rich organisms has a higher extinction coefficient than RNA from very AT rich organisms, because G and C bases do not absorb as strongly as A and U). Since most organisms have an intermediate GC content, however, this is not a problem for most users, but those of you who work with extreme GC organisms should bear it in mind. Usually, however, the aim of RNA quantification is simply to facilitate the dilution of all RNA samples to the same final concentration prior to comparative analysis of transcript levels in each sample and to adjust the concentration to be within the working range of the technique to be used. For more specialist applications, where the exact concentration is required, a method other than the use of a spectrophotometer should be chosen.

For spectrophotometric determination of RNA concentration, mix 2–5 μl of sample with 1 ml of RNase-free water in a 1 ml quartz cuvette. The best way to mix the sample in the cuvette is to pipette up and down gently five to ten times with a pipette set to 100 μl. Take care that the water is at room temperature so that condensation does not form and that no air bubbles are formed in the solution in the cuvette. Both these events will increase the apparent absorbance of the solution, and lead to an overestimation of RNA concentration. If air bubbles are present, simply tap the cuvette on the surface of the bench (gently!) a few times. It is worth thoroughly rinsing the cuvette with RNase-free water prior to its use for RNA quantification so as to minimize RNase contamination. It is not advisable to treat quartz cuvettes with DEPC and autoclave them; they have a habit of exploding! You are aiming at a spectrophotometer reading of 0.1–0.5 A_{260}, which is where the machine gives its most accurate results. The spectrophotometer is zeroed using RNase-free water. Accordingly, if the A_{260} reading is significantly less than 0.1, add more sample to the cuvette, and mix again. If the A_{260} reading is higher than 0.5, dilute the RNA sample accordingly and repeat the analysis. Spectrophotometer readings correlate with concentration of the absorbing material in an approximately linear manner, though the correlation can tail off at higher A_{260} readings, giving an underestimation of absorbance. This is why you need to repeat your spectrophotometer reading using a lower concentration of RNA if the reading given the first time is too high. Accordingly, if an A_{260} reading of 2.0 is observed, dilute the RNA sample 1:4 before repeating the test. Do not be tempted to reduce the amount of RNA sample added to the cuvette below 2 µl, since pipetting error may give a false concentration. When you finally have an A_{260} reading that falls within the appropriate range, use the following calculation to determine the concentration of RNA in your preparation.

The concentration (in µg µl^{-1}) of the RNA in the preparation is

$$A_{260} \times 40 \times \left(\frac{DF}{1000} \right)$$

where DF (dilution factor) = µl in cuvette/µl of preparation added to the cuvette.

2.9 Sources of contamination in RNA preparations and how to spot them

There are three major sources of contamination of an RNA preparation. Most common is DNA contamination, closely followed by protein contamination and finally, the often overlooked salt contamination.

DNA contamination

Large-scale DNA contamination can be a problem in array hybridization experiments (Chapter 3) because it can titrate out labeled cDNA, reducing the hybridization signal on the array. It may be possible to see this level of genomic DNA contamination as a high molecular weight smear following electrophoresis of the RNA preparation (see Section 2.10). However, this approach is not particularly sensitive to contamination of the RNA preparation with small amounts of DNA. The best approach to locate low-level DNA contamination, is to perform PCR using primers targeted to amplify a housekeeping gene and a portion of the RNA preparation as template (see Chapter 4 for details). This is appropriate because it is particularly during RT-PCR experiments that DNA contaminants become apparent, and cause the most significant problems. In this case, DNA contamination would cause you to see RT-PCR amplification products in all samples, irrespective of whether they contain transcripts or not.

If you find DNA contamination in your RNA preparation, you will need to consider a solution. It may give the best results if you perform lithium chloride re-precipitation of the RNA, though this will take many hours. So, if you are in a hurry you must turn to the use of DNase I. This enzyme is available commercially with a concentrated buffer. Both are added to the RNA solution as set out in the manufacturer's protocol. If you are making cDNA from your preparation of RNA, it is advisable to remove the DNase I enzyme, using chloroform cleaning and ethanol re-precipitation of the RNA. Some manufacturers of silica matrix RNA purification columns provide DNase I, which can be added to the RNA on-column, and then washed away along with the deoxynucleotides produced prior to RNA purification. Indeed, it is the fact that such kits give very little DNA contamination, coupled with this neat method for removing any contaminating DNA that may be present, which makes them very popular for RNA purification prior to RT-PCR or array hybridization experiments.

Protein contamination

To test for protein contamination is more straightforward than testing for DNA contamination. RNA absorbs light about half as efficiently if the wavelength is 280 nm as it does if the wavelength is 260 nm. So, absolutely pure RNA has an A_{260}/A_{280} ratio (i.e. the value obtained by dividing the absorbance reading of a sample at 260 nm by that at 280 nm) of almost exactly 2.0. Aromatic amino acids, which make up a minor but consistent component of all proteins, maximally absorb UV light with a wavelength 280 nm. Thus, the presence of protein lowers the A_{260}/A_{280} ratio of a solution. You should routinely perform A_{260}/A_{280} ratio determinations for your RNA preparations for this reason and should consider chloroform cleaning and re-precipitation to remove protein contaminations if the A_{260}/A_{280} ratio is significantly less than 2.0. It is important to remember that pH affects the A_{280} reading when proteins are present, and since water has no buffering potential, the A_{260}/A_{280} ratio should be determined by adding the RNA sample (2–5 μl) to 1 ml of TE buffer (see Section 2.7) in the cuvette. This buffer should not be used to determine the RNA concentration (i.e. the A_{260}) since the extinction coefficient of 40 mg l^{-1} per A_{260} is not valid in this buffer (Wilfinger et al., 1997). So two separate tests are required, one to determine RNA concentration, and the other to check for protein contamination.

Contamination with salts

To look for salt contamination, the A_{260}/A_{240} ratio can be determined. The ratio for pure RNA should be around 1.4; if it is significantly lower than this, you should suspect salt contamination. This is an often overlooked source of RNA contamination, and can play havoc with some downstream enzyme reactions, particularly those that are excessively fussy about ionic concentration and the types of ions in solution. To solve the problem, a further isopropanol precipitation coupled with a 70% v/v ethanol wash of the pellet should be considered to remove this salt.

A note about spectrophotometers

It goes without saying that if you are measuring absorbances using a spectrophotometer producing a beam of light at 260 nm or below, the machine must be up to the job. Very old machines, and more commonly, very old halogen bulbs produce very weak and erratic UV light beams, and can suffer from wavelength wobble. Both these phenomena result in variable and inaccurate absorbance readings making RNA quantification and contaminant determination very difficult. It is possible to purchase RNA standards in order to check and calibrate spectrophotometers, and you should consider doing so. The best way to avoid problems, however, is to invest in a good UV–visible spectrophotometer, keep it maintained and keep the bulb fresh. Make sure the halogen bulb is switched off when you are using the machine out of the UV (i.e. >340 nm on most machines), or when it is not in use at all, since some machines do not automatically turn off the bulb. If you want to avoid the time it takes for your spectrophotometer to set itself up by not switching it off at the mains between UV absorbance readings, at least reset the wavelength to 400 nm or so, and make sure the halogen bulb is turned off. If you do not feel your laboratory needs to purchase a UV–visible spectrophotometer, when all you will do with it is measure A_{260}/A_{280} ratios (see above), then simple machines can be purchased with bulbs and filter wheels that do only that. In some cases, the machines can be upgraded by the purchase of new filter wheels to enable OD_{600} readings for cell density measurements, and, for example, A_{595} readings for protein concentration determination protocols, e.g. the Bradford protocol (Chapter 6). In my experience, such machines perform as well as mid-range spectrophotometers, and for a fraction of the cost, though they are obviously much less versatile.

2.10 Separation of RNA samples using electrophoresis

Historically, RNA species have been separated using centrifugation according to their sedimentation coefficients (S). This is where the names '23S', '16S' etc. for different RNAs come from. However, in recent years, gel electrophoresis has been the staple method for separating RNA molecules according to their molecular size. The main reason for doing this is that the RNA preparation can be assessed in terms of quantity and quality (e.g. whether significant degradation or DNA contamination has occurred) prior to its use in downstream applications.

Gel electrophoresis: theory and practice

Gel matrices for electrophoresis basically consist of a cross-mesh of polymer strands that interweave. When molecules flow through these meshes, they are subject to friction. Hence gel matrices retard the progress of different molecules based on the differential amount of friction imposed on each. The average mesh size of a gel matrix can be manipulated to provide the exact separation parameters required for the specific analysis being performed (*Table 2.4*). The basic way of separating molecules through a gel is to add the molecules in solution to a 'well' cut into one end of the gel, and then to apply an electric current across the gel in such a way that the molecules flow out of the well and through the gel because of charge–charge attractions between them and the electrode at the far end of the gel. Of course, for this to work predictably, the molecules to be

Table 2.4 Useful ranges of agarose and polyacrylamide gels for nucleic acid electrophoresis

Matrix type	Percentage (w/v)	Useful range (base pairs)
Agarose	0.6%	1000–10 000
	0.8%	600–8000
	1.0%	400–7000
	1.2%	300–6000
	1.5%	200–3000
Acrylamide:bis acrylamide	4.0%:0.2%	100–1500
	5.0%:0.25%	80–500
	8.0%:0.4%	60–400
	12.0%:0.6%	40–200

separated must all be charged in the same sign. When separating proteins by electrophoresis, this is achieved by mixing the proteins with SDS, which coats their surfaces with negative charge. In the cases of DNA and RNA, the samples are generally placed in an alkaline buffer meaning their phosphodiester backbones become fully deprotonated, and, therefore, negatively charged. In both cases, the molecules are loaded onto the gel at the cathode end, and migrate towards the anode when an electric current is applied (*Figure 2.9*). The speed with which each molecule travels through the gel depends upon the amount of retardation the gel imposes on its progress (i.e. the amount of friction). In the case of proteins coated in SDS, the amount of friction is dictated by the size (molecular mass) of each molecule. The bigger the molecule is, the harder it is for it to squeeze through the gel matrix so the slower it moves through the gel. Hence if the current is applied across the gel for a set amount of time, small proteins will have moved a relatively long, and large proteins will have moved a relatively short, distance through the gel. The relationship between size and distance traveled is not linearly proportional, however, but exponential (Helling *et al.*, 1974). Thus a protein of 10 kDa will not travel twice as far as a protein of 20 kDa. The relationship between mass and distance traveled is a complex one, but the use of molecular weight size standards each time you run a gel will allow calibration of the gel and determination of the size of the specific molecule you are interested in.

The density of gel you choose will depend on the separation range required (*Table 2.4*). If you are interested in separating high molecular weight molecules, then you either need to apply the current across the gel for a long time, or you need to use a gel with a mesh size that is large. In both cases, the side effect is that small molecules will probably run so far that they fall off the anode end of the gel. To avoid this, you will need a very long gel indeed. On the other hand, if it is small molecules that you want to separate, large mesh sized gels, run even for a short period of time will probably not resolve the molecules into tightly defined bands. In this case, what is needed is a very small mesh sized gel. The side effect here, is that large molecules will hardly make it out of the wells.

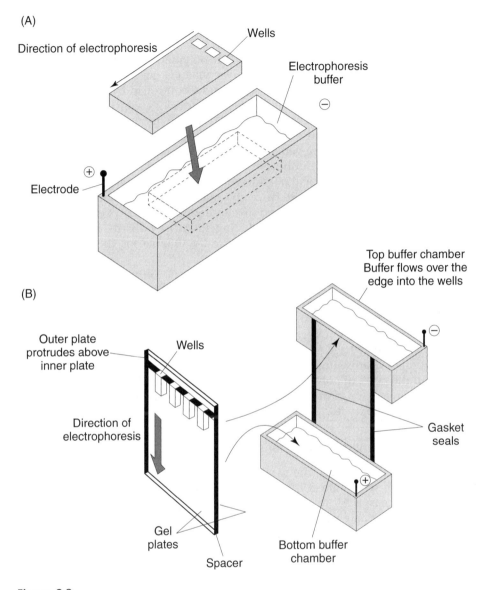

(A)

Direction of electrophoresis

Wells

Electrophoresis buffer

⊖

⊕

Electrode

(B)

Top buffer chamber
Buffer flows over the
edge into the wells

Outer plate
protrudes above
inner plate

Wells

Direction of
electrophoresis

⊖

Gasket
seals

⊕

Gel
plates

Spacer

Bottom buffer
chamber

Figure 2.9

Two types of general electrophoresis apparatus. In (A), a horizontal gel (normally an agarose gel) is made with wells cut out at one end. The gel is submerged in electrophoresis buffer in a tank. An electric current is applied across this buffer, and it also permeates the gel. Since nucleic acids are predominantly negatively charged at pH 8–9, they move towards the positive electrode. In (B) a vertical (typically polyacrylamide) gel is made between two glass plates. Wells are formed in the top. The gel is clipped into the apparatus such that the electrophoresis buffer in the top chamber fills the wells. A protuberance of the outer gel plate over the inner plate allows this to happen without buffer spilling over the side. In the case of nucleic acid electrophoresis, the same buffer is loaded into both top and bottom chambers. This time, samples are loaded at the top, and move towards the positive electrode at the bottom.

Different gel matrices

There are essentially two types of gel matrix used routinely for electrophoretic separation of biological polymers: the polyacrylamide gel and the agarose gel (*Table 2.4*). Polyacrylamide gels (*Figure 2.10*) involve very small mesh sizes, and the sizes can be controlled by adding different amounts of acrylamide and bis-acrylamide moieties, which polymerize and cross-link in the presence of TEMED (which provides the 'glue') and ammonium persulfate (an oxidizing agent that sets the glue). Polyacrylamide gels are generally used for separating proteins and we will revisit protein gel electrophoresis in Chapters 6 and 7. Many people forget, however, that nucleic acid polymers are, in general, far larger than proteins. Each nucleotide is at least twice the mass of an amino acid, and whilst proteins are generally made up of hundreds of amino acids, nucleic acids are made up of thousands (including RNA) or millions of nucleotides. Hence polyacrylamide gels are only really suitable for the separation of small DNA or RNA molecules (*Table 2.4*)

The gel matrix most commonly used to separate mixtures of nucleic acids is the agarose gel (*Figure 2.11*). These gels are also much more straightforward to make. A powder of agarose polymers is suspended in buffer and heated so that it dissolves. When the mixture cools, it solidifies into a gel with the agarose polymers crossing over each other in a random manner to generate an average pore size that is inversely proportional to the concentration of agarose used. The pores of even very thick agarose gels are larger than the pores of polyacrylamide gels, and the resolution achieved by them is lower, but for most purposes, agarose gels make excellent DNA and RNA separation matrices.

Denaturing versus non-denaturing gels

If all molecules were the same shape, then the only difference between them that would affect friction whilst passing through a gel matrix would be their masses. However, in the case of an RNA preparation, all molecules are not the same shape. All molecules will have some degree of secondary structure (this is not a property of double-stranded DNA, only of single-stranded nucleic acids) but the amount of secondary structure in each molecule is sequence-dependent as well as to some extent length-dependent. The more secondary structure present, the more compact the molecule is, and so the less friction is imposed on it by the matrix. It is best, therefore, if one wants to separate RNA species simply based on their mass, that you make sure all the molecules to be separated are devoid of secondary structure prior to loading them onto a gel, and during electrophoresis. Removing secondary structure is referred to as denaturation, and specific chemicals must be added to both the RNA sample, and to the gel matrix (a so-called 'denaturing gel') to ensure denaturation throughout the electrophoresis run (Lehrach *et al.*, 1977). Indeed, unless you are undertaking very special analysis of *in vitro* RNA/protein interactions (which would be blocked by denaturation), or if you want to experimentally estimate the amount of secondary structure in an RNA molecule (i.e. run the RNA on both denaturing and a non-denaturing gels having

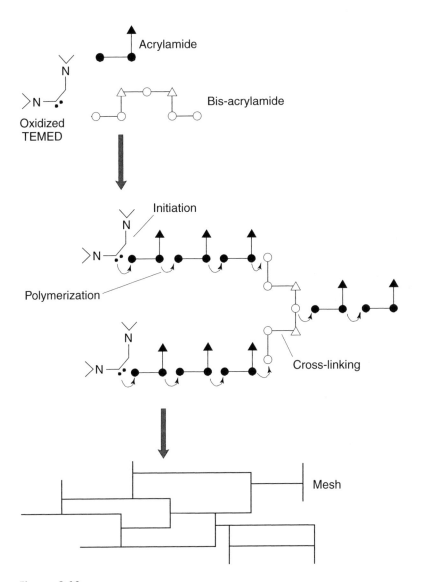

Figure 2.10

Polymerization of acrylamide and bis-acrylamide to form a polyacrylamide gel. In the presence of ammonium persulfate, the catalyst TEMED becomes oxidized. This donates a pair of electrons to an acrylamide molecule, and starts a chain reaction, with electrons being passed from acrylamide to acrylamide, polymerizing them together. If a bis-acrylamide molecule happens to be joined, it can cause cross-linking, because there are three possible sites for attachment of acrylamide strands. Overall, a mesh of strands is created, with the density of this mesh being dependent upon the acrylamide concentration, and the acrylamide:bis-acrylamide ratio.

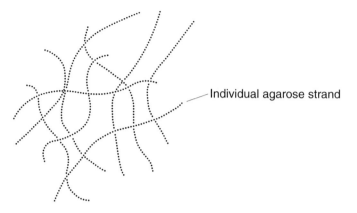

Individual agarose strand

Figure 2.11

Formation of an agarose gel. Here, there is no chemical reaction. Simply, each agarose molecule is a long strand of random length. The strands become mixed up when agarose is melted, and the strands fold around each other. When the gel cools and solidifies, the result is a mesh, with an average pore size dependent upon the concentration of agarose used to make the gel.

identical pore sizes in parallel (*Figure 2.12*), and look for a change in migration pattern) it is difficult to see why you would want to run a non-denaturing gel for RNA analysis at all. The most common denaturant used in RNA agarose gel electrophoresis is formaldehyde (Reijnders *et al.*, 1973) which works by reducing hydrogen-bond potential, meaning that intra-molecular hydrogen bonds do not form, resulting in a linear RNA molecule devoid of secondary structure. It is important to make up the RNA sample in denaturing conditions before loading it onto a denaturing gel and running the gel using buffers containing denaturant. If at any stage the denaturant concentration is allowed to dwindle, secondary structure can reform and problems will result.

Formaldehyde is a tricky chemical to work with. It is toxic, and should only be handled in a fume hood. This means that RNA-denaturing agarose gels containing formaldehyde should also be run in a fume hood, and never on an open bench surface. Formaldehyde oxidizes readily, so buy a small quantity of the chemical frequently, rather than purchasing a large bottle and hoping it will last for a long time. It won't. You should also be aware that agarose gels containing formaldehyde are considerably less rigid, i.e. are more likely to fall apart than normal agarose gels of the same matrix density. Make sure you support them from underneath at all times and never try to pick them up in your hand, they will probably disintegrate! When running formaldehyde gels, the sample buffer (for making up the RNA solution prior to loading it onto the gel) and the running buffer must also contain formaldehyde. The buffer base is usually a MOPS buffer, pH 8.0 (*Table 2.5*).

If you do choose to run a denaturing polyacrylamide gel, your denaturant of choice will be 8 M urea. This cannot be used for agarose gels, since it inhibits the setting process. When using urea gels, the RNA to be separated

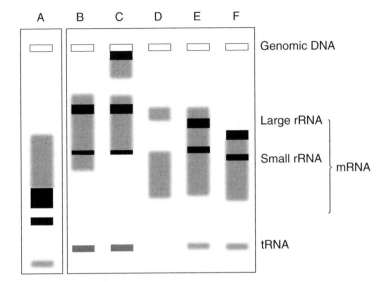

Figure 2.12

Illustration of the separation of total RNA by agarose gel electrophoresis. In lane A, a nondenaturing gel is run. This is to be compared with lane B, which is an illustration of the same sample, but in denaturing conditions. The average apparent mass of all RNA molecules is greater because secondary structure is reduced. Looking at lane B, you can see strong rRNA bands, with the large rRNA being approximately twice the intensity as the small rRNA band. The mRNA runs as a smear because it represents a heterogeneously sized population. Sometimes, but not always, a discrete tRNA band is visible at the bottom of the gel. In C, the presence of genomic DNA can easily be seen at the top of the gel. The sample should be treated with DNase I or re-precipitated with lithium chloride. In D, another common problem is shown, that of RNA degradation. The average size of mRNA is much reduced and the rRNA bands, as well as being less intense, are more smeared. This is a particularly severe example, low-grade degradation can usually be seen by a reduction in the large-to-small rRNA subunit band intensity ratio. Lanes E and F illustrate total RNA from yeast and bacteria, respectively, together showing the different rRNA band sizes, and in F, a reduction in the average mRNA size and tRNA band intensity.

Table 2.5 Recipes of buffers for RNA gel electrophoresis

Gel type	Gel/running buffer	Loading/sample buffer
Denaturing agarose gel	1× MOPS buffer, pH 8.0 (20 mM MOPS, 5 mM sodium acetate, 0.5 mM EDTA) 3% v/v formaldehyde	Running buffer containing 50% v/v glycerol, 9.2% v/v (total) formaldehyde, 0.25% w/v bromophenol blue
Denaturing polyacrylamide gel	1× TBE, pH 8.3 (89 mM Tris base, 89 mM boric acid, 2.5 mM EDTA, 0.1% SDS) 8 M urea	Running buffer containing 50% v/v glycerol, 8 M (total) urea, 0.25 % w/v bromophenol blue

by electrophoresis must be made up in urea sample buffer – the formaldehyde sample buffer used with formaldehyde agarose gels is not suitable. For the running and sample buffers, TBE (Tris-borate-EDTA), pH 8.3 is used as the base with urea being added to 8 M (Reijnders *et al.*, 1973). Often, running the gel at 60°C helps to improve the denaturant action of urea, and this should be recommended if your polyacrylamide gel electrophoresis equipment has the facility to control the temperature of the gel. You should remember that all polyacrylamide gels heat up to some extent when an electric current is applied, so additional heating may not be required. Special adhesive thermometer strips can be purchased, which, once stuck onto the surface of the gel plates, will give a reasonably accurate readout of the temperature of the gel.

A step-by-step guide to RNA electrophoresis

Gel electrophoresis (Ogden and Adams, 1987) is a seemingly simple technique, which is often performed in a very lax manner, producing far from satisfactory results. If you follow the tips set out below, you should produce excellent results every time.

Electrophoresis equipment

Electrophoresis equipment is notoriously rich in nucleases. These can degrade your sample whilst it is sitting in the well or gel, making it literally disappear before your eyes. Whilst denaturants do inhibit nucleases, if you cut RNA bands from the gel, and remove the denaturant later on, the nuclease will have been carried over, and can work its evil. It is best, therefore, to soak all gel combs, trays, plates etc. in 0.5% w/v SDS containing 50 mM EDTA in RNase-free water for 1 h, then rinse them in RNase-free water. Be tempted to use gel trays/plates that can be clipped into special casting trays with rubber seals, meaning that you don't have to use tape in order to pour the gel. The internal compartment of gel running apparatus should be soaked in 3% v/v hydrogen peroxide in RNase-free water then rinsed with RNase-free water alone.

Agarose gels

Agarose gels should have a thickness of 0.5–0.75 cm and never more than 1 cm, because this impedes the electric current. The wells should be made with combs 1–2 mm thick and between 0.5 and 1.0 cm wide depending on the volume of sample to be run, and the number of samples you need to load on each gel. Polyacrylamide gels are normally 1–2 mm thick using 0.5 cm wide combs.

To make an agarose gel, mix the appropriate mass of agarose with the gel/running buffer without formaldehyde (*Table 2.5*) in a conical flask and heat until melted. Normally, a 0.8–1.5% w/v agarose gel is suitable for RNA electrophoresis (*Table 2.4*). The conical flask should be at least twice the volume of agarose gel being melted in it. When molten, agarose is clear, and you will need to periodically swirl the solution to mix it. When all the lumps have gone leave the gel to cool before pouring. Agarose sets at around

50°C, so gels can be kept in a 60°C incubator until needed. In this case do not add the formaldehyde until you need to pour the gel. Indeed, this is good practice in all cases, or otherwise heating the gel is potentially very hazardous and must be performed in a fume hood, which can prove troublesome in some laboratories. If you are going to add formaldehyde after melting the agarose, make sure you take the volume of formaldehyde to be added into consideration when working out how much buffer to you need to melt the agarose into so that the gel is not less dense than required. When adding formaldehyde to molten agarose, this MUST be done in a fume hood. Pour the molten agarose slowly into the gel tray, which has been taped or placed into an appropriate casting tray beforehand. Pour the gel into the center of the tray, making sure the gel tray is on a level surface. It is possible to mark onto the gel try the appropriate thickness for the gel with a marker, though some gel casting trays have depth markers engraved on them. Once you have done this a few times, you will learn the volume of molten agar needed to make one gel. Once the gel has been poured, put in the comb. Make sure the teeth do not touch the bottom of the gel tray. Use tape to raise the comb up if necessary, but make sure it is level. Remove air bubbles using a pipette tip. It is not necessary to burst them, simply push them down to one corner of the gel, at the opposite end from the comb. Leave the gel to set; the agarose will turn opaque. Do not be tempted to put the gel in a fridge to cool it down quickly and do not disturb the gel whilst it is setting, and don't be tempted to prod the gel with your finger, since this could cause disaster! Half an hour should do it. When the gel has set, place it into the gel tank (remove the tape first, if it has been used to cast the gel!) and submerge the gel in running buffer containing formaldehyde, which is almost always the same buffer as used to make the gel (*Table 2.5*). Use only enough running buffer to just completely cover the gel. Make sure you look at the edges of the gel, since here the agarose is often thicker than in the middle. These ridges of agarose must be covered with running buffer. Be careful since the gel can slide off the gel tray. It is best to tilt the gel tray towards you, and support the gel with your fingers (wear gloves) to prevent it moving. When the gel has been submerged for a minute or two, remove the comb with a single steady pull with one hand gripping each end of the comb. This will minimize damage to the gel. If air bubbles appear trapped in the wells, then gently run your gloved finger over the wells, and they should be released.

Polyacrylamide gels

To make polyacrylamide gels it is best to use freshly made buffers (see *Table 2.5* for the gel buffer recipe), and certainly don't use buffers that have been stored for more than a few weeks. The buffers should be filtered and degassed through a 0.22 μm pore size filter under vacuum. Another essential point is that the gel plates must be scrubbed clean and thoroughly dried before use. As a final touch, they should be rinsed in 1:1 (v/v) ethanol:ether and dried with clean tissue. A final rinse on the gel side of each plate with a proprietary gel lining solution, e.g. SigmaCoat, can help give a nice smooth finish essential if polymerization of acrylamide gels is to be efficient. Assemble the gel plate arrangement with a 2 mm spacer on each side, and clip into the casting apparatus according to the manufacturer's instructions. It is worth checking for leaks at the bottom seal by pouring a small amount of RNase-free water between the plates. Leaks can be sealed using Vaseline. Remember to pour away the water before proceeding! If you do not have access to a proprietary casting apparatus, put together the plates, sandwiching a 2 mm spacer on each side, and tape the sides using strong plastic tape. The use of bulldog clips is recommended to hold the arrangement in place. Make up a standard 3% w/v agarose gel solution in water and melt it as above. Pour the solution into an appropriate container and clamp the gel plate sandwich vertically in place, such that the bottom is submerged up to about 0.5 cm in the agarose until it sets. Leave the gel plates in this position whilst pouring the polyacrylamide gel.

The standard polyacrylamide gel strength is 4% w/v acrylamide with 0.2% bis-acrylamide final in the gel (this is referred to in the books as a 19:1 ratio, though don't ask me why!). For 20 ml of gel, add 2 ml of a solution of 40% w/v acrylamide containing 2% w/v bis-acrylamide to a 10× solution of gel running buffer without urea (*Table 2.4*). Add 6 ml of nuclease-free water and then add solid urea to give an 8 M final solution, when made up to 20 ml total volume of gel. Make up the gel to 19.80 ml with more nuclease-free water (due to the volume of urea added, it may not take very much water) and mix thoroughly by inverting. You will know that you have the correct volume because you will have accurately measured out 19.75 ml of water into the container you are making the gel solution up in, and will have marked the meniscus of the water on the surface of the tube before pouring the water away. Some people prefer to use gels containing 5% v/v glycerol, or made using 5× TBE rather than 10× TBE. Also, different acrylamide:bis-acrylamide ratios are used, e.g. 37.5:1. All these modifications will alter resolution and performance of the gel. Only consider changing the above recipe if you are having problems achieving what you want from your gels.

However the polyacrylamide gel has been cast, once the plates are fixed vertically and all leaks have been sealed, add 1 µl of TEMED per ml of gel and 10 µl of 10% APS (ammonium persulfate) w/v in water (this must be freshly made) per milliliter of gel to the acrylamide/bis-acrylamide solution and mix gently by inverting. Apply the mixture between the gel plates using a large pipette by gently dribbling the gel mixture down one edge of the gel, allowing it to fill slowly from the bottom without leaving air bubbles. When the gel is full, apply a comb at the top and leave the gel to set in a perfectly level and upright position. It is best to leave a little gel mixture in a disposable container, since when this has solidified, you will know that the gel has solidified given that setting of polyacrylamide gels is due to a chemical reaction and is not a matter of cooling, as with the setting of agarose gels. The gel can be unclipped from the casting apparatus and the agarose plug and the bulldog clips can be removed, if they have been used. The gel is then clipped into the running apparatus and buffer added to the upper and lower reservoirs. When buffer is present in the upper reservoir, the comb can be removed. Air bubbles in both the wells, and in the bottom portion of the gel where the agarose bung resided must be removed, and this can be done using gentle pipetting of running buffer.

Samples for electrophoresis are made up in sample buffer (*Table 2.5*), which is essentially running buffer to which a high density component, e.g.

glycerol, has been added so the sample sinks to the bottom of the well, and sometimes a dye (e.g. bromophenol blue) is added so the sample is visible, both during loading, and when the sample is being resolved on the gel. Heating the sample to 60°C for a couple of minutes can be very beneficial. Loading the sample is a matter of being careful, and it is particularly important that you do not pipette air bubbles into the well, which will cause loss of sample, as well as deposition of the sample into neighboring wells. In a comparison of the amount of a particular RNA species (e.g. using northern blotting) in different samples separated on a single gel, such cross-contamination or differential loss can produce entirely uninterpretable results. If you have difficulty finding the wells, place a piece of dark paper behind or beneath the gel equipment. This will improve the contrast and you should be able to make out the edges of the wells against the buffer or gel plate surface, which can be reflective. Be careful not to touch the sides or more importantly the bottom of the well when pipetting the sample, since this can cause damage to the well, deforming the band obtained, or causing loss of sample. Also, make sure you do not overload each well. You can easily work out the maximum capacity of each well by multiplying the width (W) and the thickness (D) of the comb teeth used, and then multiplying this by 90% of the length (H) of the comb teeth (to take account of the fact that the comb does not go right into the gel). If all dimensions are used in millimeters for the calculation, the result is the capacity in microliters of each well. For example, a typical well would have dimensions ($W \times D \times H$), in millimeters, of $5 \times 1 \times 4.5$, which equals 22.5 mm^3 or 22.5 μl capacity. In this case, the amount to be loaded should not exceed 20 μl. Do not load RNA at a concentration of more than 1 μg μl^{-1}.

Running a gel

When loaded, the gel equipment should be sealed and connected to the power supply and the gel run as set out in the manufacturer's instructions that come with your equipment. The rule of thumb for agarose gels is 5 V for every centimeter length of gel, though this is often exceeded, and gels are run for around 1 h.

For polyacrylamide gels, 2.5 V per centimeter is appropriate, with gels being run for 3–4 h. Be very careful not to leave exposed wires, or to touch the equipment whilst the power supply is on, since the high voltages used can give you a nasty shock. Leakage of current is a typical problem when running gels. It is usually caused by poorly maintained connectors. Regularly sand down metal connectors and terminals, make sure wires remain firmly soldered in place and terminals are screwed tightly down.

Size markers

Size markers run on the gel alongside your standards must be used if you aim to determine the molecular size of the molecule(s) making up a particular band. Remember that DNA molecules of 1000 base pairs are twice the size of RNA molecules of 1000 nucleotides. Hence if DNA size markers are being used, this fact should be taken into consideration. There are many different ladders of DNA size markers available commercially, and some

RNA size markers; for example, Promega market an evenly spaced RNA ladder made from *in vitro* transcripts ranging from 281 to 6583 nucleotides. The position of rRNA bands within a preparation of RNA is also an excellent internal size standard if, for example, you want to locate in general terms where mRNA molecules have run. Around 1 μg of size standard should be run on each gel lane.

Visualizing RNA and DNA within gels

It is possible to stain RNA quantitatively using dyes that interact with bases. The intensity of staining of a particular nucleic acid is directly proportional to the total number of nucleotides in the molecule and the amount of the molecule present in a band. The most common nucleic acid stain is ethidium bromide, which is a planar molecule capable of intercalating (sitting between) the bases. To visualize, the stained nucleic acid must be exposed to UV (approximately 300 nm) light, which causes the ethidium bromide moieties to fluoresce, emitting orange light with a wavelength of 590 nm. This can be seen with the naked eye or recorded using Polaroid film, or digital image capture devices.

Not surprisingly given its ability to intercalate within nucleic acids, ethidium bromide is a powerful mutagen. Be very careful handling it and make sure not to contaminate surfaces. Gels stained with ethidium bromide should be disposed of according to the regulations set out where you work, and should never be placed in the general waste.

Ethidium bromide can be used to determine nucleic acid concentration by using standards of known concentration on the same gel. Many standards are commercially available for quantification of DNA in bands. Of course this is not particularly useful for RNA bands on gels, since the only discrete bands that can be seen following ethidium bromide staining are those of rRNA molecules. When measuring gene expression, it is the amounts of RNA(s) having a specific sequence that need to be quantified in the gel, so northern blotting followed by sequence specific hybridization is needed, as described in Chapter 3.

Ethidium bromide can be incorporated into the gel matrix to a concentration of 0.5 μg ml^{-1}. In this case, the same concentration must be present in the running buffer so that the stain is not washed out of the gel during electrophoresis. This procedure is normal for visualization of DNA in gels (see Chapter 4), but for analysis of RNA, it is not common, since the ethidium bromide can affect the mobility of RNA. If you are going to use the RNA gel for northern blotting (see Section 3.4) it is best to cut out the lane containing the size markers using a razor blade and stain the markers with ethidium bromide separately. To do this, the gel fragment is immersed in a solution of 0.5 μg ml^{-1} ethidium bromide dissolved in 0.5 M ammonium acetate and gently rocked back and forth for 45 min. To remove background staining of the gel (which will be significant if formaldehyde is present within the gel) a period of destaining for 10 min in RNase-free water is required. After staining, the marker lane is put back in its place in the gel before visualizing the stained size marker bands using a UV transilluminator. When the bands are visible, this will allow you to mark their positions on the gel using a razor blade. Thus, when the gel is placed onto the nitrocellulose or nylon membrane upon which the RNA is to be blotted, the positions of the size standards can be marked with a pencil. The benefit of doing this rather than staining all the RNA in the gel is that ethidium bromide intercalation may well affect both blotting and hybridization of single-stranded nucleic acids (see Section 3.4).

When RNA bands have been blotted onto membranes, it is possible to reversibly stain them using methylene blue (Herrin and Schmidt, 1988). The advantage of doing this is that there is nothing to affect electrophoresis or blotting of the RNA, and the ability to remove the stain prior to hybridization means that, unlike ethidium bromide, it does not affect that process either. Other advantages include the fact that methylene blue is not as hazardous to health as ethidium bromide. Once blotting has occurred (ethidium bromide staining of the gel after blotting verifies that blotting has occurred efficiently) the RNA must be immobilized onto the membrane (see Section 3.2) and soaked in a solution of 0.02% w/v methylene blue in 0.3 M sodium acetate, pH 5.5 for 2–3 min. The bands will be visible to the naked eye, and can be marked on the membrane using a pencil. The stain is then removed by soaking the membrane in 20% v/v ethanol in water for 5 min.

There are a large number of modern nucleic acid stains that can be used in either agarose or polyacrylamide gels. The best known of these is SYBR green (Zipper *et al.*, 2004). This is a fluorescent dye, with many properties in common with ethidium bromide, but it has a number of advantages. For example, it fluoresces very brightly under UV trans-illumination, with a very large difference between the amount of fluorescence when bound to nucleic acids compared with when not bound. This means that the dye is exquisitely sensitive to low amounts of nucleic acid, and that the background staining of gels is almost negligible. Be aware that SYBR green comes in at least two forms. SYBR green I only binds to double-stranded molecules (i.e. DNA). SYBR green II binds to double- and single-stranded molecules (i.e. DNA and RNA). Thus for visualization of RNA, only SYBR green II will do. If you spot any other type of nucleic acid stain being marketed, make sure it will work with RNA!

2.11   Analysis of RNA molecules using the Bioanalyzer

Agilent Technologies Inc. developed the Bioanalyzer systems (current model, 2100) for detailed analysis of nucleic acid and protein samples. The system uses a so-called 'chip', which is made of plastic with 16 circular sample wells cut into it. Each well is connected through the chip to a separate anode by 'micro-channels', which are filled with sieving polymer and a fluorescent dye. Each micro-channel is the same length. The machine works by inserting a cathode into each sample well, creating a circuit through the sieving polymer. Hence, the system works by separating samples according to molecular weight, following the same principles of electrophoresis described in Section 2.10, but on a very small scale. Since the micro-channels contain fluorescent dye, there is no need for separate staining, and the Bioanalyzer has a laser, which reads the electrophoretogram from each sample by scanning along each micro-channel, exciting the fluorophore on the dye and reading the intensity of emitted light. A sample electrophoretogram is shown in *Figure 2.13*. This, coupled with computer software that can interpret the electrophoretogram, allows rapid and sophisticated analysis of sample component sizes and intensities. Internal standards are provided in each well, which allow both size and concentration determination provided one sample well is used to run a commercial size marker to calibrate the machine. Of course, it is possible to scan images of stained gels to produce electrophoretograms (*Figure 2.13*), and use commercial image analysis software specifically marketed for band intensity analysis (see Chapter 4). The use of the Bioanalyzer, however, allows small-scale analysis, meaning very little sample loss, and has the prime advantage of tighter control over separation than standard gels, meaning higher levels of resolution, and more accurate electrophoretogram data. Different chips are available for different purposes.

When analyzing RNA samples using a denaturing sieving matrix in the Bioanalyzer, two distinct peaks are seen (28S and 18S rRNA for eukaryotic samples or 23S and 16S rRNA for bacterial samples) corresponding to the two bands seen upon ethidium bromide staining of denaturing agarose gels. The coding RNA molecules are present as a smear, which produces a broad peak flanking the 16/18S rRNA peak. The tRNA (plus other small structured RNAs) molecules do sometimes produce a discrete, small peak at the very

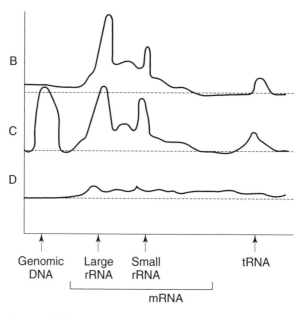

Figure 2.13

Electrophoretograms of RNA samples run on agarose gels. The figure shows band intensity measurements from top to bottom, transposed as left to right. The lanes illustrated are lanes B, C and D from Figure 2.12. They further illustrate genomic DNA contamination (lane C) and RNA degradation (lane D). These electrophoretograms can be used to determine the amount of RNA present in each sample, one relative to the other, through calculating the area under the total curve (excluding DNA contamination).

edge of the electrophoretogram dependent upon the integrity of the RNA in the sample, since breakdown products can occlude this peak (see *Figure 2.13*).

The machine can determine total RNA concentration by measuring the area under the entire electrophoretogram. It will also locate any anomalies in the electrophoretogram; for example, DNA contamination, which will produce high molecular weight smears and peaks. RNA integrity (RIN) can also be measured. The process of electrophoretogram analysis to produce an RIN number for your sample is complex, but basically involves determining the relative intensities of the rRNA bands, the overall average size of molecules in the total RNA sample and the production of very small degradation products which occlude the tRNA peak. As part of developing this RIN analysis, Agilent determined electrophoretograms for hundreds of RNA samples after they had been treated in different ways so that they degraded to different extents, and then tested the RNA in the samples for its suitability in downstream applications. Thus the RIN value obtained for your sample can give you an idea as to how well it will perform, or whether it should be binned and the RNA preparation repeated! The analysis was performed with eukaryotic total RNA samples, however, so it is not always possible to confirm the usefulness of prokaryotic RNA samples, for example in microarray analysis, but as the knowledge base improves, more information will become available. As a rule of thumb, however, RIN numbers of > 5 are needed before you can be confident of good performance of your RNA. Bioanalyzer machines are not cheap, and unless you have an unlimited budget, should only really be considered if you are running large numbers of samples, and want to perform very RNA integrity sensitive downstream applications, e.g. microarrays or real time RT-PCR. Indeed, for these applications, I would strongly recommend using this equipment, even if you have to commandeer someone else's. The Bioanalyzer can perform many other functions that will not be discussed in this book; for example protein and DNA electrophoresis and analysis, and flow cytometry in miniature, hence if your own usage level would not justify the financial outlay, it might be that a number of research groups might club together to make the purchase. The vast majority of serious transcriptomics facilities will have a Bioanalyzer already.

Further reading

Jones, P, Qui, J and Rickwood, D (1994) *RNA Isolation and Analysis*. BIOS Scientific Publishers, Oxford, UK.

Lorkawski, S and Cullen, P (Eds) (2003) *Analysing Gene Expression, A Handbook of Methods, Possibilities and Pitfalls*. Wiley VCH, Weinheim, Germany, pp. 97–116.

References

Amara, RR and Vijaya, S (1997) Specific polyadenylation and purification of total mRNA from *Escherichia coli*. *Nucleic Acids Res* 25: 3465–3470.

Aviv, H and Leder, P (1972) Purification of biologically active globin messenger RNA by chromatography on oligothymidylic acid-cellulose. *Proc Natl Acad Sci USA* 69: 1408–1412.

Bartel, DP and Unrau, PJ (1999) Constructing an RNA world. *Trends Cell Biol* 9: 9–13.

Berger, SL (1975) Diethyl pyrocarbonate: an examination of its properties in buffered solutions with a new assay technique. *Anal Biochem* 67: 428–437.

Cathala, G, Savoret, JF, Mendez, B, West, BL, Karin, M, Martial, JA and Baxter, JD (1983) Laboratory methods: a method for isolation of intact, translationally active ribonucleic acid. *DNA* 2: 329–335.

Chomczynski, P and Sacchi, H (1987) Single-step method of RNA isolation by acid guanidinium thiocyanate–phenol–chloroform extraction. *Anal Biochem* 162: 156–159.

Favaloro, J, Triesman, R and Kaman, R (1980) Transcriptional maps of polyoma virus-specific RNA: analysis by two-dimensional S1 gel mapping. *Methods Enzymol* 65: 718–749.

Glisin, V, Crkvenjakov, VR and Byus, C (1974) Ribonucleic acid purified by cesium chloride centrifugation. *Biochemistry* 13: 2633–2637.

Helling, RB, Goodman, HM and Boyer, HW (1974) Analysis of endonuclease-R EcoR1 fragments of DNA from labdoid bacteriophages and other viruses by agarose gel electrophoresis. *J Virol* 14: 1235–1244.

Herrin, DL and Schmidt, GW (1988) Rapid, reversible staining of northern blots prior to hybridization. *Biotechniques* 6: 196–200.

Kirby, KS (1968) Isolation of nucleic acids by phenolic solvents. In: *Methods in Enzymology XIIB* (eds L. Grossman and K. Moldave) Academic Press, pp. 87–99.

Lehrach, H, Diamond, D, Wozney, JM and Boedtker, H (1977) RNA molecular weight determinations by gel electrophoresis under denaturing conditions, a critical reexamination. *Biochemistry* 16: 4743–4751.

Ogden, RC and Adams, DA (1987) Electrophoresis in agarose and acrylamide gels. *Methods Enzymol* 152: 61–87.

Reijnders, L, Sloof, P, Sival, J and Borst P (1973). Gel electrophoresis of RNA under denaturing conditions. *Biochim Biophys Acta* 324: 320–33.

Wilfinger, DA, Mackey, M and Chomczynski, P (1997) Effect of pH and ionic strength on the spectrophotometric assessment of nucleic acid purity. *Biotechniques* 22: 474–481.

Zipper, H, Brunner, H, Bernhagen, J and Vitzthum, F (2004) Investigations on DNA intercalation and surface binding by SYBR Green I, its structure determination and methodological implications. *Nucleic Acids Res* 32: e103.

Protocol 2.1 Isolation of RNA from animal cells using the acid phenol method

EQUIPMENT

Stoppered glass tubes or Eppendorf tubes
A rotor–stator homogenizer or a mechanical disrupter for the cells
18-gauge needle and appropriate syringe
Bench centrifuge
Pipettes

MATERIALS AND METHODS

1. (i) Take animal tissue or pelleted animal cells (primary or tissue culture cells, with the culture medium removed) and put in an appropriately-sized tube and add 1 ml of disruption buffer (50 mM Tris-HCl, pH 7.5, containing 50 mM NaCl, 5 mM EDTA (pH 8.0), 0.5% w/v SDS and 0.1% w/v 8-hydroxyquinoline OR 200 μg ml^{-1} proteinase K) per 50 mg of tissue. If the tissue/cells have been pre-treated with RNA stabilization reagent, then keep the tube on ice until needed. If not, then it is best to keep the sample warm, and to proceed quickly. Homogenize the tissue with two 15-s bursts with a rotor–stator homogenizer such as a Polytron with a probe size appropriate for the size of sample (see the manufacturer's instructions). Alternatively, use a mechanical disruption method, e.g. a Thermo-Hybaid Ribolyser and acid-washed glass beads in 1.5-ml plastic tubes. A disruption time of 40 s at speed setting 6 should be enough.

    (ii) For tissue culture cells grown in monolayer, pour off the medium and add 1 μl per 10^4 cells of disruption reagent (above), and disrupt cells with a rubber policeman. Homogenize the sample by passing it through an 18-gauge needle, 10–15 times if the extract appears gloopy.

2. If proteinase K has been used in the disruption reagent, incubate the cell extract (however the extract has been obtained) for 15 min to digest protein (including nucleases). If not, then proceed immediately to the next stage.

3. To the extract, add an equal volume of buffer-saturated phenol:chloroform:isoamyl alcohol (25:24:1), pH 4.5 (this reagent can be purchased in

molecular biology grade from a number of different suppliers). Put the lid on the tube tightly and shake vigorously for 20 s. Then leave the solution on the bench for 10 min to allow dissociation of RNA/protein complexes.

4. Centrifuge at 12 000 rpm (14 000 × g) in a bench-top centrifuge for 10 min.
5. Take off the top aqueous phase and move to a fresh tube. Leave the interphase intact.
6. Add an equal volume of chloroform to the aqueous phase from step 5. Shake, centrifuge and remove the top phase as above. Repeat until there is no obvious cloudiness (precipitated protein) at the interphase.
7. Add one volume of ice-cold isopropanol to the aqueous phase. Mix gently by inverting the tube and put back on ice for 10 min, inverting every minute or so.
8. Centrifuge for 10 min at 14 000 × g at 4°C if possible to collect the RNA pellet. It should be obvious – white and fluffy. Be careful because the pellet will be softer and more likely to come off the side of the tube than a DNA pellet.
9. If the pellet is very small, add sodium acetate, pH 5.2 to a concentration of 0.3 M to the liquid in the tube. Incubate on ice for a further 10 min and re-centrifuge. If the pellet is a good size, this step should be missed out because sodium acetate will be difficult to completely remove from the RNA sample and may affect downstream reactions.
10. Pipette off the supernatant gently and add 1 ml of ice-cold 70% v/v ethanol in water. Centrifuge again at 14 000 × g for 5 min to wash the pellet, and gently pipette off the residual ethanol. If you have used sodium acetate, a second ethanol wash will be required. Finally, air dry the RNA pellet.
11. Make up the pellet in 50–100 µl of nuclease-free water dependent upon the size of the pellet, and with reference to Table 2.2.

Protocol 2.2 Isolation of RNA from bacterial and yeast cells using guanidine isothiocyanate and lithium chloride precipitation

EQUIPMENT

Glass tubes or Eppendorf tubes
Vortex mixer
Bench-top centrifuge
18-gauge needle and appropriate syringe
Pipettes

MATERIALS AND METHODS

1. (i) Resuspend 1×10^8 to 1×10^9 bacterial cells in 100 µl of TE buffer (10 mM Tris-HCl, 5 mM EDTA, pH 8.0) containing 0.4 mg l^{-1} (for Gram negative bacteria) or 3 mg l^{-1} (for Gram positive bacteria) lysozyme by pipetting up and down.
    (ii) Resuspend 10^8 pelleted yeast cells in 1 ml of buffer (1 M sorbitol, 0.1 M EDTA, pH 7.4 containing 0.1% v/v β-mercaptoethanol, added just before use, and 500 U ml^{-1} zymolase).
2. (i) Leave the mixture to incubate at room temperature for 5 min for Gram negative bacteria or 15 min for Gram positive bacteria.
    (ii) Leave the suspension of yeast cells to incubate for 30 min at room temperature and centrifuge it for 5 min at $300 \times \boldsymbol{g}$ to pellet spheroplast cells. Carefully discard the entire supernatant.
3. Add 350 µl of lysis buffer (5 M guanidinium isothiocyanate, 30 mM sodium citrate, pH 7.0, 0.3% w/v sarcosyl, 125 mM β-mercaptoethanol, the last added just before use of the buffer) and vortex for 10 s.
4. Homogenize by passing through an 18-gauge needle 10–15 times if the sample is very thick.
5. Centrifuge at $14\,000 \times \boldsymbol{g}$ for 10 min in a bench-top microfuge. Remove the supernatant to a fresh tube, without disturbing the pelleted protein. At this stage, you could use Protocol 2.6 to purify RNA, or continue as set out below.

6. Add an equal volume of chloroform. Shake vigorously for 20 s, centrifuge for 10 min at 14 000 × g and remove the top, aqueous phase containing DNA and RNA to a clean tube.

7. Add lithium chloride to a final concentration of 4 M to selectively precipitate RNA. Mix gently by inverting the tube and incubate at 4°C overnight.

8. Centrifuge for 10 min at 14 000 × g at 4°C, if possible, to collect the RNA pellet.

9. Pipette off the supernatant and add 1 ml of ice-cold 70% v/v ethanol in water. Centrifuge again at 14 000 × g for 5 min to wash the pellet, and gently pipette off the residual ethanol. Air dry the pellet.

10. Make up the pellet in 50–100 µl of nuclease-free water dependent upon the size of the pellet and with reference to Table 2.2.

Protocol 2.3 Isolation of RNA from plant and filamentous fungal cells by using guanidine hydrochloride

EQUIPMENT

Aluminum foil
Pestle and mortar
Glass tubes or Eppendorf tubes
Rotor–stator homogenizer
Bench-top centrifuge

MATERIALS AND METHODS

1. Snap freeze the sample in liquid nitrogen after first weighing it and wrapping it in aluminum foil.
2. Unwrap the frozen material and place it in a pestle filled with liquid nitrogen.
3. Using a mortar, grind up the material into a smooth paste, adding more liquid nitrogen as required.
4. Place the pestle onto a bed of dry ice and allow the liquid nitrogen to evaporate.
5. For each 100 mg of material add 1 ml of disruption buffer containing 8 M guanidine hydrochloride, 20 mM 2-(N-morpholino)ethanesulfonic acid (MES) at pH 7.0, 20 mM EDTA at pH 8.0, and 0.22 M β-mercaptoethanol (added immediately before use).
6. Homogenize the sample using a rotor–stator homogenizer (three 15-s bursts) or some form of mechanical disruption (two 30-s bursts).
7. Centrifuge the cell extract, $14\,000 \times g$ for 10 min in a bench-top centrifuge. Remove the supernatant and proceed to chloroform clean, then lithium chloride precipitation of RNA, as set out in Protocol 2.2 from step 6. Alternatively, you could use Protocol 2.6 to purify RNA from this supernatant.

Protocol 2.4 Rapid isolation of RNA from animal tissues and cells using guanidine isothiocyanate, lithium chloride and cesium trifluoroacetate isopycnic density centrifugation

EQUIPMENT

Glass tubes or Eppendorf tubes
Rotor–stator homogenizer or Bead-grinding machine
Bench-top centrifuge

MATERIALS AND METHODS

1. Select the appropriate tube. Up to 50 mg of tissue or 5×10^6 tissue culture cells can be processed in a 1.5-ml Eppendorf tube. The volumes set out in this protocol are for this amount of starting material. For your experiment scale the volumes up or down as necessary dependent upon the amount of starting material (do not use for less than 25 mg tissue or 1×10^6 cells). For larger amounts of material, use larger centrifuge tubes, such as a 15-ml glass Corex tubes.

2. Put 50 mg tissue or 5×10^6 pelleted tissue culture cells from which the supernatant has been poured off (if these are monolayer cells, disperse the cells into $1 \times$ PBS by pipetting up and down before pelleting) into a 1.5-ml bead grinding tube, which already has acid-washed glass beads in it, or into a 1.5-ml Eppendorf tube if you will be using a rotor–stator homogenizer. Alternatively, use a glass homogenizer tube with a Teflon-coated plunger, such as a Potter–Elvejhem homogenizer. Add 150 μl of disruption buffer (4 M guanidinium isothiocyanate, 25 mM sodium citrate, pH 7.0, containing 2% v/v β-mercaptoethanol, added immediately before use, and 0.25% w/v sarcosyl).

3. Homogenize for one 15-s burst (Polytron); one 20-s burst, (mechanical disruptor) or until an even, but thick homogenate has been produced (manual homogenizer).

4. Add 350 µl of 6 M lithium chloride and homogenize again as in step 3. It is particularly important that homogenization is complete because any debris will pellet with the RNA. This is why mechanical disruption is favored over manual homogenization. If you have a number of samples, place each one on ice after homogenization has been completed. Complete all homogenizations before moving to the next step. For samples homogenized manually in a homogenization vessel, pour the homogenate into a clean tube. For samples homogenized using a bead grinding machine, pellet the beads and any cell debris with a 2-min centrifugation step (full speed in a bench-top centrifuge) and transfer the supernatant to a fresh tube.

5. Add 500 µl of cesium trifluoroacetate (this is purchased as a liquid with a density of 1.50 g ml^{-1}) to each homogenate. Place the tubes on ice for 10 min.

6. Centrifuge the samples at $14\,000 \times g$ for 15 min at room temperature. If the centrifuge has a cooling capacity, select 15°C.

7. There will be a pellet of RNA, a liquid phase being the cesium trifluoroacetate cushion, within which will be the DNA, and a protein layer on top. Remove the liquid phase and protein layer using a vacuum aspirator or a single pipetting motion (if the volume is not too great). Pipetting off the supernatant using a number of steps will lead to protein contamination of the pellet.

8. To wash the RNA pellet, add 75 µl of disruption buffer (above), 175 µl of 6 M lithium chloride, 250 µl of cesium trifluoroacetate and 1 ml of ice-cold 70% ethanol. If there is significant protein contamination on the side of the tube (more common with large glass tubes), then add 10% of the total required volume of ethanol directly to the bottom of the tube and pipette up and down to disrupt the RNA pellet, and transfer the pellet to a fresh tube before adding the rest of the wash reagents.

9. Centrifuge again, as in step 6 but for 5 min not 15, and aspirate the supernatant as in step 7.

10. Carry out a final wash with 1 ml of ice-cold 70% ethanol, centrifuge for 5 min and aspirate the supernatant.

11. Make up the pellet in 50–100 µl of nucelase-free water.

Protocol 2.5 Separate isolation of cytoplasmic and nuclear RNA from tissue culture cells

EQUIPMENT

Bench-top centrifuge
Glass tubes or Eppendorf tubes
Pipettes

MATERIALS AND METHODS

1. Add 175 µl of refrigerated lysis buffer (50 mM Tris-HCl, pH 8.0, 140 mM NaCl, 1.5 mM MgCl$_2$, 0.5% v/v Igepal CA-630 (Sigma) containing 1000 U ml^{-1} human placental RNase inhibitor (Table 2.1)) to no more than 1×10^7 pelleted tissue culture cells. Pipette gently up and down to disrupt the pellet.
2. Leave the suspension on ice for 5 min. The cells will lyse very rapidly, causing the suspension to clarify.
3. Centrifuge the lysate at $300 \times \boldsymbol{g}$ for 2 min at 4°C to pellet nuclei and cell debris. Take off the supernatant (which contains cytoplasmic RNA only) and transfer to a fresh tube.
4. To purify cytoplasmic RNA, add 600 µl of disruption buffer (5 M guanidinium isothiocyanate, 30 mM sodium citrate, pH 7.0, 0.3% w/v sarcosyl, 125 mM β-mercaptoethanol, added just before use of the buffer, to the supernatant from step 3. Next, add 500 µl chloroform, shake vigorously for 20 s and centrifuge at $14\,000 \times \boldsymbol{g}$ for 10 min. Remove the aqueous phase to a fresh tube and precipitate RNA as set out in Protocol 2.1 from step 7. Alternatively, instead of adding chloroform and proceeding to precipitate RNA, you could use Protocol 2.6 to purify RNA after the addition of disruption buffer to the supernatant from step 3.
5. To isolate and purify nuclear RNA, proceed exactly as set out in Protocol 2.2 from step 3.

Protocol 2.6 Purification of RNA using silica beads

EQUIPMENT

Glass tubes or Eppendorf tubes
Bench-top centrifuge
Pipettes
Vortex mixer

MATERIALS AND METHODS

1. To the cell extract, usually generated using guanidine denaturation, add an equal volume of 70% ethanol. Pipette up and down to mix.
2. Centrifuge to pellet debris, 5 min, $14\,000 \times \boldsymbol{g}$, room temperature.
3. Transfer the supernatant to a fresh tube and 25 µl of a 50% w/v slurry of chromatography grade silica gel (approx. 25 µm diameter; for example, Sigma catalogue number 60734) in 70% ethanol. Mix by vortexing and leave at room temperature for 5 min.
4. Centrifuge to pellet beads, 1 min, $14\,000 \times \boldsymbol{g}$, room temperature.
5. Discard supernatant; add 1 ml 70% ethanol, vortex, centrifuge as in step 4 and again discard supernatant.
6. Add 50–100 µl of water to the beads, vortex, centrifuge as in step 4 and pipette off the supernatant containing RNA.

Hybridization-based methods for measuring transcript levels

3

3.1 The basics of nucleic acid hybridization

The natural structure of DNA is in the form of an anti-parallel double helix with each strand possessing a negatively charged phosphate backbone opposite nitrogenous bases, which are capable of donating hydrogen bonds. The precise geometry of the different bases means that these hydrogen bonds are donated to other bases placed in opposition to them in an anti-parallel fashion using the classic Watson–Crick base pairing rubric: A=T and G≡C. These pairs of bases are said to be 'complementary' to one another. Pairs of complementary nucleotides can form hydrogen bonds in solution, but if two complementary polynucleotides come together, the strength of the hydrogen-bond attraction between the molecules is equal to the sum of all the individual nucleotide interactions. Pure genomic DNA consists of two perfectly complementary strands.

Hydrogen bonds between the individual strands of a DNA molecule are usually broken in the test tube in two main ways: first, by heating the molecule so the strands vibrate faster and faster until they separate; second, removing the hydrogen bonding potential by raising the pH such that the protons are sucked out to interact with OH⁻ ions in solution. When DNA molecules are heated to separate the two strands, they are said to have been 'melted', when sodium hydroxide is used to raise the pH, the DNA is said to be 'denatured'. When the temperature or pH is lowered again and the strands come back together to reform a duplex, this is referred to as 'annealing', or perhaps more accurately, 're-annealing'.

Since it is much easier to manipulate the temperature of a solution rather than its pH, and since long-term exposure to alkaline pH can result in damage to nucleic acids (particularly to RNA) heat is the preferred choice for converting double-stranded nucleic acids into their constituent strands. The amount of heat needed to melt a DNA molecule is defined by the 'melting temperature', which is assigned to each nucleic acid duplex. Mathematically, the melting temperature of a duplex is the temperature in degrees centigrade at which half of all the double-stranded molecules in a sample are melted. The melting temperature depends on a number of factors, but essentially it is dictated by the number of hydrogen bonds that are keeping the duplex together. The more hydrogen bonds, the more heat needed to break them all. The overall length of the DNA molecule dictates the total number of hydrogen bonds present, but this is a very complex relationship, since long, single-stranded molecules can form intramolecular secondary structures, adversely affecting inter-strand annealing. The 'GC content' (i.e. the proportion of the sequence that is made up of guanine and cytosine nucleotides) of a DNA molecule affects its melting temperature because G and C bases donate three hydrogen bonds, whereas A and T bases donate two. Thus the higher the GC content, the greater the density of hydrogen bonds, and the more heat needed to break them all. Other factors that can affect annealing include Na^+ concentration (concentrations of around 1 M are required for maximal annealing). This is the case because Na^+ ions neutralize to some extent the very strong negative charge on the phosphate backbone of each nucleic acid strand. If this negative charge is present on both strands at its maximal extent, charge–charge repulsion occurs, which tends to push the strands apart, disfavoring annealing. The presence of organic solvents such as formamide or dimethyl sulfoxide (DMSO) also prevent hydrogen bond formation, as does formaldehyde, which alkylates the bases.

Hybridization of probes to target nucleic acid strands

Once a DNA molecule has been melted, it is possible to add a separate, single-stranded DNA molecule (known as a 'probe') that has a sequence which is complementary to one of the strands of the melted DNA duplex (the so called 'target strand'). In this case, when the temperature is lowered, annealing can occur either between the target strand and its original partner strand, or between the target strand and probe. In the latter case, annealing is referred to as 'hybridization'. The amount of hybridization that occurs depends on the melting temperature of the hybrid duplex compared to the melting temperature of the original duplex; this difference can be significant. For example, the probe is not always perfectly complementary to the sequence of the target strand; that is, it may have a slightly different nucleotide sequence, perhaps because the exact sequence of the target strand was not known. Incomplete complementarity results in a number of 'mismatched' bases, which reduces the number of potential hydrogen bonds that form, reducing the melting temperature of the hybrid duplex. To overcome any possible bias towards reforming the original DNA duplex, therefore, and to favor hybridization, the probe is always used in large excess. The basic principles of nucleic acid hybridization are illustrated in *Figure 3.1.*

What is the point of a probe? Well so much of what we do in molecular biology is to devise ways of observing things that cannot be seen with the naked eye. How, for example, can you see nucleic acid molecules that have a specific nucleotide sequence? Well, one way to do this is to make the molecules more visible by incorporating a label into them, usually either a colored or fluorescent dye or a radioactive isotope. The best way to achieve this is to make a labeled single-stranded probe complementary to the sequence of the target nucleic acid strand and then to hybridize the probe and target strand together, allowing the target strand to be visualized.

Another type of molecule that is essentially a probe is a very short

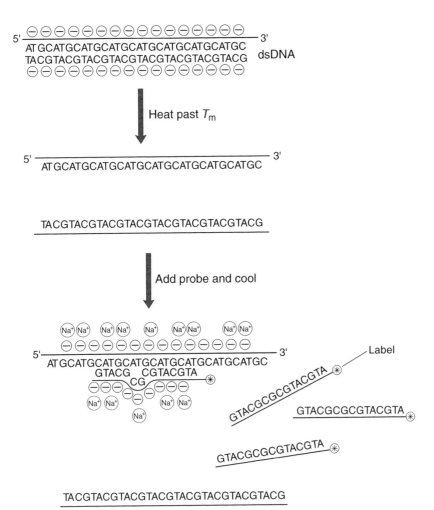

Figure 3.1

The principle of nucleic acid hybridization. When a target double-stranded nucleic acid (the figure illustrates DNA) is heated above the melting temperature (T_m), the strands dissociate. Probe sequences complementary to one of the two target strands are added in considerable excess and when the temperature cools, the probes anneal. The figure illustrates a number of points. The Na^+ ions in solution neutralize somewhat the negative charges on the phosphate backbones of the DNA strands, promoting hybridization; if the probes are labeled in some way, the hybrid product can be visualized; the probe does not have to cover the whole target DNA strand, and mismatches can be accommodated without blocking hybridization, though it will reduce the melting temperature.

complementary oligonucleotide sequence, which hybridizes to its target sequence and whose sole role is to provide a 3' hydroxyl group onto which an enzyme with DNA polymerase activity can add nucleotides. Such probes are commonly referred to as 'primers' and will be discussed later in this chapter and in Chapter 4.

3.2 Blotting

It is quite possible to perform nucleic acid hybridization when both target strand and probe are in the liquid phase (i.e. in solution in a test tube). The major reason for hybridizing a labeled probe to a target strand is, however, to visualize the hybrid product. This requires that any unbound probe must be removed after hybridization has occurred, since a free probe would mask visualization of the hybrid. The most efficient way of removing free probe is to place the target nucleic acid strand onto a solid phase surface, such as a membrane, and then to wash over the probe in solution so that it hybridizes to the target strand. When the excess probe has been rinsed away, the labeled hybrid remains bound to the membrane and can be visualized by a variety of means dependent upon the particular label used (*Figure 3.2*) as discussed in Section 3.5. Binding a nucleic acid to the surface of a membrane is referred to as 'blotting'. Other types of molecules are said to be blotted when they are bound to the surfaces of membranes; the most common example is protein and this type of blotting procedure will be dealt with in Chapter 6.

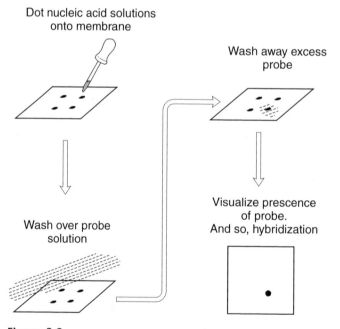

Figure 3.2

The basics of nucleic acid blotting and hybridization. In a dot nucleic acid blot, samples of total nucleic acid are spotted onto a membrane. A labeled probe, complementary to one specific nucleic acid is washed over the surface and hybridization occurs if the complementary nucleic acid is present within the spot. When the excess probe is removed any hybridization can be visualized using any number of methods described in the text, but most commonly, autoradiography. Not only does this procedure give a binary result, but the amount of hybridization is proportional to the amount of target nucleic acid, so it can be at least semiquantitative.

Types of blotting technique

The first nucleic acid blot was undertaken by Ed Southern, who immobilized DNA (Southern, 1975). Since that point, the blotting of DNA has been referred to as a Southern blot. There are many different ways of blotting, whether the type of molecule being blotted is DNA or something else. The most simple is the 'dot blot' where a sample is taken out of a test tube with a pipette and is spotted onto the membrane (*Figure 3.2*). Next comes the 'slot blot' where a liquid sample is placed onto the membrane in a plastic slot cut into a specially made manifold and a vacuum is applied, sucking the liquid through the membrane, leaving the nucleic acid bound to the surface. Then comes the 'gel blot', where nucleic acids are separated using gel electrophoresis, as set out in Section 2.10, and the membrane is placed on top of the gel. The nucleic acids are drawn through the gel until they hit the membrane and bind to its surface. Gel blots are generally performed in one of two main ways. In the most common application, the so-called 'capillary blot', transfer buffer moves from a reservoir below the gel into a dry stack of paper towels above the membrane by capillary action (*Figure 3.3*). In so doing, the buffer passes through the gel and carries the nucleic acid with it, until it hits the surface of the membrane above, where it stops. The second gel blot application is the 'electro-blot' (Bittner *et al.*, 1980). Here, an electric current is applied across the gel and membrane in such a way that the anode is above the membrane and the cathode is below the gel. Since nucleic acids have a net negative charge, they move towards the anode, out of the gel, but stop once they hit the membrane (see Chapter 6 for more details). The final form of blotting is a so-called 'colony blot' and is generally restricted to a Southern blot for analysis of recombinant bacterial colonies to find those that carry particular sequences. This technique is used extensively when searching a genomic or cDNA library for recombinants carrying the DNA sequence you want to clone. Clearly, this has nothing to do with measuring gene expression, so this technique will not be referred to any more in this book.

If hybridization of a probe to a target nucleic acid that has been blotted onto a membrane is your aim, the target nucleic acid must be made single-stranded before blotting. Of course, RNA is single-stranded in its normal state, but DNA must be denatured prior to blotting by adding sodium hydroxide or very high concentrations of salt. In the case of Southern gel blots, the entire gel is soaked in 20× SSC (*Table 3.1*) (Southern, 1975) or 0.4 M sodium hydroxide (Reed and Mann, 1985) to denature the DNA prior to transfer. Alkaline transfer is best, since denaturation is more consistent. It cannot be heated up to melt the DNA, or the gel would also melt. In the case of dot and slot blots, concentrated sodium hydroxide solution is added directly to the sample of DNA until the required final concentration is reached. RNA should not be exposed to alkaline conditions, because this can cause alkali hydrolysis of the molecule.

Once blotted onto the surface of a membrane, nucleic acid molecules must be fixed in place ('immobilized') so that they do not float away when solutions, particularly high ionic strength buffers, are washed over the surface of the membrane. This is done in different ways depending on the type of membrane the nucleic acid has been blotted on to.

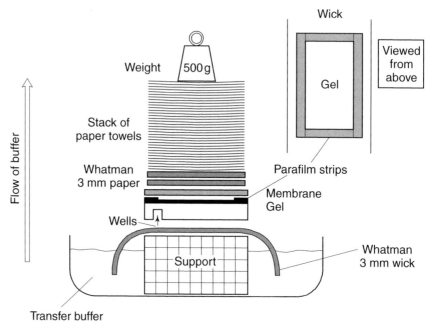

Figure 3.3

How to perform a capillary gel blot. At the start of the procedure, the membrane, the gel, the 3 mm filter wick and the two sheets of 3 mm filter paper above it are soaked in transfer buffer, and more buffer is placed in a reservoir, into which the wick is dipped. Buffer is drawn up from the filter papers into a wad of dry paper towels because of capillary action induced by the presence of the weight. Accordingly, buffer is sucked from the reservoir, through the agarose gel and membrane to replenish that which has been removed from the filter papers. The flow continues until the stack of paper towels becomes saturated. As buffer flows through the gel it carries nucleic acids with it which hit the membrane above and stick, because they are too large to pass through the pores of the membrane. The edge of the gel must be covered with Parafilm (insert) to prevent the possibility of the gel or wick touching the filter papers above, which would result in buffer flowing past the membrane and transfer would stop. The gel is placed well-side down simply to reduce the distance from the nucleic acid to the membrane and so reduce the required transfer time.

Types of blotting membrane

Two main types of membrane materials are used for nucleic acid blots: nitrocellulose and nylon. Polyvinylidine difluoride (PVDF) membranes are also sometimes used for nucleic acid blotting, but are generally reserved for protein blots (see Chapter 6), so they will not be discussed further here. All membranes have pores; 0.45 µm pore size is standard, but 0.22 µm pore size membranes should be used in preference to larger pore sizes when blotting short nucleic acids (which might otherwise pass through the pores). However, do not use small pore sizes unnecessarily, particularly when capillary blotting is being employed, since small pores will slow the rate of

Table 3.1 Buffers for optimal use in northern blotting and hybridization

Buffer	Constituents
Transfer buffer	10× SSPE (1.5 M NaCl, 100 mM NaH_2PO_4, 10 mM EDTA, pH 7.4
Hybridization buffer	50% v/v formamide in [5× SSPE] 750 mM NaCl, 50 mM NaH_2PO_4, 5 mM EDTA, pH 7.4 containing [2× Dehardt's solution] 0.4 g l^{-1} Ficol, 0.4 g l^{-1} polyvinyl pyrrolidine, 0.4 g l^{-1} bovine serum albumin, fraction V plus 0.1% (w/v) SDS and 0.1 mg ml^{-1} heterologous DNA Hybridize at 42°C
Wash buffer	[0.2× SSC] 30 mM NaCl, 3 mM sodium citrate, pH 7.0 containing 0.1% w/v SDS Wash twice at 50–68°C for 10 min depending upon experience with your probe
Stripping buffer	10 mM Tris-HCl, pH 7.5 containing 1 mM EDTA and 1% SDS Incubate for 10 min at 96°C

capillary action. It is also very important to remember that all types of membrane can be overloaded with nucleic acid, meaning that some does not bind properly and is washed away. This is an important consideration when the purpose of the blot is to quantify the amount of nucleic acid in a sample, since the assumption must be that all of it has bound.

Nitrocellulose was the first membrane to be routinely used for nucleic acid blotting. The interaction between nitrocellulose and nucleic acids is believed to be hydrophobic in nature meaning that very high ionic strength buffers (which cancel out surface charge on the nucleic acids, making hydrophobic interactions most significant) are required for binding to occur. There are several important problems with using nitrocellulose, however. It has a relatively low nucleic acid binding capacity, particularly for small molecules. Furthermore, the membranes are brittle when dry, making hybridization experiments difficult. It is not an option to dry the membrane incompletely to get over this problem of brittleness, because any remaining moisture will severely compromise the hybridization reaction. Furthermore, if moisture is present in blotches on the surface, any hybridization within those blotched regions will follow different kinetics to that in the rest of the membrane, making attempts to accurately compare hybridization levels at different points on the membrane nearly impossible. Baking the membrane at 80°C for 2 h in a vacuum oven is the only way to properly dry nitrocellulose, and this process also immobilizes nucleic acids onto the surface. Given that vacuum ovens are not standard pieces of equipment in most laboratories, this can be difficult to achieve. The final problem is that because the interaction between nitrocellulose and nucleic acids is weak, even immobilized nucleic acids can dissociate into solution, meaning that multiple rounds of probing and stripping the blots (see below) are not possible.

Nylon filters have for a long time been more popular for nucleic acid blotting than those made from nitrocellulose. This is primarily because they are stronger and more capable of binding nucleic acids. These advantageous properties are even more pronounced for the more advanced so called 'nylon+' membranes now commercially available. As a result, these types of membranes are being used more and more. Binding of nucleic acids to the surfaces of nylon membranes is due to charge–charge interactions; the membranes are positively charged and the backbones of nucleic acids are negatively charged. This means that blotting can be undertaken at medium salt concentrations. Furthermore, because binding is very strong, once bound and visualized, probes can be stripped efficiently from the target nucleic acid sequence without the fear of significantly disrupting the inter-action between the target nucleic acid strand and the membrane. Thus, sequential stripping and re-probing of the immobilized sample looking for a number of different target molecules is a real possibility with these membranes. This is particularly useful when attempting to quantify a number of different RNA molecules in the same preparation of total RNA. It negates the need to perform multiple blots, each time using up more precious sample and potentially introducing human error. It should always be remembered that even when using nylon+ membranes, the immobilized nucleic acid is not eternal. It will gradually leach out of the membrane each time the blot is stripped and re-probed. This can cause problems of repro-ducibility, and eventually, hybridization will be beyond the limit of detec-tion.

One problem caused by the high affinity between nucleic acids and nylon membranes is the potential for nonspecific interaction between the probe and the membrane, causing high background labeling of the entire membrane. This one drawback can be minimized by 'blocking' the membrane surface with nonspecific nucleic acids (see below) and the decid-ing plusses are that nylon membranes are physically very strong, and nucleic acids can be fixed far more easily to nylon than to nitrocellulose. Simply baking in a normal oven for 1 h at 70°C is sufficient. Hence, in spite of the potential problem of a raised background, I would recommend the use of nylon+ membranes for nucleic acid blotting on every occasion.

Handling blotting membranes

Membranes for blotting should only be handled whilst wearing gloves. The membranes come sandwiched between two sheets of protective paper. This should be left on whilst handling the membrane, and whilst cutting it to size, and should only be removed immediately prior to blotting. Most membranes need to be soaked in nuclease-free water, and be equilibrated in the transfer buffer of choice before use. Read the instructions that come with the specific membrane you have purchased, but in most cases, the membrane must be floated on the surface of the liquid and allowed to soak up the liquid and sink of its own accord. If you force the membrane under the surface, air bubbles can be caught within the matrix, which can be very difficult to remove, and will affect the future use of the membrane. If any part of the membrane appears white or smudged following soaking in buffer (the membrane should be an even light gray), this is indicative of air bubbles, damage to the surface, or contamination. You should seriously consider disposing of the membrane and starting again if any of these imperfections are suspected, unless the imperfection is at the extremity of the membrane, and will not form part of the area to which nucleic acids will be blotted.

3.3 Using hybridization to quantify RNA molecules

RNA molecules are single-stranded. For the purposes of hybridization, they can therefore be considered equivalent to one strand of a denatured DNA molecule. Since duplexes can be formed between nucleic acids irrespective of the nature of the backbone (RNA or DNA) and since uracil and thymine have identical hydrogen-bonding potentials, a complementary DNA probe will interact with an RNA target sequence in an almost identical way to the way it will interact with a DNA target having the same sequence. In further homage to the father of DNA:DNA hybridization technologies, Ed Southern, who invented the Southern blot, RNA:DNA interactions are referred to as northern hybridization (get it?) and if hybridization occurs on a membrane, the process of binding and immobilizing RNA onto the membrane is known as a northern blot (Alwine *et al.*, 1979). The main reason for hybridizing a probe to an RNA target is to quantify the amount of an RNA molecule having a particular sequence present in a sample. If each probe molecule is labeled at the same level of intensity, then the total amount of label incorporated into the hybrid duplex will be directly proportional to the amount of target sequence. This only works if the probe sequence is in excess of the target during hybridization, and if the unbound probe is completely removed prior to visualization.

3.4 The northern blot

It is possible to perform northern hybridization using a dot blot; that is, a mixture of RNA molecules within a single drop spotted onto the membrane and immobilized, with the probe providing specificity and labeling only molecules of a particular sequence. There are several negatives that make this approach less than adequate for RNA quantification, however. First, if the target RNA has low abundance, it will be necessary to use a lot of total RNA to be able to visualize the target:probe interaction. The use of large amounts of RNA concentrated into small dots on the surface of a membrane can lead to saturation of the binding capacity of the membrane and incomplete immobilization of the sample. Even if the RNA target has high abundance, it is difficult to control the amount of sample blotted onto a membrane due to the inherent error associated with pipetting small volumes. If RNA quantification is the aim, you need to know precisely how much total RNA is present, though these problems are less significant if you also have a probe that hybridizes with a control target RNA (see Section 3.8). Another major problem with dot northern blots is that the presence of large numbers of non target RNA molecules in the same sample will inevitably interfere with target:probe hybridization. Finally, RNA molecules carry internal secondary structure, which needs to be removed since it will interfere with hybridization. The addition of denaturants like formaldehyde to the sample before dot blotting will remove secondary structure, but by the same mechanism of blocking hydrogen bond formation, it will severely inhibit hybridization.

The way to overcome all these problems is to use gel blotting to bind your RNA onto a membrane. Running the RNA samples on a denaturing agarose gel (which removes secondary structure) will allow larger and therefore more accurate and useful amounts of RNA to be loaded. The individual molecules will be separated by size, and so will be spread out on the membrane without saturating it and reducing the possibility of interference between nontarget sequences and target:probe interactions. Details of running denaturing agarose gels for separation of RNA samples are given in Section 2.10. Once these have been run, the gel must be soaked in electrophoresis running buffer for 30 min to remove the formaldehyde, if used as a denaturant, which will affect subsequent transfer and hybridization. RNA in the gel should never be stained with ethidium bromide if it is to be used in a northern blot, since this will affect transfer onto the membrane, and is very difficult to completely destain. If you need to ensure the gel has been run far enough to separate out rRNA bands sufficiently and so to reduce interference between rRNA and target:probe interactions, then simply use two marker lanes, one at each side of the sample lanes. Cut one marker lane out of the gel and ethidium bromide stain it and visualize it using a trans-illuminator once the stained marker lane has been put back into the gel. This is good practice, because in any event, the whole RNA-containing agarose gel should be placed on a UV trans-illuminator for 5 min after it has been subjected to electrophoresis. This will cause the RNA backbones to be nicked, which will facilitate more efficient transfer. If a second marker is not being used, the positions of the stained size markers can be noted on the gel using a razor blade to cut small notches. If a second marker lane is being used, once transfer has occurred the blotted markers can be stained on the membrane, along with the RNA in the sample lanes, using methylene blue and then destained prior to hybridization as set out in Section 2.10. This is useful not only to allow the position of each size marker to be marked on the blot with a pencil before destaining, but also because it will confirm whether sufficient transfer of RNA from the gel onto the membrane has taken place.

The majority of northern blots are capillary blots onto nylon+ membranes, as set out in Section 3.2 and *Figure 3.3*, above. Capillary action is a slow method for transferring nucleic acids out of gels and onto membranes, and the transfer process must continue for at least 16 h. The thickness and density of the agarose gel affects the rate of transfer, as does the size of the nucleic acid molecule being transferred. Do not use gels that are more than 0.75 cm thick or made with an agarose density of more than 1.5% w/v. It is also important that the transfer apparatus and the gel within it, is level. It is worth using a small spirit level to ensure this is the case, and to pack the base of the transfer apparatus with tissue paper if necessary to make it level. If the RNA moves through the gel at an angle because the gel is not level, it will increase the apparent thickness of the gel, lengthening the time needed for transfer, and potentially deforming the bands when they are blotted onto the membrane. When you use a capillary blotting apparatus for the first time, it may be necessary to run a few test transfers to determine how long transfer needs to be allowed to continue in order to get a particular molecular weight of RNA onto the membrane. Once transfer has occurred, any RNA remaining in the gel can be stained with ethidium bromide, and that which has transferred onto the membrane can be stained with methylene blue, along with a transferred size marker. Take some time to do this before undertaking transfer for real, since it will avoid disappointments caused by incomplete transfer. Once transfer is stopped and the membrane moved, it will not be possible to start it again; no matter how careful you are in putting the membrane back in place, transferred bands will be severely deformed.

An alternative to capillary blotting that will increase transfer rates, and will probably produce more even transfer is the use of electro-blotting apparatus (see Section 3.2). These are commercially available from a number of manufacturers, and if you are performing a large number of blots, including western blots to detect protein levels (see Chapter 6), then you should consider investing in one of these machines.

Different transfer buffers are used for different blotting techniques. The membrane, the gel and the filter papers that sandwich them all need to be equilibrated in transfer buffer by soaking prior to transfer. The standard transfer buffer for DNA is SSC, which is used at different strengths dependent upon the type of membrane and transfer method being employed. The use of nitrocellulose membranes requires the use of very strong buffers; the use of nylon membranes does not. Capillary blots use stronger buffers; you cannot use high ionic strength buffers for transfer by electro-blotting, since the machine might explode! For Southern capillary blotting SSC is used at 20×. However, for capillary transfer of RNA onto nylon+ membranes, 10× SSPE buffer (*Table 3.1*) is very much better than 20× SSC

Once transfer has taken place, the membrane is carefully peeled away from the gel and filter papers, briefly washed in dilute transfer buffer to remove debris, and then air dried on the surface of a piece of filter paper.

Nucleic acids are usually immobilized onto the membrane by baking it. Some manuals favor the use of UV light to cross-link nucleic acids onto the surfaces of membranes, and so immobilize them. This can be tricky to get right, however, and involves a number of steps, and the end result is no more satisfactory than simply baking the membrane in an oven. Furthermore, if you attempt to UV cross-link nucleic acids onto nitrocellulose membranes, make sure you have a fire extinguisher to hand! Once baked for the appropriate amount of time (Section 3.2) membranes can be stored between two sheets of filter paper in a vacuum flask until they are needed, if hybridization is to be delayed.

3.5 Making DNA probes

It is possible to use single-stranded RNA molecules as probes in nucleic acid hybridization. RNA probes are made *in vitro* using purified RNA polymerase. This enzyme requires double-stranded DNA as a template (provided as a cloned gene) and the RNA molecule produced is an exact copy of the coding (top) strand of the gene (with thymine being replaced by uracil). Because of this, RNA probes made from a gene *in vitro* will not be able to hybridize with any coding RNA derived from the same gene present in an *in vivo* source, since the sequences of target and probe would be identical, and so not complementary. It is possible to get around this by inverting the gene within the *in vitro* transcription vector, such that an 'antisense' RNA is produced. This would be able to bind to the coding RNA produced from the gene *in vivo*, but the need for this adds an unwelcome extra layer of complexity to the process. Because of this fact, I will not discuss the production or labeling of RNA probes, though the reader should be aware that they can be very useful for detecting DNA sequences, where both the coding and non-coding strands are present, and so where there will always be a valid target.

Synthesis of DNA probes

DNA probes are single-stranded and are made by DNA polymerase enzymes, which synthesize a second, complementary strand onto a single-stranded template using deoxynucleotides as building blocks (*Figure 3.4*). The vast majority of DNA polymerases need a DNA template, but some, including reverse transcriptase, which will be discussed more in Section 3.12, use an RNA template. During the process of assembling the complementary strand, a label can be introduced into the probe. The probe must then be released by melting the double-stranded product, since the label is only found on the newly synthesized strand. Accordingly, when a DNA probe is made to be complementary to a coding RNA sequence, you should realize that the DNA coding strand must be used as the template for probe synthesis (*Figure 3.4*).

DNA polymerases do not build nucleotides opposite a template DNA strand randomly. They work in a programmed manner, adding nucleotides sequentially in a direction that is 5′–3′ with respect to the strand being synthesized. Thus, for the polymerization reaction to be established, a 3′ hydroxyl needs to be in a position from which the nucleotides can be added. The best way of providing this 3′ hydroxyl in a test tube is to add a short single-stranded oligonucleotide primer that has a sequence complementary to the template strand, which anneals in position when the temperature is below the melting temperature of the primer (*Figure 3.4*). In every case where one wants to probe a specific RNA sequence in order to quantify it, the sequence of the coding strand of the gene encoding that RNA (or an orthologue of this gene from another organism) is known. Thus, the exact sequence of the complementary oligonucleotide primer required can be determined, and the primer made in an oligonucleotide synthesis machine. Primers are available to order from many manufacturers.

Given that the synthesis of probes using DNA polymerases *in vitro* is not an amplificatory process, the amount of probe you end up with can never be more than the amount of template DNA you start with. Furthermore, if the template DNA is mixed with a lot of other DNA sequences, there is potential for the primer to anneal at more than one place, resulting in a number of probes with different sequences being produced (a so-called 'heterologous probe'). The use of a heterologous probe produces the potential complication that individual constituent probes may hybridize to different RNA bands. To be sure that neither of these problems is an issue, the template should be provided pure and at high concentration. This can be in the form of a clone (whether the whole cloning vector/insert is present, or as a purified restriction fragment) or more likely nowadays, as a PCR product. PCR will be discussed in Chapter 4. In either case, production of a probe from the template will involve a sequence specific primer,

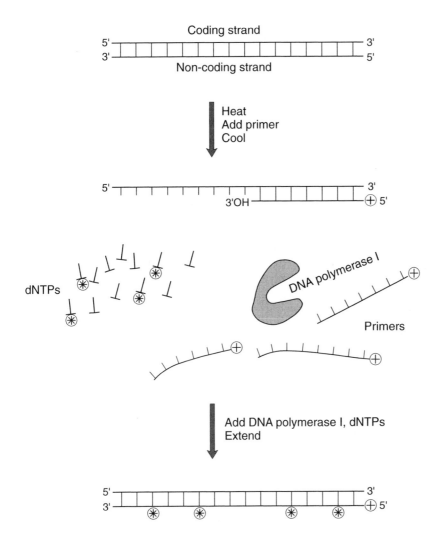

Figure 3.4

Making labeled probes. The template for northern hybridization probes is almost always a double-stranded DNA molecule. Probes are synthesized using sequence-specific primers, which provide 3' hydroxyl groups from which DNA polymerase I (the Klenov fragment is used because it works more rapidly than the holoenzyme) can build a complementary strand. Label is incorporated into this complementary strand by using either a 5' end-labeled primer or a mixture of dNTPs in which one is labeled. Primers and dNTPs are used in large excess and must be removed from the reaction mixture before the probe can be used, as must the DNA polymerase I enzyme.

though if the sequence of the DNA is not known for some reason, a mixture of random hexameric primers can be used, each mixture containing all possible hexamer sequences meaning that at least one primer molecule will bind to each melted template molecule, priming complementary strand production by DNA polymerase I.

Types of label for DNA probes

Radioactivity is the most common probe label for northern hybridization. When the probe is radioactive, its interaction with the target sequence can be visualized using X-ray film or a phosphor-imager screen. ^{32}P, ^{33}P and ^{35}S isotopes all produce beta particles, which expose X-ray film. The different isotopes produce particles with different energies, which means that different exposure times are needed. It also means that different levels of hazard are associated, but also that different decay times are found. ^{32}P gives the most energetic particles, thus is least safe, and the isotope can be stored (including probes once they have been made) for the least amount of time before it becomes useless, but the advantage of using a high-energy particle emitter is that lower amounts of label need to be incorporated into the hybridization product before the signal becomes visible than is the case with lower-energy particle emitters.

Digoxogenin (DIG) is a derivative of digitalis, and is a very commonly used alternative to radioactive labeling for DNA probes in Southern and northern hybridization (*Figure 3.5*). The DIG label is bulky, and antibodies raised against it are commercially available. These can be covalently linked (conjugated) to the enzymes alkaline phosphatase or horseradish peroxidase, allowing chemiluminescent detection of the probe. Alternatively, a secondary antibody, raised against the common region of the IgG of the animal from which the anti-DIG antibody was obtained, and conjugated to one of these enzymes, is washed over the surface, and its interaction with the anti-DIG antibody places the enzyme in contact with the DIG labeled hybrid nucleic acid (*Figure 3.5*). In either case, chromogenic alkaline phosphate or horseradish peroxidase substrates are washed over, which, when cleaved by the immobilized enzyme, release a chromophore which binds to the membrane at the point where the labeled target:probe complex is located. Many of these chromophores are fluorescent, and can be visualized using X-ray film or a phosphor-imager screen; some are stains, which produce visible bands on the membrane that can be photographed. Whilst this type of label/visualization system does not come with any of the health hazards associated with radioactive labels, it is not as sensitive. Furthermore, the chemiluminescent signals produced upon development are short-lived, meaning that long exposure times are not an option. Therefore, DIG methods are not good for detecting very-low-abundance target sequences. However, for high-abundance sequences, chemiluminescent detection gives far sharper bands than autoradiography.

Following a very similar rationale, probes can also be labeled with biotin. This interacts very strongly with streptavidin. Conjugates of streptavidin and the enzymes horseradish peroxidase or alkaline phosphatase are readily available from a variety of commercial sources. The procedure is almost identical to the use of DIG and associated enzyme-linked antibodies.

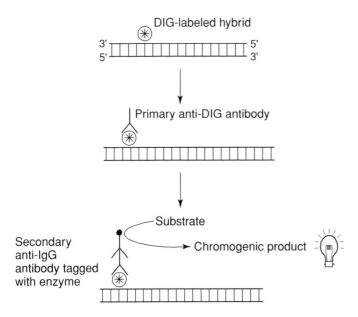

Figure 3.5

The principle of antigen-labeled hybrid visualization. The most common antigen used to label probe sequences is DIG, though often biotin is used, and rarely, other large antigens. To locate the antigen, and so hybrid nucleic acids containing the probe, a primary antibody (or streptavidin, if biotin is the label) is used. This primary antibody (or streptavidin) may be labeled with either alkaline phosphatase or horseradish peroxidase; alternatively, an enzyme-linked secondary antibody, designed to bind to the animal-specific IgG common region of the primary antibody can be used. The enzyme converts a substrate into a colored, luminous or fluorescent product, depending upon the exact substrate used and these can be seen either as a stain on the blot, or by using high-speed X-ray film.

Visualization of the label is also the same, being a case of chemiluminescent detection using X-ray film or photography.

Colored dyes can be used to label DNA probes in order to detect hybridization, but the usefulness of dye labels to visualize target:probe interactions on membranes is limited. These probes are used extensively in array-based hybridization reactions, however, and will be discussed in Section 3.12.

Incorporation of labels into DNA probes

There are two main ways in which labels can be incorporated into probes. The most common is to use a mixture of the four nucleotides as building blocks for the DNA polymerase, one of them being spiked with a labeled portion (*Figure 3.4*). Hence every time that particular nucleotide is incorporated opposite the template strand, there is a chance that the actual molecule incorporated is labeled. Since each probe strand will have multiple instances of each nucleotide, the concentration of labeled nucleotide can be manipulated so that each probe is labeled at least once. If radioactivity is being used as the label, the radioactive nucleotide used is either [^{32}P]dCTP or [^{35}S]dATP (sulfur can replace phosphate in ATP without affecting the geometry of the molecule). The radioactive atom must be in the alpha position in the triphosphate backbone (i.e. closest to the sugar) because the two outer phosphates are lost during the formation of the phosphodiester bond. If other labels are being used, the nucleotide that is modified is dUTP, which can be incorporated opposite adenine by all DNA polymerases, including reverse transcriptase.

The second, increasingly popular way of incorporating labels into probes is to label the 5′ end of the primer used to synthesize the probe prior to its use (*Figure 3.4*). If radioactivity is the label, the procedure is very simple. The enzyme required is polynucleotide kinase, and it uses [^{32}P]dATP to phosphorylate the 5′ hydroxyl of the primer. For labeling to occur, the radioactive atom must be in the gamma position (i.e. furthest away from the sugar) since this is the phosphate group donated to the 5′ hydroxyl of the primer. For other molecules (e.g. biotin) to be attached to the 5′ ends of primers it is more complex, and requires a process of chemical linkage called conjugation. Here, biotin is provided in a solid form that is activated with sulfo-*N*-hydroxysuccinimide (NHS) and is added to the solid oligonucleotide primer at a w/w ratio of 10:1 NHS–biotin to primer. The mixture is made up in 1 μl of sodium carbonate buffer (50 mM NaHCO$_3$, 100 mM Na$_2$CO$_3$, pH 9.0) per μg of oligonucleotide, and the reaction proceeds for 4 h at room temperature. The biotinylated primer then is purified using HPLC Such biotin-labeled primers can be bought from a number of commercial suppliers.

Storage of probes

Once made, probes are part of a double-stranded duplex. Free nucleotides and the DNA polymerase enzyme are removed usually via size exclusion chromatography on a Sephadex G-50 column. Radioactive probes can then be stored at −70°C until they are needed. Do not repeatedly freeze–thaw the probe or it can become degraded. DIG- or biotin-labeled probes should be stored at −20°C for up to 1 year if not defrosted in the meantime.

3.6 Northern hybridization reactions

Before hybridization

Membranes to which RNA has been immobilized are put through a pre-hybridization routine. The membrane is soaked in hybridization buffer (*Table 3.1*) containing heterologous genomic DNA (normally salmon sperm or calf thymus DNA, unless you are quantifying RNAs from salmon or calves!) that has been sheared using a syringe and needle. The heterologous DNA binds to the regions on the membrane where RNA is not already bound and 'blocks' it. This procedure is particularly important when using nylon+ membranes, because during hybridization, the probe would bind to the membrane via the negatively charged phosphate groups, producing high-level background labeling of the membrane, obliterating all hybridization events. The use of SSPE (saline-sodium phosphate-EDTA) as a hybridization buffer (*Table 3.1*) also helps prevent nonspecific interaction between probe and membrane surface because it is a phosphate buffer, and this phosphate competes with the probe's phosphate backbone for the positively charged membrane.

Hybridization apparatus

Hybridization of a probe to nucleic acids on the surface of a membrane is usually performed in a hybridization oven. These machines use glass tubes, sealed permanently at one end, and with a screw-on cap at the other. Blotting membranes are placed into the tubes so that they stick to the inner surface, with the blotted nucleic acids facing the center of the tube, to which liquid can be added. The tubes are turned by the machine, causing liquid in them to be in constant motion, flowing over the surface of the blot. This allows mechanical disruption of nonspecific interactions. The tubes are placed within the hybridization oven and the lid closed. This allows the temperature of the inner chamber, and so all of the tubes, to be controlled. Most hybridizations involve 42–50°C, but for very high stringency, the temperature is increased to around 60°C. There are a number of things that can go wrong. The most common is damage to the surface of the membrane whilst rolling it up and placing it into the tube. Use the largest diameter tubes you can so that there is more space to maneuver. Once the membrane is within the tube it can be gently slid down the sides after the addition of pre-hybridization buffer to wet the glass surface. Use a gloved finger, since it is easier to control this than some other objects. It is always worth marking the high molecular weight end of the blot, since finger marks here, if accidentally caused, are less likely to affect results. There is no need to push the blot all the way to the end of the tube, as long as it is a few centimeters inside. It is quite possible to put more than one blot in the same tube, and indeed a bit of overlap is fine, provided there is sufficient liquid to separate the membranes.

The hybridization stage

Pre-hybridization of the membranes is essential, not just to get the membrane blocked, and to equilibrate it into the right buffer, but also so that the hybridization tube and blot get up to temperature before hybridization proceeds. Once pre-hybridization of the blot is complete, the heterologous DNA solution is poured away and replaced with hybridization buffer alone. Do not be tempted to wash the membrane and so remove more DNA. The presence of a small amount of heterologous DNA in the hybridization reaction will help increase specificity.

When you are ready to use them, labeled probes could be converted into single-stranded molecules, thereby releasing the labeled complementary strand by using heat (95°C) to melt the duplex. However, some labels, e.g. DIG- or biotin-labeled probes would be damaged at this temperature. Furthermore, it is potentially dangerous to heat radioactive probes to this temperature because you could release radioactive steam. Therefore, the best and safest option for releasing labeled single-stranded probes is to add 0.1 volumes of 1 M sodium hydroxide and to incubate the mixture for 10 min at 37°C. It should not be forgotten that the unlabeled DNA strand is present in equal amounts to the labeled probe strand, and the two can re-bind in solution. The probe is applied to the surface of the membrane to which RNA has been immobilized using appropriate conditions of stringency to allow annealing of the probe to its target, whilst minimizing problems associated with non specific binding. To do this, simply add the probe to the hybridization solution that has already been poured onto the surface of the membrane. During hybridization, make sure the lid of the hybridization oven is not opened unnecessarily, because sudden drops in temperature can lead to nonspecific annealing, the hybrid products of which do not always completely melt when the temperature gets back up to optimal.

Stringency

Stringency is a term used to define the application of physical factors in order to affect hybridization of a probe with a target nucleic acid (whether DNA or RNA). High stringency conditions would include high temperature, low salt and the presence of denaturants such as formamide. High stringency conditions are used to minimize hybridization of probes to nontarget sequences due to nonspecific hybridization between sequences that have imperfect complementarity. An example where nonspecific hybridization might be a problem would be the use of a probe targeted to an mRNA encoded by a gene that has many paralogues in an organism (genes with similar sequences, but encoding different functions), because sequence similarity amongst the paralogous genes would mean the existence of a degree of complementarity between each of their mRNA products and the probe. Of course the strongest interaction would be with the true target, but since the probe would be in excess, a number of different mRNA bands might become labeled by hybridization to varying extents, ruining the possibility of obtaining accurate quantification of the true target mRNA. Low stringency conditions might be required in a situation where the sequence of the gene encoding the target mRNA is not known. An example would be where the probe is being made from an orthologous gene (i.e. a homologue in another organism encoding the same function as the target). This is more and more the case nowadays when genome sequencing projects give clues as to the genomic complement of organisms whose genomes have not been sequenced. In this case the probe will almost certainly not be perfectly complementary to the target sequence in the organism of interest and low stringency conditions would be employed, allowing sequences to hybridize even if they are not perfectly complementary. The obvious problem is that nonspecific interactions may occur. Indeed, if you need to measure the expression of an mRNA where the exact sequence is not known, a method other than northern blotting is recommended (e.g. qRT-PCR; Section 4.9) which has the distinct advantage that the RT-PCR product can be sequenced, confirming that it represents the expected target. As an added bonus, the sequence of part of the gene in your organism of interest would be determined.

Post-hybridization washes

Once the blot has been through hybridization it is washed to reduce nonspecific interactions and then can be gently removed from the tube. These washes are generally undertaken at a higher temperature and with a higher stringency buffer than the hybridization phase, and the buffer is poured off and replaced a number of times (*Table 3.1*). During hybridization and washing, the membrane often moves down the tube and can become out of reach. If this is the case use a glass rod with a smooth, rounded end, to gently coax the membrane back up the tube. Do this before pouring off the last of the wash buffer so that liquid is present and will reduce friction. Only touch the very edge of the membrane with the glass rod, and move it a small amount only before inverting the tube 90° and working another part of the blot. This will reduce the possibility of the membrane ripping. Repeat this operation until you can reach it with a finger, and then gently pull it out of the tube. Do not grab a corner and tug, since the membrane may rip in half, or the corner may come off.

Stripping and re-probing blots

Once probes have annealed to their targets, nonspecific binding has been reduced by washing and the interactions have been visualized (Section 3.7), it is possible to strip off the first probe and repeat hybridization with another probe. This is likely to be routine for gene expression analysis, since it is wasteful to run many different gels, and probe many different blots, each time probing for a different mRNA sequence. Stripping of probes from blots is undertaken back in the hybridization oven, but using very high stringency conditions, which cause the probe to dissociate from the target sequence (*Table 3.1*). It is difficult to completely strip a probe, and some residual labeling can be expected, but it will be so weak compared with the signal generated by the second probe that it need not be worried about. The primary concern when stripping probes from target sequences, is that the target sequences themselves may come away from the membrane. Even if they have been immobilized, leaching away of target nucleic acids from blots is bound to occur. The process is much quicker with nitrocellulose membranes than with nylon+ membranes, and will occur more readily at high temperatures. It is always worth hybridizing a control probe to the blot (Section 3.8) as the first probing of the blot. As well as quality control for the RNA sample and transfer, and to allow quantification of relative abundances of RNAs in each sample, if the control probe is used to hybridize the blot every five times that the blot is stripped, the signal obtained will tell you how well the RNA is holding out. If the control bands start to become faint, it may well be time to throw the blot away. In reality, it is highly unlikely that a blot will be viable after it has been stripped more than five times, but a control hybridization during probe number six will confirm this to be the case. This means that in practice, the total number of RNA species that can be quantified on each blot is four plus the essential control.

3.7 Visualization of target:probe interactions

Autoradiography

The most common method of visualizing radioactive probes bound to their targets on membranes is the use of autoradiography. X-ray films are plastic sheets coated with an emulsion that is sensitive to particles such as beta particles or photons. Most commonly, the film used is an XAR type hyper-film, and is coated on both sides, so you don't need to worry which side you place in direct contact with the membrane. When you choose a film, make sure you read the instructions. Packs of film should be stored in cool conditions (e.g. a refrigerator) and should be firmly shrouded in black plastic to prevent exposure to light.

Radioactively labeled membranes should be wrapped in cellophane, both to keep them moist (essential if stripping and re-probing is to be performed) and to prevent contamination of equipment and bench surfaces. Be very careful that creases are not formed in the membrane or in the cellophane, since this could distort the passage of particles, and affect the quality of the bands seen when the film is developed. The membrane should be placed into firm contact with the film by using a cassette. Intensifying screens, which allow more of the label's emitted energy to expose the film are an excellent feature, and should be used routinely. They work by absorbing the emitted radioactive particles onto a phosphor surface, which converts their energy into photons that expose the film more efficiently than the radioactive particles themselves. Make sure the screen is placed in the cassette in the correct orientation, since only one side is phosphor coated. Read the manufacturer's instructions. The membrane should routinely be in contact with the film overnight in a $-70°C$ freezer. Low temperatures are important for making the film respond efficiently in the presence of low photon levels. The film can then be developed in order to get an impression of the signal strength. If required, shorter or longer exposures can then be used. It is good practice to get a short and a long exposure for the same membrane. This will help with the problem of saturation during the exposure of film, as discussed in Section 3.8.

Chemiluminescence and chemifluorescence

When a probe is labeled with DIG or biotin, its location on a blot can be reported using enzyme linked anti-DIG antibody or enzyme linked streptavidin. As illustrated in *Figure 3.5*, the linked enzymes are usually alkaline phosphatase or horseradish peroxidase. These convert colorless substrates into colored products; for example, 5-bromo-4-chloro indoxyl phosphate is converted into a purple product by alkaline phosphatase and diaminobenzidine is converted into a brown product by horseradish peroxidase. These products stick to the blot, and can be used to visualize the position, and amount of target:probe hybrid. The blot can be photographed or scanned and the intensities of these bands or spots can be determined using various computer programs, as for the bands on autoradiographs. However, these reactions are sometimes difficult to develop, and involve a series of additions. It is also possible to over-develop the reaction, and if this happens, there is no possibility of stripping away the product and starting again. To get over this problem, chemiluminescent and chemifluorescent substrates have been developed for these enzymes, which, when they are broken down release light (chemiluminescence) or produce a product that fluoresces when excited by laser light of a given wavelength. The light can then be captured as an image on X-ray film. If the first exposure is too great, as with autoradiography, a shorter exposure can be performed. There are large numbers of different chemiluminescent reagents. One of the best known is the enhanced chemiluminescence (ECL) system from Amersham, which uses a two-stage reaction. First, hydrogen peroxide is broken down by horseradish peroxidase to produce water and oxygen, which rapidly oxidizes a luminol substrate, with the concomitant production of light. Light emission is increased by the presence of a phenolic enhancer compound which acts almost like a scintillant. Chemifluorescent substrates are almost exclusively broken down by alkaline phosphatase, and do not require any enhancer. Because of this, they can provide more direct readout, with perhaps more predictable exposure times. A typical example of this is ECF reagent from Amersham.

If chemiluminescence or chemifluorescence is being used to detect labeled target:probe interactions on the surfaces of membranes, each membrane should be treated according to the reagent manufacturer's instructions. High-speed X-ray film such as hyperfilm SR is used to detect emitted photons. The membrane is wrapped in cellophane and the film is placed on top. Exposure times will be far shorter than for radioactive detection and generally cassettes are not needed. Indeed, the signal degrades within an hour or so, but can be replenished by addition of more reagent, since the antibody-conjugated enzyme is very stable. During exposure of film, simply stand in the dark and hold the film onto the surface of the membrane. Make sure it is flat, and don't move! For longer exposures, place the film onto the membrane on a flat, clean surface and put a book on top. If the blot is small, you can get a number of different exposures on one film, saving money on film and developer. One way of improving exposure characteristics of film is to pre-flash it. Use a detachable SLR camera flash with a yellow filter taped over the output. A single flash on full power is sufficient to pre-expose the film, which will get over the initial lag phase for exposure, making the exposure more linear, which is particularly important when exposure times are short. You will also find that bands on films exposed to chemiluminescent or chemifluorescent detection kits are more defined that those from radioactive sources. This is because beta particles from radioactive isotopes have higher energy making them bounce around and fly off in all directions.

Phosphor imaging

Phosphor imagers are designed to record images of radioactive spots or bands as something called 'latent image formation'. A screen, containing a complex crystalline lattice is placed in contact with the radioactive source, and absorbs energy (in the form of radioactive particles), causing the crystals to move into a higher energy state. If left in the dark, the crystals would remain in this state, with the latent image being present, indefinitely. When orange light is shone onto the crystals, however, they return to ground state, releasing their stored energy in the form of blue light that can be detected. A machine is used to read latent images on phosphor-imager screens. An orange laser is shone on the image and a picture of the resultant pattern of blue light production taken. There are some considerable advantages of phosphor imaging over autoradiography. First, the response time is very rapid, meaning shorter exposure times are required, which is particularly useful for low signal strengths. Second, phosphor-imager screens tend to expose far more linearly that autoradiography film, there is much less of a lag early on in the exposure, and the 'grain' is sharper, meaning image saturation is less likely to occur with extended exposure (see Section 3.8 and *Figure 3.6*). Finally, the same screen can be blanked by exposing it to daylight for 10–20 min, and re-used time and time again. The major drawbacks are that different screens, with slightly different crystal lattices are required for different radioactive isotopes and for different fluorophores, and most particularly, the cost of buying a phosphor imager means it is probably not

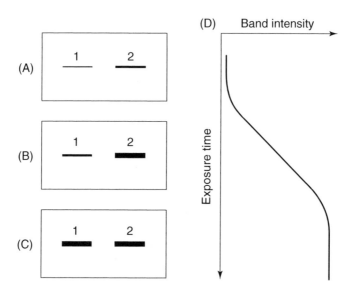

Figure 3.6

Nonlinearity in northern blot signals. Three different exposures of the same northern blot probed with a radioactive probe and visualized using autoradiography (other visualization methods suffer from a similar problem, but to differing extents). If the exposure is too low (A), the X-ray film used is poorly exposed because it takes a certain threshold of energy to expose the film at all. Therefore, whilst there is a slight difference between band intensities, it is under-represented. In (B), the exposure is for an optimal amount of time, within the linear range of the film, showing the true apparent difference in band intensities. In (C) the exposure is too long. In lane 2, the grains making up the film have become saturated and increasing exposure will not result in a further increase in band intensity. However, in lane 1, the increased length of exposure has an effect on band intensity making the apparent intensity of band 1 'catch up' with band 2, again under-representing the difference in target RNA concentration in the two samples.

worth it unless a large number of images will be taken over a long period. It is another piece of equipment that can be shared by a collective of research teams. Indeed, many phosphor-imager screen readers are capable of reading other fluorescent images, including chemifluorescent images directly from the blot, and even bands on DNA gels stained with fluorescent dyes such as ethidium bromide.

Band intensity measurements

Ultimately, in order to quantify the intensities of bands or spots, you will need some form of densitometer. In fact, nowadays, the most likely way of doing this will be to record a digital image of the gel or blot and measure the total amount of pixels of a certain color within each band or spot. To get to this point, autoradiographs and chemically stained blots will need to be scanned or photographed. Phosphor images and chemifluorescent images are usually stored as digital files anyway.

There is a large selection of different image analysis software packages that allow band or spot intensity measurements. One of the most popular is the ImageQuant system which is provided with Amersham Biosciences phosphor-imager readers. In this chapter so far, we have been dealing with spots or bands on a blot. Each imaged spot or band must be distinct from each other band (i.e. no overlap) or quantification will not be possible. The simplest way of quantifying the intensity of a band is to generate a gray-scale image of it, and to use image quantification software to define the total amount of black in a defined region of the blot. Therefore you are not simply measuring the size of the band or spot, nor are you determining the intensity of blackness within any part of the band/spot, you are defining the total amount of black associated with the band/spot. The way to do this is to find the largest spot or band and draw a shape around it. It doesn't matter if you include a bit of the unexposed background blot; a rectangle or square will be fine. Then draw identically sized shapes (you can simply use the copy and paste function) around all the other bands or spots whose intensities you want to compare with the first. The final trick is to work out the background intensity of the blot. This is not important if the background is identical throughout the blot, but this is rarely the case; there are often smears etc. So background needs to be recorded for each band/spot whose intensity is being quantified. Simply draw an identical shape as close to each band as possible, preferably in the same lane, if using a gel blot, and obviously not including other spots/bands on the same blot. Now ask the program to determine the amount of blackness in each shape, and simply subtract the background value for each spot/band from the density value for each spot/band. You will end up with a normalized intensity for each spot/band on the blot and these values can be used to determine relative signal strengths on the original blot.

The documentation of DNA bands on gels is now most commonly via a digital image acquisition machine, which consists of a UV light box and a camera. These digital image systems often come with computer controllers and image analysis software. These different packages work in a very similar way to ImageQuant (above) and, like ImageQuant, can take external files generated from any digital image source in order to quantify band intensities. Tried and tested examples of this type of system are the BioRad GelDoc XR system which comes with 'Quantity One' image analysis software, and the Kodak Image Station 4000R with the Molecular Imaging software package. There are many more available and you should shop around before purchasing them.

3.8 Limitations and design of northern blot hybridization experiments and interpretation of the results

Northern hybridization experiments need to be carefully planned, and are not suitable for all purposes. The classical experiment would be to purify total RNA from two experimental conditions (usually the same cell type grown in two different physiological states) and to measure the abundance of a particular RNA sequence in each total RNA preparation in order to learn something about the regulation of the expression of a particular gene under a particular growth condition. Such pair-wise comparisons are typical of all gene expression analysis experiments, and the same basic principle applies even if a large number of different samples are being collected. For example, if samples of cells are removed at periods over a time course of incubation with a particular chemical and gene expression is analyzed in all the samples, the data will be still be initially analyzed in a pair-wise manner. Hence, I will limit my discussions of experimental limitations to pair-wise comparisons of RNA content in two total RNA samples.

Obtaining accurate northern blot data

There are some situations where one needs to know the absolute concentration of an RNA molecule having a particular sequence in a sample of total RNA, but this is very rarely the case. Each sequence specific target:probe interaction will have its own unique kinetics, and so the amount of label incorporated into each RNA band on a northern blot is no more than correlated with the amount of RNA present, and cannot be used to absolutely measure it, unless a series of standard RNAs are loaded, each having a known concentration of the target RNA to allow plotting of a standard curve of label intensity versus concentration of RNA. For this same reason of unpredictable hybridization kinetics, it is not possible to use northern blotting to compare even the relative amounts of two RNAs having different sequences within the same preparation of total RNA, unless both RNAs are accurately quantified.

Far more commonly than accurate quantification of RNA content, one wants to know the relative abundance of an RNA species in each of a pair of total RNA preparations, when each preparation represents a different condition of growth. This may seem simple; just measure the northern hybridization signal strength in two lanes, each containing separated RNAs from one of the two RNA preparations, then divide the signal strength in preparation one by the signal strength in preparation two. If the result (x) is >1, then gene expression is x-fold higher in condition one than in condition two. If the result is <1, then gene expression is $1/x$-fold higher in condition two than in condition one. Simple? Well, yes, but there are a few obvious and a few less obvious complications to this simplistic approach.

It goes without saying (but I will say it anyway because people always forget) that the two northern hybridization signals being compared must be from two lanes of the same gel immobilized onto a membrane at the same time, and must be from the same image (i.e. solid medium such as a sheet of developed X-ray film and digital image such as a scanned autoradiograph, chemiluminescence image or phosphor imager download). You cannot validly compare the signal intensities derived from two RNA samples separated on different gels, immobilized on different membranes and probed with different probes (or any one of these differences) since electrophoresis, transfer onto the membrane and probe production are not homogenous procedures. Neither can you compare bands from two differently acquired images of the same blot, because even if the exposure time were identical, variations in the medium, the developing process, and in digital image acquisition (particularly scanning) will add unnecessary variability to the signal intensity data. Thus always do your pair-wise comparisons between two RNA preparations absolutely in parallel, treating both identically at every stage of the process.

Even when using a pair-wise analysis of two samples run absolutely in parallel, there are problems that can affect the accuracy of the analysis of signal strength. The key problem is one of linearity and saturation of the medium used to visualize the hybridization signal (*Figure 3.6*). Autoradiography film and phosphor imager screens do not become exposed in an entirely linear manner, but as an 'S'-shaped curve. This problem is considerably less of an issue for phosphor imaging than for autoradiography/chemiluminescence. The existence of this nonlinearity of exposure means that at very high and very low signal strengths, the medium will under-represent the actual signal strength. At the low end, this is because the medium has a threshold sensitivity that must be crossed before significant exposure occurs. At the high end (and this is more commonly a problem) the medium becomes saturated. There are only so many pixels or grains of silver that can become exposed on the image which is being used to quantify signal intensity. Between these two plateau phases, most media are pretty much linear in their response to different signal intensities. Thus, if one of the pair of samples has low levels of the RNA being picked up by the hybridization, and the other sample has very high levels of the RNA, it is perfectly possible that the signal associated with the second preparation will saturate the visualization medium, whilst the first preparation produces a signal in the linear range of the medium, and so gives a true reflection of RNA levels. The end result would be underestimation of the difference between RNA levels in the two samples. To get over this, you might be tempted to reduce the exposure time so that the second sample gives a signal in the linear range of the medium, but if you do, you may find that the small amount of RNA in the first sample now barely registers at all. The result of this would be dramatic over-estimation of the difference in RNA complement between the samples, even up to the extreme situation where the difference appears to be infinity!

To prove to yourself that you are getting an accurate representation of relative RNA levels in the two samples, you will have to use at least two different exposure times. If the relative difference in signal strengths seen in both exposures is more or less the same, you can be confident you have reported a real relative difference in RNA amount. If you cannot get at least two exposures with the same relative signal strengths, however, then you might want to repeat the whole experiment, reducing the amount of total RNA used for the sample with the highest signal strength, or increasing the amount of total RNA used for the sample with a lowest signal. This is a valid approach, provided of course that the increase or decrease in total RNA loading for one of the pair of samples is taken into consideration later on when determining relative differences in RNA levels!

It takes approximately 5 pg of an RNA molecule with a given sequence to produce a visible signal following overnight exposure onto autoradiography film of a northern blot hybridized with a probe having the specific activity of radioactivity typically used. If you assume that there are 10 000 genes in a genome, and that 1% of total RNA is mRNA (which is what we generally look for in gene expression analysis experiments) you can estimate that an averagely expressed mRNA will make up one millionth of the total RNA in the sample. Therefore, to get 5 pg of an averagely expressed mRNA you will need 5 mg of total RNA in a lane of a northern blot. If you happen to be trying to measure the level an mRNA in a growth condition where its level is considerably below average, then you can see why no signal is more commonly a problem than too much signal in northern hybridization experiments. To overcome this, you will need tens of milligrams of total RNA, just to enable visualization of a hybridization product. However, when performing northern blotting experiments, the use of tens of milligrams of total RNA in each lane is just going to cause problems, including inefficient transfer, saturation of the membrane, nonspecific hybridization product formation and interference with specific target:probe hybridization. So the only real way of increasing mRNA levels ten-fold or more, without increasing total RNA levels on the blot so that they become unmanageable, is to purify mRNA from the total RNA sample prior to electrophoresis and transfer onto the membrane. This is only really possible if the mRNA you are interested in is polyA$^+$ since this is the property typically used to purify mRNA (Section 2.6). Clearly, not all mRNA in a eukaryotic cell is polyA$^+$, and in prokaryotes, none is. Another problem comes from the fact that even if it is possible for you to purify mRNA, and that your chosen target mRNA is represented in this purified preparation of total mRNA, it is not always practical to obtain the tens of milligrams of total RNA needed as a raw material for polyA$^+$ mRNA purification, particularly if the tissue sizes or numbers of cells available in each preparation is small. Do not be tempted to pool a number of total RNA preparations to get over the problem of low abundance of an mRNA in each preparation, since this will inevitably increase biological variability in the experiment and may mean that the differences in mRNA levels measured are statistically insignificant (see Chapter 8). Alternatively, you might turn to a method of amplifying the apparent RNA level prior to performing a northern blot. This might include the use of SIP-PCR amplification of cDNAs produced from polyA$^+$ mRNAs which is the starting point for a so-called 'virtual northern blot', to be discussed in Section 4.10.

Another alternative to increasing signal strength without increasing RNA levels is to make the northern hybridization probe to a very high specific activity, though this may well result in high background labeling of the blot. Alternatively, simply use very long exposure times, though this can also result in an elevated background, and is certainly not helpful when dealing with impatient supervisors. So, in the end, if it turns out to be the case that you simply cannot get enough mRNA to give a decent northern hybridization signal, then it is likely you will have to abandon northern blotting and try a technique such as qRT-PCR, which requires orders-of-magnitude less RNA to give an interpretable signal. It is for this reason that qRT-PCR is more commonly turned to as a first-choice tool for measurement of mRNA levels, though this technique has its own share of problems, as will be discussed in Section 4.9.

Controls for northern blot analysis

Let us take forward our example of a typical experiment where levels of a specific RNA species are compared in two total RNA preparations obtained from cells growing under different conditions. You have collected accurate band intensity data for northern blot hybridizations, and can work out the relative intensity difference between bands representative of the two total RNA samples. This is not enough, however. How do you know that the preparation of RNA that appears to have more of the specific target RNA doesn't just have more total RNA? Well, first, you would have measured the concentration of total RNA in both samples spectrophotometrically, and would have diluted them both so that they represent the same concentration. Then you would have loaded exactly the same amount of each sample onto the agarose gel. But this is not sufficient, since spectrophotometric measurements of RNA levels are simply measurements of the amount of nucleotides; there is nothing to tell you whether the RNA in both samples is intact. It could be that RNA from one sample has degraded more than the other, meaning that there is an apparent difference in gene expression between the two growth conditions but this is simply artifactual. Again, you can check for RNA integrity using agarose gel electrophoresis, but this will only give you a rough idea of average RNA integrity. The consequences of RNA degradation for some RNAs in a sample will be more pronounced than for others; you have no way of knowing whether your target RNA is more or less prone to degradation, and so differences in average degradation between samples, than the norm.

So we need some more appropriate internal standard of RNA loading. Here we usually turn to the so-called 'housekeeping gene'. A few years ago it was very easy to define a housekeeping gene as being one that is essential for growth under all possible growth conditions. Examples might be genes encoding subunits of the core DNA polymerase enzyme, metabolic enzymes whose roles appear ubiquitous or proteins involved in essential and consistent sub-cellular architecture. In those simple times, it was thought obvious that the expression of housekeeping genes would be at constant levels all the time, irrespective of growth conditions (so called 'constitutive expression'). It follows therefore that the RNA products of constitutively expressed genes will be present at a constant proportion of total RNA under all conditions of growth. Thus it was believed that the products of housekeeping genes would be excellent northern blot controls to normalize total RNA loading.

For bacterial systems, the housekeeping gene that is commonly used for normalization for northern blot experiments is *gyrA*, encoding DNA gyrase A, an essential protein involved in maintaining correct DNA topology. It is also common to use probes against the 16S rRNA. For eukaryotic cells, housekeeping RNAs used for controls in northern blot analysis include those encoding glycerol-3-phosphate dehydrogenase (GAP-DH) and β-actin.

It is now generally accepted that there is no single perfect control for any gene expression study. This is for two main reasons. First, there is no such thing as a constitutively expressed gene; even housekeeping genes are expressed at different levels under different conditions. This is particularly the case for bacteria, where expression of housekeeping genes is at a level directly proportional to growth rate. In eukaryotic cells, the effects are more sporadic, particularly when using genes encoding metabolic enzymes, since there expression is under feedback and feed-forward control. The second potential pitfall when using housekeeping controls is that the RNA products of housekeeping genes are generally abundant, meaning that differential amounts of degradation between RNA samples may have a smaller impact on housekeeping mRNA levels than on the target mRNA you are hoping to quantify.

The combination of the above potential problems encountered when using controls in gene expression analysis is that it is not possible to validate small apparent differences in test gene expression using one control. The only way around this issue is to use more than one housekeeping control for each measurement of test gene expression. Pick a couple of housekeeping genes encoding quite unrelated functions, so the possibility of coordinate regulation of their expression is minimized, and if one control looks nonconstitutive under the growth conditions used, the other control will show this up. How do you know which is correct? Well, it would be unlikely for you to produce two samples of total RNA with orders of magnitude differences in concentration, so a constitutively expressed control will produce only subtle differences in RNA levels between the two samples; a nonconstitutive gene will produce larger differences in RNA levels. If you are in doubt, or if you need absolutely cast-iron validation of very small differences in gene expression levels, you will have to turn to a third control. However, to be honest, if you are trying to measure two-fold differences in gene expression between growth conditions, you are best not to choose northern blotting and hybridization as the technique for doing so. Experimental variability associated with northern blotting (i.e. possible variation from experiment to experiment due simply to the complexity of the experimental technique) makes accurate expression level calculations very difficult.

A typical controlled northern blot experiment would involve:

- Immobilization of two RNA samples, each from cells growing in a different growth condition, which have been electrophoretically and spectrophotometrically analyzed and diluted to the same nucleotide concentration.
- Hybridization of a test DNA probe (i.e. designed to anneal to the RNA whose abundance is being measured) to the immobilized RNA samples.

- Visualization of the intensities of the target:probe hybridization products in the two immobilized RNA samples in parallel using at least two exposure times and calculation of the apparent relative difference in test RNA level in samples of RNA from the two growth conditions. More exposures will be required until at least two separate exposures report the same difference in RNA between the two samples.
- Stripping the test probe from the blot and re-probing with a probe targeting a control, housekeeping RNA.
- Visualization of the intensities of the control target:probe hybridization products in the two immobilized RNA samples and calculation of the relative difference in control RNA level in samples of RNA from the two growth conditions. If the control RNA appears to be at different levels in the two conditions, then a second control probe will be required. If this does not reveal the same apparent difference in RNA loading as the first control probe, then a third control will be required. In this case, the majority apparent difference in RNA loading amongst the three control reactions should be used.
- Division of the apparent relative difference in test RNA level by the control RNA level in order to calculate the actual difference in test RNA level between the two RNA preparations, and so, by extrapolation, the relative difference in test gene expression between the two growth conditions.

Biological and experimental replicates for northern blot analysis of gene expression

I cannot overstate the value of multiple replicates, coupled with properly devised statistical tests in overcoming potential variability in gene expression analysis. No fewer than three biological replicates, each using a different pair of RNA preparation should be used, and it is best to perform two separate northern blots with each RNA preparation (i.e. two experimental replicates). This applies even when large changes in gene expression are apparent. If the apparent differences in RNA levels seen, allowing for properly implemented controls, as described above, are so low that they are still not at the desired level of statistical significance (see below) more replicates will be required. Given the high levels of experimental complexity associated with northern blotting and hybridization techniques, experimental variability is likely to be almost as high as biological variability, so it will be useful to increase the number of experimental replicates first, and only if statistical significance is still not at the desired level should you invest in the time required for preparing more RNA.

3.9   Northern hybridization meets ELISA

There are now a number of commercial kits designed to measure specific RNA levels in a sample, even from a cell lysate. For example, see Allawi *et al.*, 2004. They work on the basis of enzyme-linked immunosorbance assays (ELISA). ELISA requires that the target molecule is linked to a surface so that a number of different washing routines can be carried out without loss of target. In the case of ELISA-based northern hybridizations (*Figure 3.7*), immobilization of the target RNA to a plastic surface is achieved by providing a short oligonucleotide whose sequence is complementary to the target RNA, and whose 5′ end is conjugated with biotin. Hence the use of plastic 96-well plates whose wells are coated with streptavidin will specifically immobilize the target RNA sequence whilst all other sequences in the cell extract are washed away. To be detected using ELISA, a target molecule must carry an epitope to which an alkaline phosphatase-conjugated antibody or primary antibody against which an enzyme-conjugated secondary antibody has been raised, binds. Raising antibodies against RNA sequences is not practical, so the target RNA molecule must be labeled with an epitope to which an antibody has already been raised. This is done by providing a

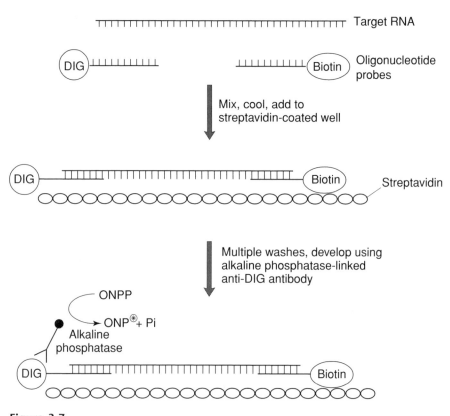

Figure 3.7

ELISA-detected northern hybridization events. The RNA being quantified is targeted to the surface of an ELISA plate well coated in streptavidin through a biotin-tagged specific complementary probe. All other RNAs are washed away. A second probe is included which binds to another point on the target RNA, and carries with it a DIG or other antigenic label, which can be used to bind a primary antibody attached to alkaline phosphatase, which as with a standard ELISA reaction can be used to develop a color from a colorless substrate, e.g. *ortho*-nitrophenyl phosphate (ONPP) in solution present in the well. The beauty of this approach is that the alkaline phosphatase enhances the signal, making this an excellent way of detecting and quantifying very low abundance RNAs.

second complementary oligonucleotide (ideally one that binds at the opposite end of the RNA molecule to the biotin-tagged oligonucleotide) that is conjugated to DIG, or some other commonly used epitope. Hybridization between the epitope-tagged oligonucleotide and the target RNA can be detected using alkaline phosphatase-linked anti-DIG (or whatever epitope is chosen) antibodies and chromogenic alkaline phosphatase substrates (e.g. *ortho*-nitrophenyl phosphate, which turns yellow when attacked by alkaline phosphatase). Commercial kits for this type of procedure use more advanced alkaline phosphatase substrates, with a variety of amplification reagents that give quantifiable signals even for low abundance target RNAs (Allawi *et al.*, 2004).

ELISA-based northern hybridization are currently only used for some specific purposes. However, they have the potential to revolutionize gene expression measurement, when only a small number of RNAs need to be quantified. If relative amounts of an RNA in two different samples (each normalized to a constitutively expressed control RNA, of course) is all that is needed, there is no reason why this approach will not give accurate results. Furthermore, it relies on equipment and reagents which are generally available in most research laboratories and DIG- and biotin-conjugated oligonucleotides can be purchased from a number of suppliers. Seriously consider having a go with this approach, it may be just the thing you are looking for.

3.10   Array-based hybridization methods

Northern blotting can be used to measure the expression of < 10 target genes and a control from each RNA sample. Up until the last 10 years, this was perfectly adequate, but now we are in the post-genomic age. When the first complete genome sequence was published, that of *Haemophilus influenzae* (Fleischmann *et al.*, 1995), suddenly, for the first time, the sequence of every gene within a self-replicating organism was known. The obvious question to ask was 'What do they all do'? Indeed, getting the genomic sequence is the easy part. Now we have to start wading through the masses of sequence data, trying to make some sense of it all. Using classical genetics, selecting mutants with particular phenotypes and then finding the gene that is mutated (Avison and Bennett, 2005), together with more sophisticated post-genomic approaches to knocking out gene functions, such as RNA interference (Shearwin *et al.*, 2005) has provided us with an insight into the functions encoded by many different genes in a number of model organisms. Through comparative genomics (Avison, 2004), it is possible to locate orthologues of these well characterized genes in other less well studied organisms, allowing predictions about their functions to be made, which themselves can be tested by specific functional genomics experiments. However, genome sequencing revealed a whole plethora of genes whose functions have never been considered before, and therefore whose orthologues (often called 'conserved hypothetical open reading frames') are really just pieces of uncharacterized DNA. These 'hypothetical' open reading frames could be disrupted by mutagenesis or RNA interference to prevent a protein product from being produced, with a view to looking for phenotypic changes, allowing a hypothesis to be generated about the functions of the products of these genes. But what phenotypic change do you look for? There are so many possibilities, meaning that the chances of hitting upon the right test to see a change is very slim. As such, we need methods for developing hypotheses concerning what areas of physiology the products of hypothetical genes take part in. Once these hypotheses are generated, it narrows down the possible phenotypes to look for when knocking out the gene, making it more likely that the true function will be found. An excellent way of developing a hypothesis concerning the function of a gene product is to find which physiological conditions the gene is expressed under. This has produced a subtle change in emphasis. Researchers are now more interested in physiological states, than in individual genes and their products. They aim to characterize all the genes that are responsible for a given physiological state. In order to find genes to study, measuring the expression of all genes in the genome when cells are in the physiological state that interests you is a very good starting point. If the number of genes that might be expressed is in the thousands, then even if northern blots were stable so that unlimited numbers of stripping and re-probing cycles could be performed, it would take a very long time indeed to make thousands of different probes, and to undertake thousands of blot hybridization, stripping, hybridization, stripping; on and on. And each time you carried out an experiment, all the probes would have to be made again, because the label would probably have degraded. It's just not feasible.

So what to do? Well the whole northern blot process needed to be turned on its head. The target RNA in a northern blot is immobilized on the solid phase surface and the probe is washed over. Thus the probe has to be the one labeled, so that unbound probe can be washed away. This gives the problem that thousands of different probes need to be made and labeled individually. But the label is only there to detect probe/target hybridization. It doesn't matter which single-stranded molecule is actually labeled. Therefore, why not immobilize the probe and label the target? Indeed, if many probes are immobilized onto one membrane, as parts of a dot blot, perhaps, then washing over a labeled heterologous solution containing all the target RNA molecules in a cell would simultaneously allow hybridization of all the RNA molecules for which probes are immobilized. Thus a large number of RNAs could be quantified in one go without stripping and re-probing (*Figure 3.8*). This arrangement of multiple spots of unlabeled probe DNA on a solid phase surface is known as an 'array'.

3.11 Types of array

The original arrays were large nylon membranes with probes spotted onto them by hand. Perhaps a few dozen probes were used at a time. There is no reason why you cannot use this approach yourself to create your own small scale array. Remember, you cannot put more than about 100 mg of DNA per cm^2 onto a nylon membrane and it is better to use larger volumes of dilute samples and so give larger spot sizes. This will allow you to control the accuracy of the spot size. However, don't get hung up about exact spot sizes and amounts of probe DNA immobilized, since the probe will always be in large excess to the labeled target. What is more important is that the DNA

(A)

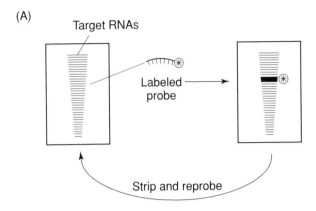

(B)

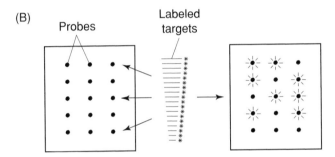

Figure 3.8

Northern blotting versus array hybridization. In (A) a classical northern gel blot is illustrated. A ladder of total RNA is run on a gel, blotted onto a membrane and a particular RNA target is probed using a labeled complementary DNA and the hybrid product quantified. If another RNA needs to be quantified, the first label must be stripped away. This causes leaching of the target total RNA pool, and is very time consuming. In (B) the array-based approach is illustrated. Here, the probes are unlabeled and spotted onto an array. The total target RNA pool (usually as a cDNA copy – see text) is labeled and this mixture is washed over the array. If a target has a probe on the array it will hybridize and the amount of hybridization, and hence the concentration of target RNA, can be determined. All unbound target RNAs are washed away.

in each spot is evenly distributed, since concentrations of probe can affect hybridization kinetics.

If you want to do real transcriptomics (i.e. measure the relative levels of all RNAs in a cell at any one time compared with any other time), you will need a micro-array (Duggan *et al.*, 1999). There is a sliding scale of micro-array spot density, but a true micro-array should have more than about 100 probe spots per square centimeter. These low density arrays (<250 spots per cm²) are commonly produced in the laboratory on nylon membranes, to be hybridized with radioactively labeled target molecules and visualized by autoradiography. The probes are dispensed using micropipetting robots, which are commonly used for a whole variety of purposes nowadays. These robots are quite expensive, but cheaper to run than a research technician, and they give more accurate volume dispensation, and, more importantly, spot position on the array. Make sure there is no possibility of carry-over between the probes by using negative pressure pipetting systems and wash solutions.

High density micro-arrays (500–50 000 probe spots per cm^2) are also available, but their production is more typically on an industrial scale. The probes are spotted onto glass or less commonly, silica plates, and the size of the plate gets smaller as the technology of spotting probes onto them gets more advanced, increasing spot density. Very high density arrays are referred to as 'DNA chips'. Glass is, in fact, a very poor surface for binding nucleic acids given its negative charge density. To assist in DNA binding, the plate is coated with polylysine, a positively charged polypeptide, which binds to silicon dioxide surfaces and the phosphate backbones of nucleic acids, bringing to two together. The fine details of producing micro-arrays will not be discussed in this chapter because the chance of you doing it yourself is minimal. On most occasions you will take your samples to your institute's micro-array facility, or send them to a micro-array analysis company. The most common probe transfer technique, however, is the 'contact-tip deposition printing' approach. This works by using a metal tip with a fine vertical slit cut into it. The tip is submerged in the probe solution for sufficient time that the appropriate volume moves into the tip by capillary action. The tip is then moved to the slide and touched onto the surface so that the sample moves out by capillary action onto the surface. Affymetrix arrays are produced by a variation of this approach where the metal pin has a cylindrical hole drilled into it. Once dipped into the probe solution and moved onto the glass surface the hole defines the size of the spot and surface tension draws the liquid onto the glass slide.

You should be aware that the smaller you go in terms of micro-array slides, the more complex the technology needed to deal with the slides, particularly in terms of array production, but also for hybridization and visualization procedures. This complexity, and the requirement for sophisticated and expensive pieces of equipment takes micro-arrays out of the reach of most laboratories. However, if you can get your hands on good commercially produced array slides, hybridization and visualization can be performed in most laboratories with a modest outlay. Given the cost of commercial array hybridization and data analysis, and accepting the extra amount of control over the experimental approach that doing it yourself gives you, if you are anticipating undertaking a lot of analysis, it is well worth investing in hybridization and visualization equipment, perhaps as a resource to be used by a number of labs.

3.12 Labeled targets for array hybridization

How can the target RNA molecules be labeled? Well, it is possible to label the 3' ends of RNA molecules with radioactive phosphate from [γ^{32}P]dATP using the enzyme polynucleotide kinase, or using [α^{32}P]dCTP, which is added to the 5' end using the enzyme terminal transferase. However, directly labeling RNA is tricky to get right, and RNA floating around in solution is inherently unstable, and folds into secondary structures which can only be inhibited by raising the stringency of the hybridization conditions to such an extent that hybridization with the immobilized probe is very unlikely to occur.

The method chosen to both label the target RNA and make it more stable at the same time, is to use the DNA polymerase enzyme, reverse transcriptase. This enzyme is programmed to use a single-stranded RNA template, and to synthesize a complementary DNA (cDNA) strand using deoxynucleotide building blocks from a 3' hydroxyl provided, *in vitro*, by an oligonucleotide primer. The reaction can be spiked with $[\alpha^{32}P]dCTP$ or labeled dUTP, or can use 5' end-labeled primers, as set out above for making DNA probes for northern blotting (Section 3.5 and *Figure 3.4*). The result would be a labeled cDNA molecule complementary to the target RNA species that you want to quantify. Quantification is a possibility, because the number of labeled target cDNA molecules produced is equal to the number of RNA molecules that have been copied, provided the reverse transcriptase reaction goes to completion, all reagents are in excess and each RNA molecule is copied ony once.

Types of reverse transcriptase enzyme

There are a bewildering number of different reverse transcriptase enzymes available commercially. They represent enzymes from different natural sources, many of which have been mutated to provide particularly useful properties. They should be used in a 'horses for courses' approach depending on the application. RNA-dependent DNA polymerase activity is something that is essential for the lifestyles of all RNA viruses, such as retroviruses, because before replicating in eukaryotic cells, the RNA must be converted into double-stranded DNA. The role of the viral reverse transcriptase is, in fact, to produce a single-stranded complementary DNA copy onto an existing RNA template (a so-called first-strand cDNA). So, the two most common reverse transcriptase enzymes come from well-characterized retroviruses: Maloney murine leukemia virus (M-MuLV) and avian myeloblastosis virus (AMV).

M-MuLV reverse transcriptase was the first to be generally available. It is a single subunit enzyme provided in recombinant form, and is very cheap. As well as being an RNA-dependent DNA polymerase working in a 5'–3' direction, M-MuLV reverse transcriptase possesses significant 3'–5' RNase H activity. When making first-strand cDNA targets for micro-arrays, RNase H is useful. It significantly degrades the RNA portion of the cDNA–RNA duplex (it does not touch single-stranded RNA) formed by reverse transcriptase, which means that each RNA molecule cannot be used as a template for reverse transcriptase more than once. Thus, there will be no amplification during reverse transcription, and so no introduction of possible quantification errors if the amplification were at different levels in different samples. The presence of RNase H within the reverse transcriptase reaction is also useful for second-strand cDNA synthesis, because the nicks produced by RNase H can be filled in using DNA polymerase I, generating the second strand and pushing off the remaining RNA fragments using the 5'–3' exonuclease activity of DNA polymerase I

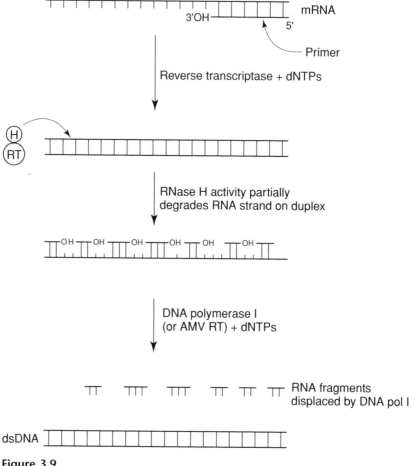

Figure 3.9

First- and second-strand cDNA synthesis. First-strand cDNA synthesis involves an RNA template and the enzyme reverse transcriptase, which is targeted to a particular RNA (or all RNAs) by the use of sequence specific (or random) complementary primers. Once the first-strand cDNA has been synthesized, the reverse transcriptase-associated RNase H activity removes pieces of the RNA strand of the cDNA/RNA duplex. If this is allowed to proceed for a long period, all the RNA will be degraded. However, when sufficient cleavage has been achieved, DNA polymerase I is added (or when using AMV reverse transcriptase, no polymerase need be added) and the second-strand cDNA is synthesized, using the fragments of RNA as primers. The polymerase pushes off the RNA fragments, meaning that the entire second strand, and so a blunt-ended, double-stranded cDNA is produced.

(*Figure 3.9*). RNase H minus derivatives of M-MuLV reverse transcriptase are available, but their usefulness in first-strand cDNA production for micro-arrays is limited. They can be used to amplify the effective signal of an mRNA, because each mRNA can be converted into a first-strand cDNA probe multiple times. However, this approach is not to be encouraged because it may introduce an extra element of experimental variability into the reaction. For example, the degree of amplification may be different each time the procedure is carried out and is likely to be dependent on the starting mRNA concentration. The most serious problem with M-MuLV reverse transcriptase, however, is that reactions must be performed at 30–37°C. This is particularly problematic when using RNA templates with significant secondary structure within this temperature range. Point mutants of M-MuLV reverse transcriptase are now available, however, with improved thermostability, and better reproducibility in general, but for micro-array analysis where all the mRNA must be converted into cDNA efficiently, this enzyme cannot be recommended.

AMV reverse transcriptase is a much more appropriate enzyme for cDNA template production for micro-array analysis. It carries high level RNase H activity, but its real benefit is its thermostability. Reactions can be performed at 55°C, at which temperature, secondary structure in all but the most obstinate RNA molecules will have melted. There are two negatives to using AMV reverse transcriptase. The first is that it has substantial DNA-dependent DNA polymerase activity, so any genomic DNA contamination of the RNA preparation may well become labeled. These labeled genomic DNA fragments would hybridize onto the array, causing virtually every probe to light up. Therefore, if you choose to use AMV reverse transcriptase for first-strand cDNA production, be very sure there is no DNA contamination. The second problem with AMV is its cost. This is not simply because of market forces – the enzyme is better than M-MuLV, so manufacturers can charge more – but, equally, because purification of the AMV enzyme is more difficult. It has to be purified from virus particles, because in a recombinant form it is lethal to the cells making it, due to its DNA-dependent DNA polymerase activity. Despite this, AMV is the reverse transcriptase of choice for qRT-PCR (see Section 4.9) and cDNA library production (partly because of the ability to synthesize a second strand onto the cDNA using the DNA-dependent DNA polymerase activity not present in the M-MuLV enzyme, where DNA polymerase I has to be added after first-strand cDNA production (*Figure 3.9*).

A new wave of enzymes with reverse transcriptase activity have recently become available from a variety of different manufacturers. They are highly thermostable, so they can be used to generate cDNAs at up to 70°C, where even the most stubborn secondary structure in the RNA template will have melted. Essentially, however, they are thermostable DNA polymerases that happen to have reverse transcriptase activity, so you are paying for two functions and using only one. Furthermore, except for use with really problematic templates which require high reaction temperatures, they are similar in performance to AMV, but for an increased cost. Accordingly, for most first-strand cDNA synthesis procedures, their use is a bit of a luxury. If you have a very precious RNA sample, however, and want maximum first-strand cDNA production, labeling, and a more forgiving enzyme in terms of variation in reaction conditions, then I would recommend investing in one of these new enzymes. They will be discussed in more detail in Section 4.9, where reverse transcription PCR is described, because it is during this procedure that thermostable DNA polymerases with significant reverse transcriptase activity come into their own. Which one is best, though, will be a matter of what works for you.

Primers for labeled first-strand cDNA target synthesis

Like all DNA polymerases, reverse transcriptase needs a 3′ hydroxyl group to build from, and so a primer oligonucleotide is required. If you want to convert a single RNA molecule with a known sequence into a first-strand cDNA copy, then a sequence-specific negative (complementary) primer can be synthesized. Using this primer, however, only one type of RNA molecule will be converted to first-strand cDNA. This is the basis of the first stage of qRT-PCR (see Section 4.9), but is useless if you need to quantify a large number of different RNAs in the same sample using an array. For this, all the RNA molecules need to be converted into labeled first-strand cDNA at the same time. The only way to do this is to use a primer that binds to all RNA molecules. If you work with eukaryotic cells, then this is relatively straightforward, since the vast majority of coding RNAs in these cells have a polyadenine tract at their 3′ ends. Thus a polythymine primer will anneal to all polyA$^+$ RNA molecules, and will prime all of their conversions into cDNA using reverse transcriptase. It should be remembered, however, that there are a number of RNA molecules that are not polyadenylated in eukaryotic cells, and particularly that organellar RNAs are not polyA$^+$ at all. Furthermore, if you work with prokaryotes, where polyadenine tracts are not added to mRNA molecules, the use of polythymine primers is not an option.

Random primers are more generally used now to prime total first-strand cDNA synthesis, because they do not suffer from any of the problems associated with polythymine primers. Most commonly, random hexamers are used. These primers are really a mixture of all possible 6 bp sequences. The rationale behind this approach is that at least one of the individual sequences will bind to every RNA molecule due to chance complementarity with six nucleotide runs within the RNA. Binding of each hexamer will prime cDNA synthesis. There are a number of problems with this approach. Normally, the reverse transcriptase enzyme will be able to extend around 1000–2000 nucleotides from the primer, and if it reaches the 5′ end of the mRNA, it will literally fall off. Since you cannot guarantee where within the RNA molecule the primer will bind, however, you cannot tell how long the first-strand cDNA produced will be. Indeed, each RNA may be converted into a diverse population of first-strand cDNAs with different lengths. Provided the average length obtained is identical each time you make cDNA, then this is a not a problem, but if different RNA samples from cells in different physiological states are converted into first-strand cDNAs with different lengths, then the first-strand cDNAs in each sample can have different hybridization kinetics with the immobilized probe. Furthermore, if the cDNAs are labeled by the random incorporation of labeled nucleotides, the amount of label in each cDNA will be directly proportional to its length, compounding any differences in average length seen with different cDNA productions. The way to get over this is to end label the primers in the random primer mixture (on their 5′ hydroxyl groups) prior to using them, as described in Section 3.6. This will mean each cDNA will be labeled with one moiety, and will give the same signal strength when hybridization is visualized. Second, it is very important that first-strand cDNA production is undertaken in very tightly controlled conditions. The amounts of RNA, primer, enzyme and buffer, and the length of time the reaction is allowed to proceed for must be the same for each cDNA preparation. Many people find that the used of random decamer primers (10 nucleotide primers will bind less frequently within RNA molecules) provides more controllable cDNA lengths than random hexamers, and will dramatically increase the average cDNA length in a sample.

Amplification of cDNA targets prior to hybridization

Sometimes, if you are using very low RNA concentrations, perhaps from very small and precious tissue samples, the amount of labeled first-strand cDNA produced is not enough to produce a hybridization signal that can be accurately quantified. For this, cDNA amplification may be required and can be achieved using a PCR technique known as sequence independent primed PCR (SIP-PCR), whose method is described in Section 4.10. This is not always satisfactory, because different cDNA molecules may be amplified with different PCR kinetics to others, meaning that they will be amplified to different amounts. If all you want to do is quantify the levels of an RNA molecule in two samples, however, it is likely that a few rounds of PCR will not dramatically affect the integrity of the hybridization results.

Another cDNA amplification method is to use RNA polymerase to transcribe the cDNA into RNA a large number of times in the test tube, and then repeating cDNA production using the amplified RNA sample as template. To do so, first-strand cDNAs are made double-stranded via second-strand cDNA synthesis using a DNA polymerase such as DNA polymerase I. Because these enzymes require 3' hydroxyls, the best way of achieving this is to use an RNase H-containing reverse transcriptase for first-strand cDNA synthesis, followed by the addition of DNA polymerase I (alternatively, just use AMV reverse transcriptase which has both RNase H and inherent DNA-dependent DNA polymerase activity inbuilt). Thus, small degradations of the RNA strand of the first-strand cDNA/RNA duplex are filled in by DNA polymerase, degrading the remaining RNA using its inherent 5'–3' exonuclease activity (*Figure 3.9*). Once double-stranded, blunt-ended cDNA has been produced, a double-stranded oligonucleotide linker containing the sequence of the bacteriophage T3 RNA polymerase binding site (a T3 promoter) is linked onto both ends of all the double-stranded cDNA molecules using DNA ligase. Next, purified T3 RNA polymerase is added in order to transcribe large numbers of RNA copies of the cDNA, which are finally reverse transcribed back into cDNA (*Figure 3.10*). This is one of those molecular biology techniques that sounds simple, but is incredibly difficult to get right. The control you have over the amplification process is poor, and overall, RNA polymerase-based amplifications are less satisfactory than SIP-PCR amplification of cDNAs direct.

Labels used in first-strand cDNA target production

Large, low-density arrays are generally hybridized with radioactive or DIG-labeled target cDNAs. These labels are incorporated randomly into the molecule as labeled nucleotides, or are conjugated onto the 5' ends of primers, as set out above (Section 3.6). However, large numbers of colored

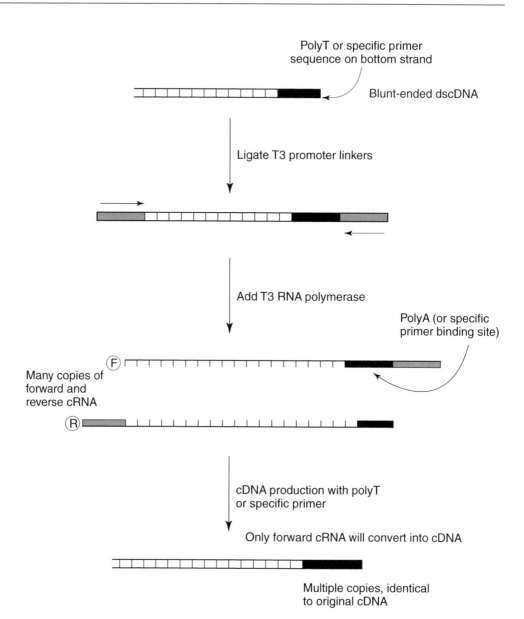

Figure 3.10

Amplification of cDNA concentration. Blunt-ended cDNA, produced as described in Figure 3.9 is modified by the addition of T3 RNA polymerase binding sites (T3 promoters) to each end (because it is not possible to only put the linker on the 5′ end with respect to the coding strand, which is all that is actually needed). In the presence of recombinant T3 RNA polymerase, both cDNA strands are converted into cRNA many times, i.e. there is amplification. Finally the forward cRNAs are converted back to cDNA using a strand-specific complementary primer, though the concentration of this cDNA will be dramatically more than the concentration of the original cDNA. If polyA+ mRNA is used to start with, this method can be used to simultaneously amplify all cDNAs from an organism, if a polythymine primer is used for cDNA production.

dye labels are available which are particularly useful for visualizing hybridization of target first-strand cDNAs to probes immobilized onto glass slides as part of high density arrays. These probes are available chemically linked to nucleotides (normally, dUTP) for random incorporation into cDNAs by reverse transcriptase, or can also be conjugated onto the 5′ hydroxyl ends of primers.

The most commonly used colored dyes for target cDNA labeling prior to micro-array hybridization are Cy3 and Cy5. These are trademarked commercial names for the water soluble cyanine dyes: 2-[(1*E*,3*E*)-5-(1-{6-[2,5-dioxo-1-pyrrolidinyl-oxy]-6-oxohexyl}-3,3-dimethyl-5-sulfo-1,3-dihydro-2*H*-indol-2-ylidene)-1,3-**propa**dienyl]-1-ethyl-3,3,dimethyl-3*H*-indolium-5-sulfonate and 2-[(1*E*,3*E*)-5-(1-{6-[2,5-dioxo-1-pyrrolidinyl-oxy]-6-oxohexyl}-3,3-dimethyl-5-sulfo-1,3-dihydro-2*H*-indol-2-ylidene)-1,3-**penta**dienyl]-1-ethyl-3,3,dimethyl-3*H*-indolium-5-sulfonate (!). The bold, underlined portion of each chemical name represents the only difference between these two molecules and yet their chromogenic properties are remarkably different. Cy3 and Cy5 absorb light at 552 and 650 nm, respectively, and this causes their fluorophore to become excited. When the fluorophore returns to its basal state the molecules emit light at 568 and 667 nm, respectively. Cy3 and Cy5 are available as adducts of dUTP nucleotides. They do not greatly inhibit incorporation of dUTP by reverse transcriptase opposite adenine in the target RNA sequences, unlike some other dyes that might be chosen. Furthermore, they possess excellent photostability and brightness, and because the excitation and emission spectra of each are far apart, there is little chance of the excitation laser being visible by the emission detector. The fact that the two dyes are chemically almost identical, means that the kinetics of their incorporation into cDNA probes is, likewise, identical. This fact, coupled with their dramatically different emission spectra means that the use of these two dyes together in one experiment is quite possible, and increasingly common. Examples of why this would be useful are given in Section 3.14.

A major drawback of using fluorescent labels is that a large amount of labeled first-strand cDNA (and so target RNA) is required to give label signals that are bright enough to be seen above background levels following hybridization of the labeled cDNA to probes on the array. Typically 50–250 µg of RNA is needed, which for very small samples (e.g. micro-dissection samples or rare blood cells) is not achievable. For these cases, cDNA amplification may be essential (see above).

Another way of visualizing hybridization of target to immobilized probe is the use of gold or silver staining techniques (*Figure 3.11*). It is possible to directly label cDNA with gold, but this is not to be recommended for the generation of array targets because gold is known to affect hybridization of nucleic acids. So, to load hybridization products with gold or silver, the target RNA must be converted to cDNA that is labeled with DIG, biotin or some other large molecular weight epitope by the use of 5′ conjugated primers or labeled nucleotides in the cDNA synthesis reaction. The basis of silver or gold loading of hybrid duplexes is that the epitope label on the target cDNA interacts with a silver- or gold-conjugated antibody (or in the case of biotin labels, gold- or silver-conjugated streptavidin). Thus once hybridization has occurred and excess labeled cDNA has been washed away,

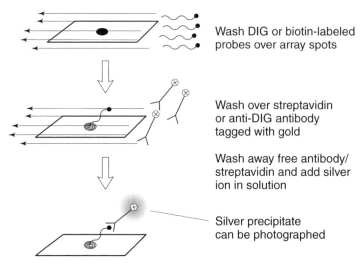

Wash DIG or biotin-labeled
probes over array spots

Wash over streptavidin
or anti-DIG antibody
tagged with gold

Wash away free antibody/
streptavidin and add silver
ion in solution

Silver precipitate
can be photographed

Figure 3.11

Silver precipitation to visualize hybridization products. Here, the labeled probe (or target in the case of array hybridization) is visualized using a primary antibody raised against the label (e.g. DIG), which is linked to a gold particle. This gold particle forms a focus for silver precipitation onto the blot, and so the hybrid product can be visualized. The more hybrid product, the more silver precipitates, and so the stronger the intensity of the spot.

the gold- or silver-conjugated antibody/streptavidin is washed over the surface and loads target/probe duplexes with gold or silver. The more duplex that has formed, the more gold or silver is loaded. Once the excess conjugated antibody etc. has been removed, the signal is dramatically enhanced to one that is easily visible by the addition of silver ions in solution, which precipitate around the gold- or silver-loaded probe/target duplexes with reproducible kinetics. Thus the amount of precipitation is directly proportional to the amount of gold or silver loading, which is directly proportional to the amount of hybridization, and so target cDNA. The silver precipitate on an array can be visualized using a camera attached to a standard laboratory microscope and be quantified using densitometry.

Silver precipitate staining is an excellent way of obtaining very accurate quantification data, since the dynamic range and signal-to-noise ratio are both far greater than for fluorescent labels. The spots are very sharply resolved, allowing high density micro-arrays that are refractory to the use of some fluorescent labels to be visualized. Finally, silver precipitates are visualized far more simply than are fluorescent labels, which require sophisticated laser emission and filtered detection systems.

When using fluorescent labels, if the signal in parts of the array is not strong enough, hybridization can be allowed to occur for a longer time. This will help with the potential problem of some hybridization interactions becoming saturated, whilst others remain virtually invisible (i.e. is similar to the use of a number of different exposure times in autoradiography). The major drawback to the use of silver precipitation-based visualization procedures in that once the silver has been precipitated, there is no going back. One final note of caution if you are considering using silver precipitation is that it is a tricky technique. The amount of time the silver solution is incubated with the array is critical, because slightly too long and nonspecific precipitation can occur, both in solution and onto the surface of the slide. Since different slides behave differently, it is not always possible to predict what will happen. Thus, if you want a quick, cheap, but dirty experiment that will allow qualitative visualization of target/probe interactions, then gold/silver conjugation coupled with silver precipitation is to be recommended, but if you want reliable visualization of probe/target interactions in real time, fluorescent labels are better.

3.13   Probes for array-based hybridization

The aim of array hybridization approaches is to analyze the expression of a number of genes at once. Hence efficient hybridization of labeled cDNA targets derived from RNA molecules to an unlabeled single-stranded DNA probe representing the sequence of the gene whose expression is being measured is the key to success. For this, both partners in the hybridization reaction must be in an optimal state. We have considered cDNA target production, above, so here we must discuss the problems of probes, and how they can be minimized. There are basically three different types of probes that are commonly used in array hybridization experiments. Genomic DNA probes, cDNA probes and oligonucleotide probes. This produces the different terms 'DNA array', 'cDNA array' or 'oligonucleotide array', depending upon what probe types are used.

Genomic clones as probes for arrays

The production of large amounts of double-stranded DNA probes for spotting onto arrays is performed using the natural DNA copying enzyme, DNA polymerase, since this is by far the most efficient way of making multiple copies of a DNA sequence. The original way of copying a DNA sequence was to manipulate the sequence so that it becomes part of a recombinant, autonomously replicating plasmid vector (i.e. to 'clone' the gene). Once the recombinant vector is placed within a host permissive of vector replication, the entire plasmid is copied *in vivo*. Large amounts of cells can be harvested, and the plasmid DNA purified. The entire recombinant plasmid can be used as a probe for array hybridization, and this method has been used many times in the past for cheap and dirty approaches to array-based gene expression analysis. The use of recombinant plasmids as array probes is popular since the recombinants are by-products of automated genome sequence projects. These projects use random shotgun cloning of genomic sequences, and then sequencing of inserts within individual purified recombinants. The use of robots to pick off recombinants, allows recombinants to be stored in individual wells with computer-defined unique clone identification numbers. Hence, when the sequence for all of the recombinants has been analyzed, a library of non-overlapping recombinants can be generated, each carrying the sequence of a different region of DNA. These libraries can then be spotted onto an array, with each recombinant forming a different spot. This approach is illustrated in *Figure 3.12*.

There are a number of problems, of course. First, there is no guarantee that only one gene will be present on each recombinant, it could equally well represent the ends of two genes that are adjacent. Thus two labeled first-strand cDNA targets could hybridize to some probes. The second

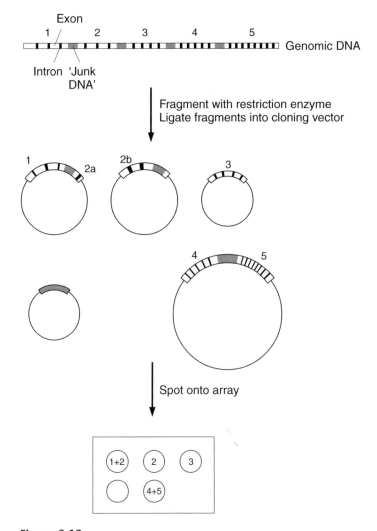

Figure 3.12

Production of genomic libraries. This is simply the digestion of an entire genome into fragments and the ligation of each of these fragments into a cloning vector. The recombinant vectors are sorted by sequencing, and each individual recombinant can be used as a probe on an array. Problems include the presence of junk and intron sequences within probes, which can affect hybridization, and the presence of sequences for more than one gene in some probes, complicating the hybridization results. Both of these problems are illustrated in the figure.

problem is particular to eukaryotic genomes, and involves the presence of introns within the genomic sequence, meaning that there might be very little coding sequence in some recombinants, making the use of recombinants with large insert sizes essential. The final problem is exacerbated by the use of large inserts, and it is the presence of significant nonprobe sequence (i.e. a sequence which does not form the coding sequence of the gene whose expression is being measured). The more 'junk' sequence (including vector sequence) present in the spot, the greater the chances of nonspecific hybridization events occurring. Hence the use of genomic library probes for array hybridization is only really viable for prokaryotic organisms, where gene density is very much higher, and introns do not exist.

Extended sequence tag cDNA probes

With the use of the phrase 'cDNA array', things can get very confusing, since all arrays, no matter what the origins of the probes used, are hybridized with labeled first-strand cDNA targets. The phrase 'cDNA array' comes from the fact that the probes are made originally from RNA sequences. Because the labeled target is first-strand cDNA, cDNA array probes must represent double-stranded cDNAs, so that the second strand is available for the first-strand cDNA target to hybridize with (*Figure 3.13*).

The use of cDNA probes is one way of overcoming the problem of intron and other non-coding sequence contamination, and overlap between genes when making eukaryotic arrays. cDNA libraries are made from total genomic mRNA using reverse transcriptase and a polythymine primer (hence they are only made from eukaryotes). Once the first-strand cDNA copy has been made, RNA in the RNA cDNA duplex is degraded with RNase H, and the second-strand cDNA built using DNA polymerase I (*Figure 3.9*). Double-stranded DNA linkers are then attached to each end, forming restriction enzyme sites, and the double stranded DNA molecules are then ligated into a cloning vector (*Figure 3.13*). Automated separation of different recombinant vectors is coupled with so-called 'extended sequence tag' (EST) analysis. Basically, the sequence across one or both ends of the cloned cDNA insert is determined, allowing different cDNA recombinants to be identified and separated, and so spotted onto an array. EST-based arrays are very popular, but it should be remembered that you only get as many different cDNAs in your library as different genes expressed in the tissue you used to purify the mRNA. Since this is highly variable, it is never possible to cover all the genes in the genome in a single cDNA library. Indeed, it is likely that a combination of all the cDNA libraries ever made for any organism does not cover all the genes that might possibly be expressed in cells from that organism. Hence the possibility exists that in any experiment involving cDNA target hybridization to an EST array, the expression of genes that do not have probes on the array, will be missed. Notwithstanding this potential problem, EST-based cDNA arrays are very popular for the analysis of global gene expression patterns in cells from eukaryotic organisms. Another problem that comes with any method where a number of complete gene sequences are being represented on the array is the possibility of cross-hybridization of a single-labeled cDNA target molecule with more than one

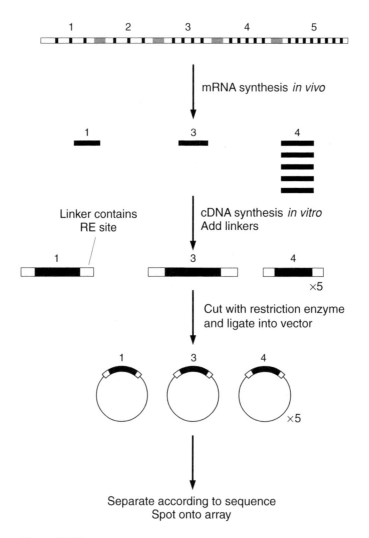

Figure 3.13

Production of cDNA libraries. RNA isolated from a cell will have no intron or junk sequences. However, no cells will produce RNAs for all genes, so a cDNA library may exclude many genes. The RNAs are isolated and converted into double-stranded cDNA (see Figure 3.9). Linkers are ligated to each end so that cutting with a restriction enzyme will allow cloning of the inserts. Sequencing across the ends of each recombinant insert (so-called EST analysis) will allow separation of different recombinants, which can be used as probes in array analysis.

probe. This could be due to the fact that two paralogues exist in the genome having high levels of sequence identity. Or it could be due to smaller-scale regions of sequence identity between two probe sequences, meaning that weak, but significant, cross-hybridization occurs. Finally, when using EST cDNA probes or cloned genomic fragments, there will be much heterogeneity in terms of melting temperature, secondary structure, length and other factors that will affect target hybridization kinetics. This can mean that when fixed hybridization conditions are used, some probes are saturated, possibly by a number of nonspecific target cDNAs, and others might not hybridize to their designated targets at all.

PCR products as probes

Another way of producing probes for DNA arrays entirely *in vitro*, is to use the polymerase chain reaction (PCR), which is an *in vitro* tool for copying double-stranded DNA sequences. The details of PCR will be discussed in Chapter 4, but the key point is that a pair of oligonucleotide primers is essential for PCR; one bounds the 5′ end and the other bounds the 3′ end of the region of DNA to be amplified. Thus, manipulating the sequences of the primers will alter the length and exact sequence of the probe to be amplified for each gene you want to provide as a spot on an array. Working from a genome sequence, therefore, one can pick regions of the genome that represent fragments of genes, and are free from contaminating non-coding sequence. In this way, every gene in the genome can be covered on an array. Furthermore, this approach can help to minimize the problem of cross-hybridization between one target first-strand cDNA and a number of DNA probes, seen with the use of genomic DNA probes, above. The regions that are amplified to make the probes can be run through a computer program to find out if any of them are similar enough in sequence to produce the possibility of cross-hybridization with the same target cDNA. At the same time, regions being amplified can be normalized in terms of melting temperature, and other factors that might affect hybridization kinetics. Possible problems can be flagged up at this stage and, if necessary, some pairs of primers can be redesigned to amplify different regions of some of the genes. The fundamental problem with using PCR to amplify sequences is that it occasionally goes wrong. Whilst primers can be designed to target one DNA sequence, in some cases they will anneal to another part of the genome, particularly if the annealing temperature used is too low. More problematically is where primer pairs, or individual primers anneal close together and in opposite orientations within the genome, because this will result in the amplification of an entirely unwanted sequence. Furthermore, PCR primers anneal to their targets with specific hybridization kinetics. If these are not allowed for in the PCR reaction, then it is possible that both primers will not anneal at the same time; for example, if the annealing temperature used is too high. In this case, no PCR product will be forthcoming. Thus PCR reactions with new primer pairs need to be optimized, and the products assessed in terms of size, whether there are multiple products, and ultimately need to be sequenced, before one is happy that the PCR reaction has worked. To scale this sort of optimization up to cover the thousands, if not tens of thousands of PCR primer pairs and PCR reactions needed to generate probes to cover an entire organism's genome, is simply not practical. Automated analysis of PCR product size and integrity (e.g. running an electrophoresis gel robotically and analyzing the gel products by computer) can be performed, but the possible artifacts of PCR mean that the size can never be determined exactly, even by a robot and a computer, and if you are unlucky enough to get a PCR product that is within a few percent of the expected size of the true product, then you are likely to be none the wiser. If such false probes were spotted onto an array, then they would hybridize with entirely spurious target cDNAs, giving some real artifactual headaches in the data interpretation part of the experiment. It is estimated that for the first PCR array of *E. coli*, which was made commercially available, 10% of the probes represented the wrong or mixed sequences.

Oligonucleotide arrays

PCR products are an excellent way of getting over all of the problems of cross-hybridization, nonprobe sequence interference and heterogeneity of probe hybridization kinetics, but the production of PCR products is fraught with potential problems, which can be very difficult to spot. This is due to the PCR process itself, which is prone to introducing errors. Hence the synthesis of DNA products with all the advantages that the exact probe sequence can be chosen for optimum hybridization and low cross-reactivity properties, but by a mechanism that is almost free of error is the 'holy grail' of array design.

The best approach so far is to synthesize the probes in a machine as single-stranded oligonucleotides. The sequences are designed by computer to have excellent hybridization properties, and to have minimal potential of cross-hybridization with nondesignated target cDNA sequences. Synthesis in a machine is the best way we have of producing the correct sequence. Indeed, if errors do occur, the results of these errors tend to be represented by a very low abundance contamination of the oligonucleotide sample, and so their influence upon hybridization of target cDNAs is very limited.

The standard size for oligonucleotide array probes is 70 nucleotides. This makes them expensive to produce, but this length is sufficient to allow specificity of hybridization. If the probes were much shorter they might well hybridize with nontarget sequences, if they were made much longer, this would improve specificity, but not to the extent that the extra cost would be justified. In most arrays, a number (sometimes as many as 10) of different oligonucleotide probes are made for each target cDNA. This redundancy allows for potential problems with the hybridization of specific probes and their targets; enough correct target:probe interactions will occur to give interpretable data.

There are a number of companies who produce oligonucleotide arrays covering all the predicted protein coding genes of many completed genome sequences, in some cases prior to the publication of those sequences. These arrays are printed to order, with an initial cost to the consumer for the first printing, and then a reduced cost per array slide thereafter. The slides can be purchased and hybridized to labeled target cDNA in house, or the companies that make the slides are able to perform the array hybridizations for you; all you provide is purified total RNA, or polyA$^+$ mRNA. When it comes to data interpretation, however, you are likely to be on your own.

3.14   Experimental and data analysis approaches for use with array hybridization

As discussed above, the simplest array is a collection of probes spotted onto a nylon membrane being hybridized to radioactively labeled target first-strand cDNAs. The hybridization protocol and conditions, including methods for handling the membrane, immobilizing probes and pre-hybridization, the stringency issues associated with washing the membrane and finally, visualization of positive hybridizations and dealing with the data generated are all identical to those described for northern blots (Section 3.8). Experiments using glass micro-array slides or chips, however, are more complex. Not in terms of the basis of the methods, since the principle of blocking the surface of the slide/chip, and the conditions of hybridization and variable stringency washing are identical, in principle, to those used for northern blots. However, the equipment needed to perform these techniques is considerably more advanced than a glass tube rolling around in an oven! Sometimes, the cost of slide/chip hybridization equipment seems unreasonable. In some ways, manufacturers of array slides can force you into choosing a particular piece of equipment because it is the only one into which their array slides fit. There is very little you can do, however, and you should take comfort in the fact that hybridization devices allow very accurate control of stringency issues such as temperature and flow rate over the surface of the array.

Once the slide has been placed in the hybridization apparatus, labeled target has been hybridized onto it, and then nonspecific interactions have been minimized by stringent washes, the slide is taken out of the machine and moved to a visualization device. For colored dye labels, the visualization device has a laser light source and a very accurate wavelength filter attached to a charged coupled device (CCD) which converts photons into electrical impulses, storing a monochromatic digital image of the array. More advanced visualization machines can work with many different dye labels, because they can alter the laser light emitted to excite the chromophore, and can alter the wavelength of light that the CCD responds to.

In some experiments, two different colored cDNA target preparations, each made from RNA isolated from cells in a different physiological state can be hybridized to a single array. In the typical example, one preparation of cDNA is labeled with Cy3, and another with Cy5. If the cDNA representing a given gene is equally abundant in both preparations, then one would expect equal amounts of Cy3 and Cy5 labeling of the spot/s on the array representing that gene. Thus the amount of red light and green light being returned from the array at that point in a visualization machine would be equal, and the computer can report the ratio of red to green light as approximately one. However, if a gene is differentially expressed in the two physiological states, one preparation of cDNA would have an over-abundance of a particular cDNA compared with the other. In this case, there would be a preponderance of red or green light being returned from the array in the visualization machine, and the computer analyzing the image can report that as well (*Figure 3.14*). At first, this approach appears highly beneficial. It means that only one array slide is required for each replicate of an experiment, making the experiment cheaper, easier and quicker. However, many

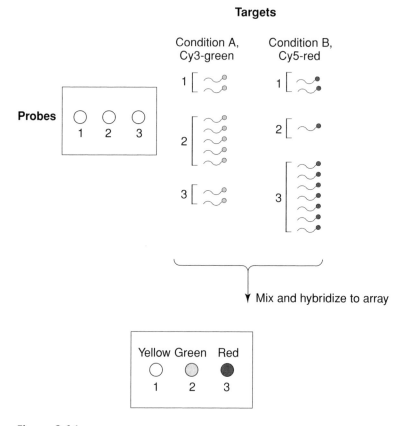

Figure 3.14

Dual dye-labeled array hybridization experiments. cDNA is made from two preparations of RNA: one is labeled with the green dye Cy3, and the other with the red dye Cy5. The two preparations are mixed and hybridized onto the same array. If the amount of a given RNA target in RNA preparation A and B is identical, the amount of red or green label incorporated onto a particular spot will be identical. This results in a yellow color. However, if a particular RNA target is present predominantly in one of the total RNA preparation, its color will predominantly label the array.

array facilities have reverted to using individual array hybridizations, one for each cDNA preparation. The advantage here is that it is easier to obtain accurate variability data for each physiological state. The presence of two different preparations of cDNA in a sample has the potential to complicate matters, particularly if one preparation out-competes the other for access to the immobilized probe, as might be the case with high-abundance cDNAs. Therefore, it is recommended that data from dual-hybridization experiments is treated with caution, and small apparent differences observed in the expression of normally highly expressed genes should be treated with downright skepticism.

For the detection of gold/silver labels, the hybridization apparatus can be used for the various steps of developing the array, and then it is moved to a more simple setup of microscope and digital camera to record the spots. Once digital images of spots have been generated, from whatever source, they can be quantified and compared using simple pixel density analysis software; the same principle is used to quantify labeled bands on northern blots (Section 3.8). This sort of software application is provided with all commercially available array visualization machinery. The same problems of image saturation exist with images of hybridized arrays as with northern blots, so multiple exposure times of each image will be required. There is also a requirement to standardize RNA loading between experiments, though with arrays this tends to be easier than with northern hybridization experiments given that housekeeping genes are very likely to be represented as probes on the array. A statistical analysis of the data is identical to the procedures suggested for analysis of northern hybridization data, and represents pair-wise comparisons of individual target:probe hybridization product intensities using cDNA probes made from RNA isolated from cells grown in two physiological states: experimental and control.

When spot intensity is converted to a numerical scale, then statistical analysis can be performed on the intensities seen for each spot under different growth conditions. Most of these statistical tests are built into the software supplied with the array visualization machinery. A description of the ways these statistical tests work is given in Chapter 8.

3.15 Limitations and design of micro-array transcriptomics experiments

It is beyond the scope of this book to discuss the different array platforms, and the equipment used to perform hybridizations of labeled target cDNA to probes on each type of array that is currently available. It is an ever-increasing market, and more and more manufacturers are introducing products into it. In fact, it is highly unlikely that you would be left to proceed with micro-array hybridization and visualization yourself. You would probably take your labeled cDNA target mixture and your commercially produced micro-array slides to a micro-array facility, and hand them over to an expert, who, at the very least will guide you through the use of the hybridization and visualization equipment and at best will do the experiment for you. Indeed, many micro-array facility managers are loath to allow outsiders to touch their valuable equipment and in some places, to even enter the room where the equipment is kept!

Micro-array facility managers, however, will generally know nothing (nor will they have time to learn) about the biology behind the specific experiment you are trying to perform. So designing the experiment will be down to you. I can't possibly address all the possible experiments you might do, but I want to spend some time discussing experimental design, because it is at this point that a number of problems can occur which will manifest themselves as poor data and long nights in front of statistics packages, working out what can be salvaged.

On a descending scale, sources of experimental variability that will affect the reproducibility of the experiments you perform are: biological variability (i.e. not getting the samples in exactly the same physiological state every time); sample variability (i.e. variation in RNA isolation due to sporadic problems during total RNA purification and first-strand cDNA production); and system variability (i.e. variation in performance of the equipment used from experiment to experiment).

System variability

The use of modern, sophisticated micro-array printing robots, generating highly homogeneous slides, which are passed through the safe, microchip-controlled hands of a variety of hybridization and visualization equipment removes so much potential system variability that you can almost ignore it. The only caveat to this is that machines do occasionally malfunction, and if a sensor breaks, this can often go unnoticed. Hence in micro-array facilities the same quality control experiments are periodically run to make sure the results obtained are within the required limits, and so to pick up on possible system variability. Of course, at the other end of the scale, if you print your own arrays on a nylon membrane with a multi-channel pipette, hybridize the probes to target cDNAs labeled with inconsistent amounts of radioactive specific activity in Pyrex dishes in an incubator that other people are constantly delving into for their own experiments, and then visualize hybridization using old X-ray film in a freezer that is renowned for temperature fluctuations and using an intensifying screen held together with sticky tape, then system variability is going to be a big problem. Be very clear in your mind, therefore, what the aim of your experiment is. If you want to get very accurate relative gene expression data, so that you can be confident of expression differences of five-fold and below, then minimal overall variability is essential, so the use of techniques that add in system variability is going to 'hole your experiment below the water line' before you even consider other possible sources of variability. However, if you want to confirm 100-fold changes in gene expression in a semi-quantitative manner, then why go to the expense and hassle of using more sophisticated equipment and protocols? Indeed, nylon arrays can be made less variable simply by using good hybridization ovens that are kept exclusively for your use during a specific experiment so the door is not being opened and closed all the time.

Sample variability

Variation in the integrity and purity of different RNA samples, even produced using the same method and from the same cell or tissue types can be surprisingly large. Sections 2.9 to 2.11 give insights into the methods used to assess RNA integrity and purity. At the very least, RNA for labeled target cDNA production should be run on an agarose gel and stained with ethidium bromide to view the rRNA bands. These should be sharp, and their intensities should reveal a 2:1 ratio between the intensities of the large rRNA subunit (23S in bacteria or 28S in eukaryotes) and the small rRNA subunit (16S in bacteria or 18S in eukaryotes). Of course, a better measure of RNA integrity is the use of a Bioanalyzer (Section 2.11) and these machines are used routinely for analysis of RNA to be used for micro-array hybridization. Purity and quantity of RNA can be assessed using spectrophotometric measurements, e.g. A_{260}/A_{280} ratio, which locates significant protein contamination and A_{260}/A_{240} ratio which locates salt contamination (which is frequently overlooked, and can significantly affect cDNA production).

Reduction in variability associated with enzymatic incubations, for example cDNA synthesis and labeling reactions, is also essential for very accurate micro-array hybridization results. Commonly, this is achieved by all RNAs being processed on the same day, using reagents from the same batch and preparing reagent master mixes. On no account should you open a new batch of anything halfway through a group of labeling reactions. In this case, 'batch' would mean not only the batch number the manufacturer stamps on a tube, but also how you have treated the contents. Once one tube out of a batch has been thawed out, some of the contents have been removed and the tube is put back in the freezer, it quite clearly is no longer of the same batch as its colleagues that have been left in the freezer all along. Make sure you have enough of each reagent in a single batch to complete all the labeling reactions needed for one particular experiment. If you are doing repeats of the same experiment (as you must do!) preparing the labels for each set of repeats is a good point to punctuate and open a new batch – as long as all of the samples for each experiment are labeled using the same batch of reagents. A final point of contention is the need for a single scientist to be responsible for all samples from each experiment. It is well known that some scientists exude more RNases than others, and some pipette with a consistent error. You know who they are, and don't let them 'help' you with your experiment! If it is very difficult to achieve this, for example, if you have just far too many samples to be processed, then divide the samples into sensible groups, and each of you should take one group. However, have a single pair of samples that every scientist processes. Using array hybridization data for this pair from all the scientists involved will help to control all the other sample preparations, by showing up the serial erroneous pipetters and RNase exuders. Another way of finding the cDNA production 'weakest links' is to use the 'signal probe' provided with most micro-arrays. The signal probe consists of total genomic DNA, or a very abundant fraction thereof, such that many labeled cDNA targets will bind and the resultant hybridization signal depends on the amount of RNA purified, the amount of cDNA produced and the amount of label incorporated. Each scientist should be able to produce labeled total cDNAs from two different batches of cell or tissue samples from the same physiological state in parallel which give signal probe hybridization intensities that are less than 2% different from each other. Workers who cannot achieve this level of reproducibility can be investigated. The classic mistakes, apart from pipetting error, are being inconsistent with incubation times or with how the samples are treated between manipulations (e.g. leaving some of them on the bench instead of on ice; leaving them for variable periods of time between manipulations). Once these errors are pointed out to people, they generally improve the reproducibility of their experiments – and not just for micro-array analysis.

Biological variability

In the majority of cases when using micro-arrays to measure global changes in gene expression, a single biological variable is changed between two comparator samples, and the array is designed to tell you what changes in gene expression occur in response to the changed biological variable. These are often referred to as the 'control sample', and the 'experimental sample'. It is obvious to most people that all other biological variables should remain unchanged, or a true picture of gene expression changes specific to one biological variable is impossible to achieve. Within a single experiment, where both members of a pair of sample cells or tissues are treated in parallel, controlling biological variability is relatively easy. It is difficult, however, if you don't know what changes might be affecting the samples. The classical example of this is the *Arabidopsis* leaf, where a few drips of water onto a leaf can lead to a stress response and changes in gene expression. Thus if you didn't notice the water dripping onto the leaf you chose for RNA isolation from the control plant, and the leaf you chose from the experimental plant did not get a drip of water, then you might never know what caused the changes in gene expression seen. The moral of the story is 'minimize all possible biological variability'. If growing your plants in sealed containers is essential then accept that condensation might form in the containers and drip onto some of the leaves. You could take a number of different leaves from each plant and extract RNA from all of them in one tube in the hope that, on average, the same number of leaves will have been dripped on in each plant. But this is not really solving the problem, it just puts artifacts into both data sets. Yes, the artifacts in the control sample will cancel the artifacts in the experimental set. But this can only work if the changes due to dripping water are entirely separate from the changes due to the experimental biological variable. If there is overlap between the changes then the whole experiment can turn into a bit of a mess. The best way of dealing with this specific example would be to pick several leaves from each plant and extract RNA from each of them separately, and perform separate array hybridization with all the cDNAs products. If you see two distinct expression profiles, one from each plant, then you can be sure no biological variability has crept into the system. However, if you see four patterns; control, control + dripping; experimental, experimental + dripping, then you know the dripping has caused a problem. In this way, if you did not know about the drips of water, you would still know that unwanted biological variation had entered the system and could search it out. Of course, what if the biological variation didn't result from a change in the leaf, but from a fundamental change in the root systems of the two plants, say one had a fungal infection and the other one didn't? Now analyzing gene expression in individual leaves from the two plants would give only two gene expression profiles, but there would be potential biological variability that you didn't know about. This would be discovered when using entirely new plants, when you repeated the experiment and obtained different results, which highlights the need for multiple experimental and biological replicates.

If this scenario makes you lie awake at night, then think about the following. Do you have to work on plants? Perhaps a change of research area would be better? The biological variation possible between two cultures of bacteria grown in an identical batch of medium in identical flasks at the same time in the same incubator is minimal, and that's why I work with bacteria! But seriously, the point to make is that the experiment you perform should be made as simple as possible so that as few possible biological variables are involved.

For those of you who want to perform micro-array analysis to measure changes of gene expression in cells or tissues from animals, inherent biological variability will almost certainly be a problem. It is not possible to control all the variables such as nutritional status (you can't really force an animal to eat when you want it to do so!) and hormonal status. Those who work with animal tissues or primary cells know that biological variability comes with the territory. Micro-array analyses will amplify this inherent variability, not reduce it. Of course, those with experience will make sure that the animals are kept to a strict routine, that procedures should be performed on them, and samples taken at the same time each day, and that animals which are genetically identical and physiologically (age, body mass etc.) as similar as possible should be used.

It is tempting to try to overcome inherent biological variability between individuals by purifying RNA from cells/tissues taken from different animals and pooling the RNA samples. The hope is to average out the effects of outliers producing an arithmetic mean hybridization intensity for each probe spotted onto the array. It should always be remembered when doing this, however, that relative gene expression data is typically measured on a log scale. Hence, biological outliers where the level of a particular cDNA is unusually high will affect the arithmetic mean of log data more than outliers where the level of a cDNA is unusually low, biasing the mean data upwards.

If you cannot avoid doing experiments on a large scale and/or using a highly biologically variable system, then you must accept that biological variability may well be a significant problem when trying to interpret the data you obtain. To get over the problem you will have to perform many more micro-array hybridizations involving more samples from each experimental condition than if the experiment and/or the system being studied were simple. An important final point to make concerning biological variability is that there may be more variability in samples taken from one experimental situation than from another. To return to the plant leaf example, the use of hormones in the experimental plant as part of an experiment to determine what transcriptome changes result from hormone treatment might exacerbate the effects of water dripping on leaves if, for example, one effect of hormone treatment was to make leaves more sensitive to water drips than those on untreated plants.

Pilot studies

It makes sense to estimate the level of variability seen in a biological system by running a pilot experiment where a set number of replicate RNA samples are taken from cells or tissues in one (usually the control) experimental condition, converted into labeled cDNA and hybridized separately onto arrays. When doing this, you may find that samples which are biological outliers contain RNAs from some genes that are under-expressed and others that are over-expressed. Hence each probe should be taken separately when considering the effect of biological outliers. The pilot study will provide crucial information concerning the statistical variance of the data, which can be calculated as described in Chapter 8. From this, you can decide how many replicates are required to make the data statistically significant. This test can be applied to every gene on an array, or can be applied to a northern blot or other gene expression experiment. First consider what threshold of fold change in expression level you want to be able to resolve between the two conditions in the experiment. Then take the control data you generate in the pilot experiment and simply multiply the data by the fold change you want to be able to resolve. Next, undertake a t-test (Section 8.4) using the two sets of sample data, to give a t-value for the prospective experiment. Next choose a statistical significance critical value (a so-called p-value) that you want to be able to apply to the experiment and simply read off the t-values associated with that particular p-value from a statistical table until the closest approximation to the actual t-value you got in your pilot experiment is observed. Each t-value for a particular p-value is dependent upon a particular number of degrees of freedom in the experiment. This essentially means the number of experimental replicates required to give that level of significance at that t-value. Thus, simply reading from the graph you can work out the number of experimental replicates required in the full experiment. Of course, this is only an estimate, but is likely to be an over-estimate of the number of biological replicates needed.

A micro-array can have tens of thousands of different probes so you will have to use a micro-array pilot experiment to generate t-test data for each probe/target interaction. Theoretically, the number of experimental replicates needed is that dictated by the most variable target:probe data set. However, for some genes, it may simply not be practical to perform enough array hybridizations to give statistically significant values for comparisons of their expression in two growth conditions. This is a call that the individual scientist will have to make.

Separate pilot experiments may also be needed in order to optimize the particular experimental growth conditions. For example, the length of time that cells or tissues are incubated with a drug, or the concentration of drug needed. Decisions concerning concentration and time should be informed by prior biological knowledge. It may be that you know the drug will lead to a change in the expression of at least one gene. In this case, the pilot experiment would involve measuring the expression of that gene (using northern blotting or qRT-PCR) following treatment of cells for various concentrations of drug for various amounts of time, looking for the minimal amount of time and concentration that gives significant changes in expression of the control gene. It is advisable not to go beyond these concentration or incubation times, since the possibility of knock-on and stress-induced artifacts increases, particularly with length of exposure time. If you have no idea as to which genes might be differentially expressed under conditions of the drug, then you may have to look at some phenotypic change. If there is no known phenotype to look for, then you will need to go straight to a micro-array, but be careful to perform lots of experiments, whilst varying time of exposure and drug concentration independently. From this, complex multivariate statistics, such as analysis of variance (ANOVA) will be required to dissect out the different effects of incubation time and concentration.

If financial constraints mean that large full-array pilot studies to determine biological variability are not possible, it is just about reasonable to use northern blotting or qRT-PCR analysis to measure the expression of a small number of genes in multiple samples of control and experimentally treated cells/tissues. From variance data obtained in this minimal pilot study and associated *t*-test results, it will be possible to estimate the number of arrays needed in the global transcriptome experiment. Of course, in doing this, you run the risk of under-estimating the number of arrays required, leaving you with a considerable amount of statistically insignificant data. If this is the case, it is possible to hybridize additional arrays after the number crunching. To protect against this eventuality, it is always advisable to make more RNA samples than you anticipate needing, so that the extra samples can be used if the statistics make additional hybridizations essential. This is preferable to going back and making more RNA when it becomes apparent it will be needed, since sample variability is more likely to be increased if RNA samples are produced on different dates.

3.16 Nuclear run-off assays

The vast majority of the techniques described in this book are designed to measure steady-state transcript levels or protein levels at any given time in a cell or population of cells. Steady-state levels are brought about through a complex interplay of factors, including rate of synthesis and rate of degradation. For proteins, this is even more complex, because rates of transcription and translation, coupled with RNA and protein stability issues all come together. In eukaryotes, there is also the complexity of whether you are measuring mRNA or hnRNA levels, since only the former can be used to produce protein, and is made from the latter via a splicing reaction with its own complex control. Furthermore, steady-state measurements of transcript levels using hybridization or polymerase chain reaction protocols tend to muddle up mRNA and hnRNA levels giving a combined estimate of RNA levels relating to the expression of a given gene.

  In some situations, particularly when your research aims to understand the control of transcription, what really matters to you is the absolute rate of synthesis of a particular mRNA (or hnRNA in eukaryotes) rather than the steady state. It is likely that, particularly in eukaryotes, the rate of transcription and the steady-state transcript level is only loosely correlated, particularly for transcripts whose genes are under tight control. The way to measure total nascent RNA synthesis in a prokaryotic cell or nucleus of a eukaryotic cell is to perform a so-called 'nuclear run-off assay' (also called the 'nuclear run-on assay' (Profous-Juchelka *et al.*, 1983), just to confuse matters). *Figure 3.15* illustrates this assay. Essentially, bacterial cells or eukaryotic nuclei (purification of nuclei is described in Section 2.5) are incubated with all four ribonucleic acids, one of which (usually UTP so that there is no labeling of DNA, which would complicate the results) is spiked with a radioactive form (i.e. $\alpha[^{32}P]UTP$). The RNA polymerase enzymes present in the bacterium or nucleus will do their job and make transcript at a rate dictated by the transcriptional control signals present within the isolated bacterium or nucleus, and will incorporate radioactivity into the newly synthesized transcripts so they can be visualized and quantified. Thus if the bacteria or nuclei are incubated with radioactive UTP for a set

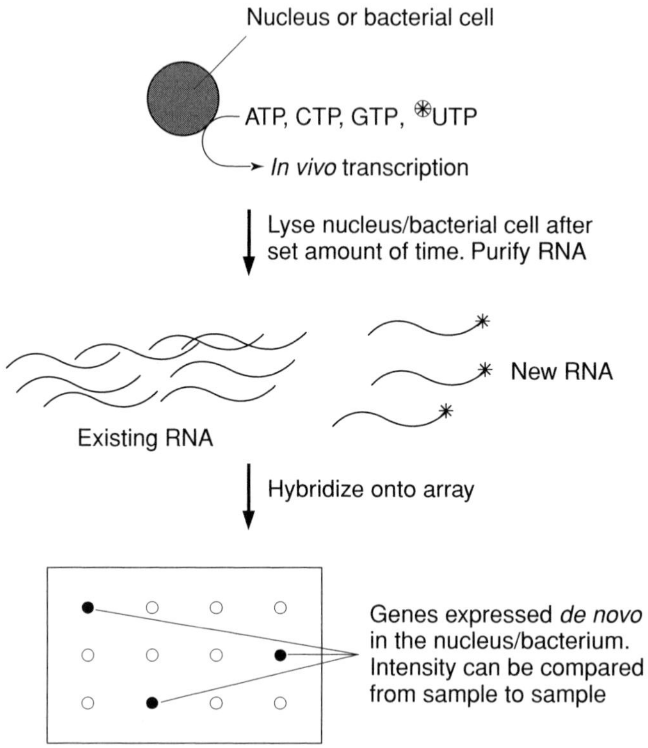

Figure 3.15

The nuclear run-off assay. RNA is synthesized in isolated nuclei or bacterial cells in the presence of a labeled UTP. This means any newly synthesized RNA will be labeled. After synthesis has been allowed to proceed for a set amount of time, RNA is extracted and hybridized to an array. The amount of label incorporated into an array probe spot will inform of the rate of transcription of the RNA target designed to hybridize with it. If the experiment is performed using cells/nuclei treated in different ways, it will be possible to determine the true differences in transcription rate caused by the different treatments.

amount of time, and then lysed into a denaturant solution, stopping transcription, and free nucleotides are removed, usually via chromatography, then the total amount of incorporation of radioactivity into bacterial/ nuclear extract in that set incubation time can be calculated and used to determine a relative rate of transcription in different sets of bacteria or nuclei under different growth conditions.

An extension of this more general method allows one to measure the rate of incorporation of radioactivity into one or a number of transcripts with specific sequences (and so by inference, their rates of transcription). The specificity is provided by using a DNA array, over which the radioactive total RNA product from the lysed cell or nuclear extract is washed. The target RNA having a specific sequence will hybridize to its immobilized

DNA probe and the target:probe interaction can be quantified using autoradiography of the hybridized array.

An alternative form of the nuclear run-off assay involves the use of biotinylated UTP instead of the radioactively labeled form, such that following incubation of the cell or nucleus with the labeled nucleotide for a set amount of time, all nascent transcripts are biotinylated and, following removal of unincorporated UTP can be purified using streptavidin-coated beads. To determine the amount of a particular transcript in this population, qRT-PCR (Section 4.9) can be used with a pair of primers specific for the transcript of interest.

References

Allawi, HT, Dahlberg, JE, Olson, S, Lund, E, Olson, M, Ma, WP, Takova, T, Neri, BP and Lyamichev, VI (2004) Quantitation of microRNAs using a modified Invader assay. *RNA* 10: 1153–1161.

Alwine, JC, Kemp, DJ, Parker, BA, Reiser, J, Renart, J, Stark, GR and Wahl, GM (1979) Detection of specific RNAs or specific fragments of DNA by fractionation in gels and transfer to diazobenzyloxymethyl paper. *Methods Enzymol* 68: 220–242.

Avison, MB (2004) Comparative genomics, digging for data. *Methods Mol Biol* 266: 46–69.

Avison, MB and Bennett, PM (2005) Bacterial genetics. In: *Topley & Wilson's Microbiology and Microbial Infection – Bacteriology,* 10th Edn (eds S.P. Borrellio, P.R. Murray, G. Funke). Hodder Arnold, London, pp. 80–135.

Bittner, M, Kupferer, P and Morris, CF (1980) Electrophoretic transfer of proteins and nucleic acids from slab gels to diazobenzyloxymethyl cellulose or nitrocelluloase. *Anal Biochem* 102: 459–471.

Duggan, DJ, Bittner, M, Chen, Y, Meltzer, P and Trent, J (1999) Expression profiling using cDNA microarrays. *Nature Genetics* 21(S1): 10–14.

Fleischmann, RD, Adams, MD, White, O, Clayton, RA, Kirkness, EF, Kerlavage, AR, Bult, CJ, Tomb, JF, Dougherty, BA, Merrick, JM, *et al.* (1995) Whole-genome random sequencing and assembly of *Haemophilus influenzae Rd*. *Science* 269: 496–512.

Profous-Juchelka, HR, Reuben, RC, Marks, PA and Rifkind, RA (1983) Transcriptional and post-transcriptional regulation of globin gene accumulation in murine erythroleukemia cells. *Mol Cell Biol* 3: 229–232.

Reed, KC and Mann, DA (1985) Rapid transfer of DNA from agarose gels to nylon membranes. *Nucleic Acids Res* 13: 7207–7221.

Shearwin, KE, Callen, BP and Egan, JB (2005) Transcriptional interference – a crash course. *Trends Genet* 21: 339–345.

Southern, EM (1975) Detection of specific sequences among DNA fragments separated by gel electrophoresis. *J Mol Biol* 98: 503–517.

Protocol 3.1 Production of a DNA probe

EQUIPMENT

Water bath or heating block (x2)
Eppendorf tubes

MATERIALS AND METHODS

1. Take 11 μl of dsDNA template (around 50 ng) and heat for 5 min at 100°C.
2. Place the mixture on ice for 3 min: this actually prevents annealing.
3. Add 14 μl of 0.4 M Hepes, pH 7.0 containing 40 μM dATP, dCTP (could be radioactively labeled dCTP to approx. 200 TBq mmol^{-1} instead) dGTP, dTTP (could be DIG or biotin-labeled UTP instead) and 100 ng of primer (could be 5′ end-labeled primer) plus 5 U of Klenov DNA polymerase I.
4. Incubate at 37°C for 1 h.
5. Stop the reaction by adding 5 μl of 50 mM EDTA containing 1 mg ml^{-1} calf thymus DNA and 0.1% w/v dextran.
6. Purify the labeled DNA using a Sepharose CL6B column (Amersham) according to the manufacturer's instructions.

Protocol 3.2
5′ end-labeling of oligonucleotides

EQUIPMENT

Eppendorf tubes
Thermostatically controlled water bath or heating block (x2)

MATERIALS AND METHODS

1. Mix 50 pmol of oligonucleotide in 50 μl of 70 mM Tris-HCl, pH 7.6, containing 10 mM MgCl$_2$, 5 mM DTT and 50 pmol [γ^{32}P]dATP (5 × 10^6 cpm μmol^{-1}) and 20 U of polynucleotide kinase.
2. Incubate for 30 min at 37°C.
3. Heat inactivate for 20 min at 65°C.

Protocol 3.3 Synthesis of first-strand cDNA

EQUIPMENT

Eppendorf tubes
Thermostatically controlled heating block

MATERIALS AND METHODS

1. Add 5 µg of total RNA to a total volume of 40 µl of 50 mM Tris-HCl, pH 8.3, 40 mM KCl, 6 mM MgCl$_2$, 50 µM of each dNTP (can use labeled UTP or radioactive dCTP, as for Protocol 3.1), 5 mM DTT, 1 µg oligo dT primer or 100 ng of sequence specific primer (primers can be 5′ end-labeled as in Protocol 3.2).
2. Add 40 U of AMV reverse transcriptase.
3. Incubate for 45 min at 42–50°C.

Protocol 3.4 Synthesis of second-strand cDNA

EQUIPMENT

As for Protocol 3.3

METHODS

If you used AMV reverse transcriptase to make first-strand cDNA, simply reduce the reaction temperature to 37°C after first-strand cDNA synthesis is complete (Protocol 3.3) and incubate for a further 30 min. This will synthesize second-strand cDNA.

Protocol 3.5 Ligation of linker sequences to blunt-ended cDNA

EQUIPMENT

Eppendorf tubes
Bench-top centrifuge

MATERIALS AND METHODS

1. Take the double-stranded cDNA product (Protocol 3.4) and make up to 250 μl with molecular biology grade water. Add 300 μl of chloroform, shake and centrifuge at $14\,000 \times \textbf{\textit{g}}$ in a bench-top centrifuge for 10 min.
2. Take off the aqueous phase and add an equal volume of ice-cold, 90% ethanol. Mix by inverting and place at −20°C overnight. Pellet the precipitated DNA by centrifugation, pour off the ethanol and air dry the pellet.
3. Make up the pellet to a total of 20 μl in 25 mM Tris-acetate, pH 7.7, containing 100 mM potassium acetate and 1 mM DTT, 100 μM each dNTP.
4. Add 3 U of T4 DNA polymerase and incubate at 39°C for 30 min to make the cDNA completely blunt ended and phosphorylated.
5. Repeat steps 1 and 2 to purify the DNA. Make up the pellet in 10 μl total volume of 30 mM Tris-HCl, pH 7.8 containing 10 mM $MgCl_2$, 10 mM DTT and 1 mM ATP plus 1 μg (total) double-stranded linker DNA.
6. Add 1 U of T4 DNA ligase and incubate at 15°C overnight.
7. Repeat steps 1 and 2 to purify the DNA produced.

Protocol 3.6 RNA production *in vitro* for cDNA amplification

EQUIPMENT

Eppendorf tubes
Thermostatically controlled heating block

MATERIALS AND METHODS

1. To the pellet generated at the end of Protocol 3.5 (provided you have ligated T3 promoter sequences to the cDNA!), add 100 μl of 40 mM Tris-HCl, pH 7.9, 10 mM NaCl, 10 mM DTT, 6 mM MgCl$_2$, 2 mM spermidine, 0.05% v/v Tween-20 and 0.5 mM of NTPs.
2. Add 20 U of T3 RNA polymerase and incubate for 60 min at 37°C.

Protocol 3.7 Nuclear run-off assay

EQUIPMENT

Eppendorf tubes
Vortex mixer
Bench-top centrifuge
Heating block

MATERIALS AND METHODS

1. Use nuclei isolated from about 10^8 cells (see Protocol 2.5) made up to 250 µl with 20 mM Tris-HCl, pH 8.0, 100 mM KCl, 6 mM magnesium acetate, 10 mM ammonium chloride, 10% v/v glycerol 0.3 mM EDTA 1 mM DTT.
2. Add creatine phosphokinase to a final concentration of 10 µg ml^{-1}, creatine phosphate to 10 mM, combined NTPs to 220 µM total NTPs, 25 µl of [α^{32}P]UTP (10^6 cpm mmol^{-1}) or otherwise labeled UTP.
3. Make up to 305 µl and mix gently.
4. Incubate at 30°C for 30 min. Vortex four or five times during the incubation.
5. Centrifuge the samples at 2000 × g for 3 min. Discard the supernatant.
6. Purify the RNA from the pelleted nuclei according to Protocol 2.5.
7. Use the labeled RNA instead of a DNA probe in northern hybridization, or instead of cDNA targets in array-based hybridization experiments.

PCR-based methods for measuring transcript levels

4

4.1 Introduction

As discussed in previous chapters, one of the recurring aims of molecular biology experiments is to visualize what is not ordinarily visible. In this book we are concerned with visualizing the products of gene expression: RNA and proteins. In visualizing them, we hope to be able to quantify them, if not absolutely, then at least relatively between amounts of a particular product in cell extracts from two experimental states. The example in the previous chapter was to label specific RNA or DNA sequences using hybridization so that they stand out from the background of other RNA molecules. The label is highly visible, so that even if the target RNA is low abundance, it will be seen. Furthermore, label incorporation is proportional to the amount of target RNA, so can be used to estimate its concentration. An alternative approach to the same problem of making a particular RNA molecule visible against a background of many other RNAs is to specifically increase the concentration of (amplify) the target RNA molecule to a level that becomes visible using gel electrophoresis and simple nucleic acid staining. If the concentration of amplification product were directly proportional to the amount of starting material, then the observed concentration of the product could be used to estimate the concentration of the original RNA species.

This chapter will set out the use of the polymerase chain reaction (PCR) to specifically amplify nucleic acid sequences, and how this technique can be adapted to amplify and quantify RNA species. It will deal with all the potential problems of PCR, and how best the PCR product can be visualized and quantified.

4.2 The basics of PCR

Amplification of a specific DNA sequence is very important for a whole host of molecular biology experiments. Until relatively recently, DNA cloning was the sole method for routine amplification of DNA sequences. This works by incorporating the sequence to be amplified into a cloning vector which is able to be replicated *in vivo* by a cell's DNA polymerase enzyme(s) such that large amounts of cloned insert can be recovered for sequencing or other procedures (e.g. subcloning into a protein expression vector). Now, *in vitro* approaches are more commonly used than cloning to amplify DNA sequences. Here, as with *in vivo* amplification, DNA polymerases are used but in a purified form and in a test tube. The approach that has revolution- ized the amplification of DNA sequences *in vitro*, and has therefore provided a massive boost to molecular biology research is PCR.

Besides production of large amounts of DNA for cloning and sequencing, there is one additional benefit of using PCR to amplify DNA sequences. This is that the amount of amplified product is proportional to the amount of template DNA sequence. So, by analyzing the amount of amplification product (e.g. as an ethidium bromide stained band on an agarose gel) you can estimate the quantity of DNA before amplification. This can be easily used to determine relative concentrations of starting DNA in two prepara- tions of DNA, and given the right experimental approach can even be used to calculate an accurate quantity for the template DNA.

The three P's of DNA polymerases

DNA polymerase enzymes have been discussed in the previous chapter. To recap, they have a 5′ to 3′ strand extension activity, which incorporates deoxynucleotides sequentially opposite a single-stranded template DNA (and/or in some cases, RNA) strand and catalyze the formation of phospho- diester bonds between the incorporated nucleotides to form a complemen- tary strand.

There is a vocabulary associated with dealing with DNA polymerases, which I will use in this chapter and beyond, and of which you should there- fore be aware. I refer to this as the 'three P's': processivity, proofreading and primers. Descriptions of what is meant by these terms and their broader implications for your experiments are set out below.

Processivity

DNA polymerase enzymes catalyze the chemical reaction that they were designed for at different rates. How rapidly a particular DNA polymerase synthesizes a complementary strand is referred to as its extension rate. A fast extension rate is an advantage in the test tube, since shorter reaction times will be required to produce a given complementary strand than when using enzyme with low extension rates. The processivity of an enzyme is a measure of how strongly the polymerase binds to template DNA. This property does not really affect the rate of extension, but has a strong influ- ence on the absolute length of complementary DNA strand that can be made. DNA polymerases, like all enzymes that perform sequential, linked reactions, each involving the addition of a substrate, do not carry on for ever. At some point in time, extension of the complementary strand termi- nates, and the polymerase falls off to start again somewhere else. This could be at a previously terminated complementary strand, and does not have to be back at a primer, but the point in time that termination occurs is random, and so the sizes of complementary strands follows an almost perfectly normal distribution, assuming a very long template strand. Different DNA polymerase enzymes give different average complementary strand lengths because their propensities to terminate strand synthesis differ. Enzymes with high processivity bind more strongly to the template, and are therefore less likely to terminate extension. This means that if you want to copy a very long template strand in the test tube, some DNA polymerases (so-called 'long-range' polymerases) with high processivity will be better for this task than others. Highly processive enzymes are not always polymerases with high extension rates. Indeed, a rapid rate of catalysis, can actually mean a higher propensity for termination of strand extension.

Proofreading

Different DNA polymerases have different roles to play in nature. DNA polymerases that are involved in replication of an organism's genome must be very accurate. This accuracy is facilitated by the presence of one or more proofreading activities alongside the DNA polymerase activity. This means that the multi-functional enzyme adds a deoxynucleotide, then goes back and re-checks the template to complement interaction in order to make sure the correct nucleotide has been added. In some cases, this so-called proofreading activity is present in the same protein that carries the DNA polymerase active site. In many cases, however, additional proofreading activities are provided by separate protein subunits. Indeed, in some cases, the polymerase itself is a cluster of several proteins, each essential for the enzyme to work.

Because proofreading polymerases are multimeric, they are more difficult to produce, purify, store and use, and are therefore generally more expensive than non-proofreading polymerases. The presence of proofreading activity in the test tube might be very important for your experiments, however, and particularly if you want to clone the resultant double-stranded DNA product and express the encoded protein, where mutations could ruin your future experiments; or if you want to sequence the double-stranded product and interpret anything sophisticated from the sequence obtained. On the other hand, proofreading, by its very nature, slows down the extension phase of complementary strand synthesis, and makes proofreading polymerases far less rapid than non-proofreading polymerases.

Another point to make at this stage is that all DNA polymerases have inherent terminal transferase activity. When they reach the end of a template strand the enzyme falls off. Just beforehand, however, it adds an adenine deoxynucleotide to the 3' end of the complementary strand. If proofreading activity is present, however, this additional A residue is removed again. Thus, proofreading polymerases produce perfectly blunt-ended (at the 5' end relative to the template strand) double-stranded molecules, but non-proofreading polymerases produce a single A overhang at the 3' end of the complementary strand, known as a 'sticky end'. This overhang can be used to improve the likelihood of insertion of the double-stranded DNA polymerase product into a cloning vector cut such that it carries a single thymine overhang at each end.

Primers

All DNA polymerases require a 3' hydroxyl group from which to build the complementary strand *in vitro*. This 3' hydroxyl is provided by an oligonucleotide primer, which anneals to the template strand using complementary base pairing. DNA polymerases are not affected by whether or not primers are long and thus highly specific to one target, or whether they are short random hexamers that bind multiple targets. However, if the primer is highly specific, it is far more likely that only one target DNA will be copied by the polymerase, which could be an important consideration for your experiments if you are using a mixture of template strands.

Primers do not have to be perfectly complementary to the target sequence. Providing the melting temperature of the primer/target duplex is not significantly lower than the temperature used to initiate extension by the polymerase, and that the primer is in large excess of the template, the primer will be in place on sufficient numbers of template to get enough complementary strand synthesis for most purposes. It is critical that the 3′ terminal nucleotide on the primer is complementary to the target DNA strand, however, because if the 3′ hydroxyl is not fixed in place, the DNA polymerase will not initiate complementary strand synthesis.

Amplification of a DNA sequence using DNA polymerases

When provided with a single-stranded template, a primer and sufficient deoxynucleotides, a DNA polymerase enzyme will catalyze formation of a complementary strand opposite the template strand resulting in a double-stranded duplex (*Figure 4.1*). If the template strand is very long, the duplex will only be partial in most cases. The 5′ end of the complementary strand is defined by the 5′ end of the primer. Hence, given that the 3′ end of the template strand will almost certainly continue beyond the primer binding site, the duplex will have an overhang (which may be very large) at the 3′ end of the template strand. Furthermore, since termination of extension by the DNA polymerase occurs after a normal distribution of nucleotides have been added, the 3′ end of the complement strand will be of uncertain length, and the 5′ end of the template strand will overhang the complementary strand in most products (*Figure 4.1*). However, given that only a portion of the template strand located 5′ proximal to the primer binding site need be amplified, the synthesis of a complementary strand might appear on the face of it to be doubling the concentration of the original template. This isn't really the case, though, because DNA is always thought of as a double-stranded molecule, and in reality the single-stranded template would have been provided as a double-stranded molecule (made up of so-called 'coding' and 'non-coding' strands) which would have been heated to melt it and release the template strand. To truly double the concentration of a double-stranded DNA molecule, therefore, both the coding and non-coding strands of the original DNA duplex need to be used as templates for complementary strand synthesis, resulting in two double-stranded DNA amplification products for each of the duplexes in the starting material. This is achieved by designing two primers, one complementary to the coding strand, and one complementary to the non-coding strands (*Figure 4.2*). Each is designed to anneal at a position that defines the region of the starting double-stranded DNA molecule that is to be amplified. Using two primers that are oppositely oriented on the starting duplex in this way, one round of melting the original duplex, annealing the

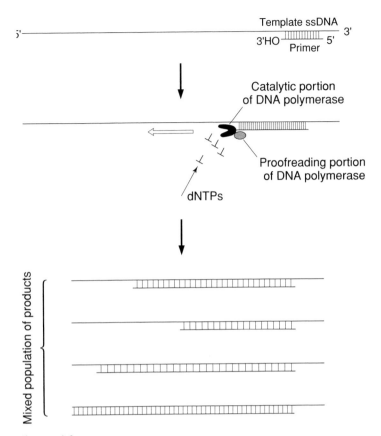

Figure 4.1

DNA polymerase mediated DNA copying. A double-stranded DNA molecule is melted by heating, and a sequence specific or random oligonucleotide anneals to one of the template strands. DNA polymerase extends the 3' hydroxyl of the primer, checking that the correct nucleotide has been inserted using a proofreading subunit (if present). The polymerase will continue until it randomly dissociates from the template, or until the end of the template is reached. Therefore, with long templates, a mixed population of products results, with the average length dependent upon the processivity of the polymerase and the time of the reaction.

primers and extending from the primers will apparently double the concentration of the region to be amplified. In fact, however, the region that has been amplified will have a template strand that is from the starting DNA, and that will probable overhang the newly synthesized complementary strand, which itself will be bounded at its 5' end by the primer, with the 3' end having a semi-normal distribution of distances away from the primer (*Figure 4.2*).

Products obtained by using a single DNA polymerase catalyzed amplification are no use at all for the downstream application for which *in vitro* DNA amplification is, in theory, so applicable: cloning, sequencing and quantification. This is for a number of reasons: the varying sizes and inconsistencies in the sorts of ends of the duplexes (occasionally blunt ends, but more likely

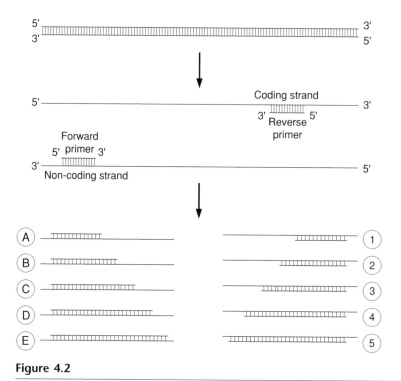

Figure 4.2

Amplification of template DNA using DNA polymerase I. The mechanism of DNA template strand copying is the same as that described in Figure 4.1, but in this case, a pair of oppositely oriented primers are used, meaning that both template strands are copied, effectively duplicating the concentration of the template DNA between the primer binding sites. However, as in Figure 4.1, a mixed population of products is produced.

template strand overhangs of different lengths at both ends) make the products almost useless for cloning (where a blunt end, or at least a predictable overhang is needed); the amount of amplification is only two-fold making it insufficient for sequencing; the sizes of the products are of varying lengths, meaning that they will not be resolved as defined ethidium bromide stained bands on agarose gels, so cannot be quantified.

The only way to amplify a defined region of DNA such that the product can be used in cloning, sequencing and band quantification, is to perform a number of rounds of DNA polymerase induced amplification (*Figure 4.3*).

• The first round is as described above, where DNA polymerase extends each of two primers, copying the coding and non-coding template strands and so effectively doubling the concentration of the required region of the original DNA duplex, but without absolutely defining the amplified region as a product of one size.

• In the second round, the products of the first round are melted to produce single-stranded templates, the primers are annealed to their targets and DNA polymerase extends from the primers. Not only does

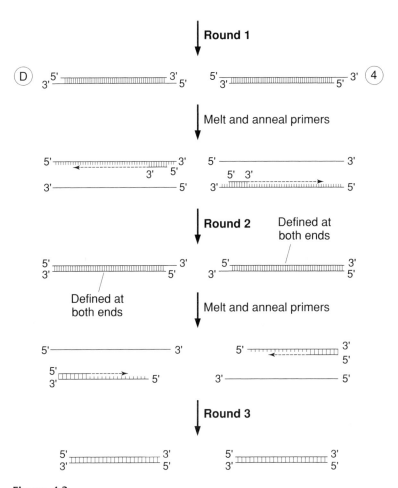

Figure 4.3

The polymerase chain reaction. Following amplification as set out in Figure 4.2, the products are used to further amplify the DNA. Only products D and 4 are shown (though E and 5 would work). First round products 1–3 and A–C are too short to be further amplified (i.e. terminate before the binding site for the second primer). After the second amplification, one of the product strands has both ends defined by primer binding sites, and from this point, the defined PCR amplicon doubles in concentration after every round.

this second round again double the concentration of DNA, but it produces product strands where both ends are defined by the pair of primers, meaning that they are a single size.

- In the third and subsequent rounds of amplification, melting, annealing of primers and extension occurs. Each time, the concentration of amplified DNA doubles and the defined size of the amplification product remains the same.

- The process of doubling the concentration of amplified DNA can continue, round upon round, until one of the reagents becomes limiting (*Figure 4.4*).

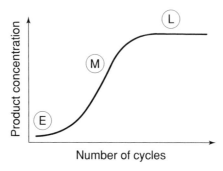

Figure 4.4

The dynamics of a PCR reaction. In the early (E) portion of the reaction, the affinity of template DNA for the polymerase is limiting (i.e. the template concentration is significantly below K_m, meaning that the rate of reaction is sub-optimal, and amplification is slow. However, once a threshold template concentration is reached, the enzyme works maximally, and in the middle (M) phase of the reaction, the DNA template concentration increase is exponential. Finally, in the late cycles (L) even if only one reagent becomes depleted, it will be enough to slow down the reaction rate of the polymerase up until a point where there can be no further amplification.

The development of PCR

This method of DNA polymerase-catalyzed amplification by multiple rounds of melting, annealing primers and extension is known as the polymerase chain reaction and the originally proscribed method for PCR is very simple (Saiki *et al.*, 1985). DNA template is provided as purified genomic or plasmid DNA, which is dissolved in a buffer appropriate for maximal DNA polymerase activity and mixed with the two PCR primers and all four deoxynucleotides in a test tube. The tube is then immersed in boiling water for a few minutes to melt the template DNA. The tube is next removed from the boiling bath and placed in another water bath set at a lower temperature to allow annealing of the primers to the template DNA strands. DNA polymerase is added at this point and extension is allowed to proceed for a few minutes. At the end of round one, the tube is moved to the boiling bath again, and the whole process repeated for as many rounds as is deemed necessary.

This original PCR approach never really took off as a widely used method for DNA amplification. There was (and still is) no better way of moving a test tube between water baths than by hand, so the PCR process involved once scientist being at the beck and call of a stop clock, moving tubes about every few minutes for several hours. This was thought of as tedious and too labor intensive. Other major problems were due to the fact that the only DNA polymerase available at the time was *E. coli* DNA polymerase I, which is used for probe synthesis in hybridization experiments (see Section 3.5). The use of *E. coli* DNA polymerase I for PCR means that when the first round of amplification has occurred and the tube has been returned to the boiling water bath, the enzyme becomes completely denatured. As such, fresh enzyme has to be added prior to each extension step. This makes the original PCR method very expensive. The second problem with using *E. coli* DNA polymerase is that it works best at 37°C, so primers must be annealed and the mixture cooled to 37°C before addition of the enzyme and extension of the primers. Given that the melting temperature of most oligonucleotide primers is around 50–70°C, incubation at 37°C gives the potential for significant nonspecific annealing of primers to the template DNA. If two primers happen to anneal nonspecifically (or one primer anneals twice) on opposite strands of the DNA at a distance of up to 3 kb apart, then this will probably result in the amplification of an unwanted product, which can make downstream experiments troublesome and complicated.

Two developments have made PCR the workhorse of molecular biology that it is today. The first was the discovery of thermostable DNA polymerases. These enzymes are collectively known as *taq* polymerases, because the original thermostable enzyme was purified from the thermophilic bacterium, *Thermus aquaticus* (Chien *et al.*, 1976). In fact this enzyme was discovered before PCR was invented, but its use in PCR allows annealing/extension temperatures to be elevated such that nonspecific primer annealing is less of a problem (Saiki, *et al.*, 1988). Furthermore, *taq* polymerases are not denatured during the melting phase of each round of PCR, meaning that enzyme does not need to be added over and over again. The second development was the invention of thermal cycler machines that can rapidly and reproducibly alter the temperature of a heating block over and over again automatically, leaving the scientist free to do something more interesting.

4.3   Methodological aspects of PCR experiments

PCR experiments often appear so simple and are thought of as so routine that people get lazy when designing them. This results in problems, and dealing with these problems will be discussed below. It is best to avoid problems occurring in the first place, however, so take a little bit of time to set up the experiment so it has the maximal chance of working.

Design of PCR primers

Oligonucleotide primers are available commercially from an increasing range of different manufacturers. They are synthesized as strings of nucleotides anchored onto resin beads using simple repetitive chemistry. The machines are programmed with the sequence you want to achieve and the error rate is very low. It is possible to induce errors, however, and usually this happens if the previous nucleotide is not removed properly from the reaction vessel prior to the next nucleotide solution being injected. The result is a very small amount of incorrect primer in any mixture, though this would cause few problems.

The amount of primer you receive from the manufacturer depends on what optical density (OD) scale you ask for. When primers are synthesized, they are run through a UV spectrophotometer cell to measure the absorbance of the sample, and so determine the amount that has been made. The OD scale system means you get a particular ballpark of absorbance. Normally, the 0.1 OD scale would give you around 200 μg of oligonucleotide. This is sufficient for many different PCR experiments, and in gene expression analysis, where each RNA will only be quantified in a limited number of conditions, this is all that will be required.

Primers are provided lyophilized – indeed it may appear that there is nothing in the tube! You will be provided with basic information about the primer such as mass (calculated from the OD of the synthesized primer solution). Primers should be made up to a stock concentration of 100 pmol μl^{-1} using TE buffer and are very stable at –20°C. It is advisable to aliquot the stock solution so that if one sample becomes contaminated, or is suspected of being contaminated, it can be discarded and another aliquot opened.

The melting temperature (see Section 3.1) of the primer will be provided by the manufacturer, who may also highlight possible secondary structures in primers, and whether a primer might readily form homodimers (i.e. will not then bind to the target sequence) or will result in primer dimer formation due to small overlaps at the 3' ends of primers that are extended by DNA polymerase. Manufacturers will make these predictions about secondary structure and will calculate the primer's melting temperature by reference to the output of a computer program that assesses the sequence. In truth, you should be checking your own primer sequences before you order them using such a computer program, or the problems flagged up by the manufacturer may mean the primer is useless when it arrives. They will not tell you about these problems before the primer is made. There are many computer programs that can be used to help design primers; see http://www.bioinformatics.vg/biolinks/bioinformatics/PCR%2520and%25 20Primer%2520Design.shtml for more details.

If you want to fly solo when it comes to primer design, you will not go too far wrong if you use the following rules:

- The primer should be 18–25 nt long (or longer if very rich in adenines and thymines to increase the melting temperature). This will allow the required specificity and allow reasonable stringency in the annealing temperature you can use.
- The primer should contain approximately equal numbers of each nucleotide, but if this is not possible, then an equal number of G/C and A/T nucleotides.
- The primer should not have any repetitive sequences. For example, runs of a single nucleotide, more than one instance of a doublet (particularly bad is GCGCnn...nnGCGC (where n is any nucleotide) since this can cause secondary structure and, if a palindrome is present at the 3' end of the primer, then the formation of primer dimers will occur.
- The primer should be absolutely complementary at its 3' end to the proposed target sequence. The 5' end can be modified, either in sequence terms, for example to insert a restriction site or other modification, or chemically, for example to introduce a label, but don't mess with the 3' end. Indeed it is always wise to have a G or C base at the very 3' end of a primer so that this base is as strongly annealed to its template as possible.
- When making a primer, consider the sequence of the primer that you want to use it alongside. The two primers should have almost identical

percent GC and melting temperature characteristics, and should not be complementary, particularly at their 3' ends.

To calculate the melting temperature (T_m) in °C of a primer with a given sequence use the formula below (Suggs *et al.*, 1981). This will give you a result that is roughly the true T_m of the primer, but computer programs use a much more complex formula, and their results are more accurate.

$$T_m = (N_{G+C} \times 4) + (N_{A+T} \times 2)$$

where N_{G+C} is the number of G and C bases in the primer and N_{A+T} is the number of A and T bases in the primer.

Reagent concentrations and PCR buffers

The concentrations of both primers should be identical, and generally around 0.1–1.0 µM, which corresponds to around 5–50 pmol of each primer in a 50 µl PCR reaction. The primer stock is normally 10 times greater strength than required. First, dilute an aliquot of the stock to 10 µg µl^{-1} and use 0.5–5 µl of this working dilution in each reaction. In a typical PCR reaction, more than 95% of the primers remain unused because the primer-to-template ratio is around 10^7:1.

Nucleotides are usually provided commercially as individual tubes, and not as a mixture. The concentration of each is around 100 mM. Make a working mixture where each nucleotide is present at 5 mM using nuclease-free water; this can be stored at –20°C for weeks. The final concentration of nucleotides used in the reaction will be 100 µM (25 µM of each individual nucleotide), or 1 µl of working stock in a 50 µl reaction. Theoretically, this will allow the synthesis of around 5 µg of product. More nucleotides will be needed for PCR using proofreading polymerases (which have a high rate of nonspecifically degrading nucleotides), and if reverse transcription and PCR is to be performed in a single tube.

PCR is performed in a standard buffer of 10 mM Tris-HCl, pH 8.0 (which will decrease dramatically during the reaction, but will remain within the buffering range of Tris-HCl). The buffer will contain at least 50 mM potassium chloride, which is essential for reducing phosphate backbone charges on the primer and template. These negative charges would repel and prevent proper annealing if the positively charged potassium ions were not there. Do not increase the concentration too much, however, as this can cause DNA to precipitate, or at least promote nonspecific annealing of primers. Magnesium chloride is provided at 1.5 mM. Mg^{2+} ions are essential for DNA polymerase activity. Remember that Mg^{2+} ions are mopped up by nucleotides in solution at an equimolar level. This means that if you use 100 µM nucleotides, you will actually have 1.4 mM free Mg^{2+} rather than 1.5 mM, though this is quite sufficient to get maximal DNA polymerase activity. Most commercial buffers also have various non-ionic detergents such as TWEEN-20 0.01% v/v.

Various denaturants are sometimes added to PCR reaction mixes to help avoid secondary structure in primers. The classical example is dimethyl sulfoxide (DMSO), which is generally used at 2% v/v.

Increasingly, it is easier and cheaper to buy PCR master mixes. These are tubes containing buffer, *taq* polymerase and nucleotides. Some even carry a colored dye, which not only proves you have added the master mix to each reaction, but can be used as a sample dye when loading the PCR product onto an agarose gel. Usually these master mixes are provided as a 2× concentrate solution. All you will have to do is remove enough master mix to go around all the template samples you want to use in PCR, add primers and sufficient water to dilute it to a working concentration (remember to take account of the volume of template).

Even if you do not choose one of these commercially available master mixes it is always worth creating a master mix of your own from the individual reagents. This is done by adding enough buffer (provided with the *taq* polymerase), nucleotides and primers, all diluted with an appropriate amount of nuclease-free water to cover all the reactions you want to carry out (again allowing for the volume of template). The *taq* polymerase should be added before all the other reagents that make up the master mix, to minimize the possibility of contaminating the enzyme stock. Once you have made a master mix, all that you need to do is aliquot it into tubes to which template has already been added. If you do this, allow for inevitable pipetting error by providing enough master mix for at least two more reactions than you are actually going to set up. You will be amazed that even if you do this, there will be very little master mix left over! Better this slight wastage than to run out of master mix before you reach the last tube of template, though.

Types of thermostable polymerase

The use of thermostable DNA polymerases makes automated PCR possible because they can withstand the high-temperature denaturation phases of the PCR reaction. There are many different thermostable polymerases now marketed with subtly different properties. It can be confusing to decide which enzyme to choose. Below are listed some of the general functional groups of thermostable polymerase and subtle variations thereof.

Taq polymerase

DNA polymerases used for PCR are often generically referred to as *taq* polymerases. However, the original *taq* polymerase, a 94 kDa enzyme is purified from *Thermus aquaticus* (Chien *et al.*, 1976). This bacterium lives in thermal vents at the bottom of the ocean, and its optimal growth temperature is around 70°C. *Taq* DNA polymerase has specific properties. The enzyme has no 3' to 5' proofreading activity, and has a moderate processivity and extension rate, with an optimal temperature of 74°C. It is stable at 95°C with a half life of around 40 min, which is suitable for 30–40 cycles of PCR. It is a workhorse polymerase for amplification of up to 3 kb products in standard buffer conditions. The error rate for *taq* is around 1:500 nucleotides, which can cause significant problems if the aim of the PCR reaction is to clone or sequence the product, though it does not matter for DNA quantification experiments. *Taq* polymerase is available from a variety of commercial sources, usually as a recombinant enzyme, over-expressed in *E. coli*. It is easier to purify the enzyme from *E. coli* than from its native source, since a simple heat treatment will denature most of the *E. coli* proteins, but would not affect *T. aquaticus* proteins. Because of this, recombinant *taq* enzymes are cheaper than native enzymes. But be aware, however, that recombinant *taq* enzymes can be contaminated with *E. coli* genomic DNA.

Taq DNA polymerase carries a 5' to 3' exonuclease activity. This removes nucleotides in front of the extending strand, and is not relevant to PCR reactions, except that its presence means that nonspecific primer annealing cannot impede extension. Variants of *taq* are available where the protein is truncated such that the 5' to 3' exonuclease activity has been lost (Merkens *et al.*, 1995). For example, 'aTaq' from Promega. This improves the thermostability of the enzyme still further, making the enzyme ideal for use in PCR with high temperature annealing and extension temperatures. An example where this is useful is when dealing with GC-rich templates, that melt at higher temperatures. Loss of the exonuclease activity reduces processivity of the enzyme, though, so the maximal PCR product size when using this enzyme is < 1 kb.

Another modification of *taq* polymerase causes the enzyme to fold up and occlude the active site. For example, 'Thermo-Start' polymerase from Abgene and 'HotStarTaq' from Qiagen. This allows so-called 'hot-start' PCR, where the polymerase is inactive whilst the reaction is being set up, but upon the initial heating phase, the enzyme unfolds, and the active site becomes available. The advantage of hot-start PCR is that nonspecific primer annealing can occur during reaction set-up, which is generally undertaken at room temperature. If the *taq* polymerase is active when it is added to the reaction, it can start to extend some of these nonspecific primer/template interactions, which can result in nonspecific PCR products being formed (Kermekchiev *et al.*, 2003) Other manufacturers provide the *taq* polymerase linked to a paraffin bead, so that it is unavailable for PCR until the initial heating step. For example, 'TaqBead' from Promega.

Many other mutant derivatives of *taq* polymerases are marketed by different manufacturers. These mutants have beneficial properties such as very rapid extension rate (allowing shorter extension times) greater temperature stability (allowing higher annealing temperatures to be used) and greater tolerance of PCR additives, buffer and magnesium concentrations (making the enzyme more forgiving). Manufacturers also provide a variety of proprietary PCR buffers and make up their enzymes in special storage buffers. These have been optimized to give maximal stability and activity of the enzyme. Be aware that if in most situations you use a commercial polymerase, you will have to use the buffer provided, since the ingredients in the storage buffer are sometimes incompatible with some reaction buffers. This is particularly the case if the storage buffer contains Triton X-100, the reaction buffer will have to contain this detergent, or the polymerase will become inactivated.

The use of *taq* polymerase results in a product with a single adenine overhang at each 3' end. This overhang facilitates high-efficiency cloning of the PCR product using linearized vectors with single thymine overhangs at each 5' end. Annealing of the T and A overhangs of vector and insert, respectively, stabilizes the insert/vector complex making ligation of the phosphodiester backbones more likely to occur (Marchuk *et al.*, 1991).

Proofreading thermostable DNA polymerases

The classic thermostable DNA polymerase with 3′ to 5′ proofreading activity is the 92 kDa *Pfu* polymerase from *Pyrococcus furiosus* (Lundberg *et al.*, 1991). This is available from a variety of manufacturers and exhibits the lowest error rate of any known thermostable polymerase. Proofreading comes at a cost, however, because constantly checking its own activity slows the reaction. This means that longer extension times are generally required for proofreading polymerases. *Pfu* has a preference for magnesium sulfate rather than magnesium chloride, which is favored by most other thermostable DNA polymerases.

Other proofreading polymerases include *Tli* (Vent) a 90 kDa protein from *Thermococcus litoralis* available in a recombinant form from Promega. This enzyme has a faster extension rate than *Pfu*, meaning it is better for amplification of longer PCR products without the need for long extension steps (Cariello *et al.*, 1991).

When using proofreading polymerases, it should be borne in mind that these polymerases produce blunt-ended PCR products. Thus if you want to insert the PCR product into a cloning vector, low-efficiency blunt-end ligation is the way to go.

Polymerase mixes for long-range PCR

One of the most common reasons for termination of *taq* polymerase extension is that the polymerase inserts the wrong base. This mis-pairing often results in the 3′ end of the complementary strand peeling away from the template strand, which causes the polymerase to stall, and dissociate from the template, and it can never restart. The use of proofreading polymerases minimizes insertion of the incorrect base, thus increasing the potential length of the PCR product, but the low extension rate of these polymerases can mean that reaction times are prohibitively long. The solution is to provide a mixture of polymerases, with *taq* polymerase being in the majority together with a small amount of proofreading polymerase (Barnes, 1994). In this situation, when the *taq* polymerase makes an error and dissociates, the proofreading polymerase comes in and corrects the error, allowing the *taq* polymerase to re-initiate extension. Such long-range polymerase mixes are suitable for products of > 25 kb. Long PCR mixes are also an excellent way of getting high-fidelity PCR whilst retaining the 3′ adenine overhang to facilitate T/A cloning of the product.

DNA polymerases with significant reverse transcriptase activity

A number of thermostable DNA polymerases have the capacity to use RNA as a template and so synthesize a cDNA strand. This reverse transcriptase activity is considerably less efficient than true reverse transcriptase enzymes, such as those described in Section 3.12, but for synthesis of short cDNAs, they are more than adequate. The great advantage of using these enzymes, rather than reverse transcriptase enzymes for cDNA synthesis, is that the reaction can take place at elevated temperatures (e.g. 70°C). This reduces the possibility of secondary structure in the RNA template inhibiting cDNA synthesis. The additional benefit is that it allows cDNA production and amplification of the cDNA by PCR (so called reverse transcription PCR; RT-PCR) in a single tube, which makes these enzymes ideal for cloning and sequencing cDNAs from very low abundance RNAs of known size in a sample.

The first thermostable DNA polymerase to be used for reverse transcription was the *Tth* polymerase, a 94 kDa enzyme from *Thermus thermophilus* (Myers and Gelfand, 1991) This unusual enzyme works as a typical, *taq*-like DNA polymerase in the presence of magnesium ions, but as a thermostable reverse transcriptase in the presence of manganese ions. In the presence of both magnesium and manganese ions, the enzyme has both reverse transcriptase and DNA-dependent DNA polymerase activity, but the latter becomes highly error prone. *Tth* does not have a proofreading activity. Therefore, if you want to clone and sequence a cDNA from an RNA sample, you need to be aware that PCR errors in the sequence can occur. One way around this is to start the reverse transcription in a buffer containing manganese only, then add EGTA (ethylene glycol bis(2-aminoethyl ether)-N,N,N'N'-tetraacetic acid) to chelate the manganese, and finally add magnesium ions before starting PCR. Under these conditions, fidelity is similar to that of *taq* polymerase.

Another DNA polymerase with reverse transcriptase activity is the *Ctherm* polymerase from *Carboxydothermus hydrogenoformans*. The advantage of this enzyme over *Tth* is that its reverse transcriptase activity is not manganese dependent, and that it is a proofreading enzyme, increasing the fidelity of the product. The disadvantage of this is that extension is slow, and so the enzyme is often provided as a mixture with *taq* polymerase, allowing longer products to be amplified, whilst retaining the high fidelity. This combination makes *Ctherm* ideal for RT-PCR cloning and sequencing. For some reason, this enzyme does not appear to be currently supplied by any manufacturer, though it used to be supplied by Roche.

Typical temperatures for PCR

We've all done it. We program the thermal cycler with a particular set of temperatures to create a PCR method and then we never change the method. We use it time and time again for each primer pair, each template and each new batch of *taq* polymerase. It's not surprising that it often gives us a headache. It can take quite a while to program a PCR machine, so that is why we often don't bother. However, generally all that needs to be changed is the annealing temperature (which is primer dependent) and the number of cycles (which is template concentration and/or complexity dependent). In some situations, however, you may need to change the extension temperature and/or the length of extension time. This is because different *taq* polymerases have different temperature optima.

A typical PCR protocol would be 95°C for 5 min to fully melt the template, followed by 30 cycles of:

95°C for 30 s to 1 min
55°C for 30 s to 1 min
72°C for 30 s to 1 min

Followed by a final extension step of 72°C for 5 min.

The optimal annealing temperature (T_p) for a primer of a known sequence can be calculated using the following formula (Wu *et al.*, 1991):

$$T_p = 22 + 1.46 \times [(2 \times N_{G+C}) + (N_{A+T})].$$

Clearly, primers are most often used in pairs, and so a slightly different T_p might be obtained for each primer. If you find that this is the case, use the lower T_p value as the annealing temperature for the primer pair. This is because using the higher value as the annealing temperature may result in the primer with the lower T_p not annealing properly and so prevent proper amplification. On the other hand if the T_p values of the primers are very different, and the lower value is chosen for the annealing temperature, the primer with the higher T_p might bind at nonspecific targets, causing multiple amplification products. This possible problem highlights the need to design the primers in such a way that their T_p values are as similar to each other as possible. Unlike the situation when the aim of PCR is to clone and/or sequence regions of DNA, if you are carrying out gene expression analysis, it should always be possible to design primers with identical T_p values, because for gene expression studies, there is no need to amplify an entire gene. Any sized fragment of the gene whose expression is being measured, as long as it is bounded by recognition sites for primers with identical T_p values, will be suitable.

Types of thermal cycler and a discussion of PCR consumables

There are many types of thermal cycler for use in PCR. Essentially, all contain a metal block with holes for PCR tubes to be inserted. The block is heated and cooled according to the method programmed into the machine. Not all blocks are heated and cooled in the same way, however. The simplest of these involves a heat source such as a bulb, which heats the metal block and a fan that cools it. Heat conducts through the block and a thermostat in the block feeds back to the microchip controlling the required heating and cooling. The block thermostat is embedded within the block, though often, there is an additional thermostat within a probe that is placed into the block within a simulated PCR tube, for more accurate in-tube temperature calibration. There are a number of problems associated with this simple type of heating block. The first is that temperature distribution is not equal across the block, so different samples in different positions within the block may be exposed to different maximal and minimal temperatures. For quantitative PCR, where the kinetics of different reactions must be solely dependent upon template concentration, this can be a significant problem. The second problem is the length of time it takes to heat up and cool down the block. This makes overall PCR reactions longer, because the timer for each phase of each PCR cycle only starts to count down when the desired temperature is reached. Since cooling is generally slower than heating, there is no real problem with nonspecific primer annealing, but there can be a problem of *taq* polymerase and PCR product stability, particularly when long products are being amplified.

More modern, and so expensive, thermal cyclers use the Peltier effect to very rapidly heat and cool the heating block. This is a physical effect when an electric current is passed across certain alloys. The alloy heats or cools dependent upon the direction and ampage of the current applied. Other advances include the block being broken up into strips, with each having a separate temperature monitoring and heating/cooling device. These heating blocks provide more accurate temperature distribution across the entire block, but also provide an important advantage that a gradient of temperatures can be run across the block. The gradient is generally applied to the annealing phase of the PCR reaction, and is used to optimize the reaction when a new primer set is being used for the first time. Identical reactions are loaded across the block, and run at different annealing temperatures. One would expect that at lower than optimal temperatures, multiple products would be produced due to nonspecific primer binding, and at higher than optimal temperatures, no product will be produced due to inhibition of primer binding.

Some PCR machines use air heating and cooling, and do not have a metal block. They use glass capillary tubes which expose a large surface area to volume ratio of the sample, making heat conductivity far less of an issue.

Different PCR machines of different models, and particularly from different manufacturers have unique characteristics. This means that PCR protocols defined for optimal amplification of a strong, single product in one machine may not be optimal for use in another. Using published protocols for a particular PCR experiment may not give the published results in your PCR machine. Try re-optimizing the temperatures before you start demanding retractions from the authors!

Heated lids or overlay oils?

Heating a water-based reaction to 95°C causes evaporation. Whilst PCR reactions are carried out in sealed plastic tubes, so sample is not lost, the top of each tube is usually colder than the base, which is embedded in the heating block. This means that the evaporated sample condenses at the top of the tube, and its temperature is beyond control of the heating block. This will cause the reaction to fail. The original way around this was to add a couple of drops of mineral or silicone oil onto the surface of each reaction. This floats on the surface and does not itself evaporate significantly at 95°C. The presence of a layer of oil prevents evaporation of the PCR reaction underneath. There are problems with using mineral or silicone oils, however. They are potential sources of contamination, either with DNA or with nucleases (though they can be purchased certified free from contamination, storage in the laboratory and repeated use in PCR is likely to result in their contamination), and their presence prevents maximal recovery of the PCR product. A pipette tip must be pushed through the mineral oil, and it is important not to contaminate the product with oil, since this might inhibit downstream applications, so some sample will inevitably be retained in the tube. For this reason, mineral or silicone oils are only really suitable for use with 50 µl reactions or above. The final problem with using overlay oils is that it is tedious to pipette a couple of drops into each reaction, especially if a large number of reactions are being set up at once.

A far better option to reduce sample condensation within the tops of the tubes is the use of PCR machines with heated lids. The inside of the removable lid covering the heating block has a spring loaded metal plate fixed to it. This plate fits snugly onto the tops of the PCR tubes, and is heated up to > 95°C during the reaction. If the top of the tube is hotter than the bottom, there is no possibility of significant sample condensation, and no oil is required. The only real problem when using thermal cyclers with heated lids is that if you touch the lid, it may burn you!

PCR tubes

PCR consumables are almost as important as the PCR machine for optimal PCR. There is a myriad of different types of PCR tubes available. As a rule of thumb, however, smaller and thinner is better, because thermal conductivity is an issue. It is all very well if the thermal cycler block temperature is very well controlled, but if the temperature within the reaction mixture in the tube does not go to the same extremes, this will affect the reproducibility and quality of the results. Tubes of 0.2 ml are best, with reaction volumes of 25 µl. Thin-walled polypropylene tubes are available. They are more expensive than traditional 0.2 ml tubes, but they present a smaller barrier to heat conductivity and so are highly desirable. Even when using a heated lid, the region between the heated lid and the top of the heating block can become cooler than the sample below. Therefore, low profile PCR tubes are a good idea. They are shorter than normal, meaning a lower available space for samples, but they do not protrude above the top of the heating block, meaning there is limited potential for condensation.

There are many different types of bulk tubes available now for particular manufacturer's thermal cyclers. Strips of tubes can be obtained, being either eight or twelve 0.2 ml tubes long (i.e. they fit widthways or lengthways into the thermal cycler heating block). The tubes come with a strip of caps, which click into place after reagents have been added. The advantage of strips rather than individual tubes is that it is quicker to move them around and place them into racks and heating blocks.

The ultimate in minimizing the time it takes to transfer tubes around is the use of PCR plates, which are low profile polypropylene plates having 96 0.2-ml wells, each of which slots into a well on the thermal cycler heating block. These can be purchased with PCR master mixes ready aliquotted into them, and even with sample dye, so the products can be immediately run on an agarose gel. Normally, the plates are provided sealed with a thin plastic or foil membrane which can be peeled off, and primers and template added (the master mix contains all the other reagents required). The plate can then be resealed using a plastic film or foil sealing machine, and slotted into the heating block. Once the reaction has been run, samples can be pipetted out of the wells by pushing the pipette tip through the plastic or foil membrane. Other 96-well plates are available which allow fitting of strips of caps.

4.4 Analysis of PCR products using agarose gel electrophoresis

The basics of agarose gel electrophoresis have been dealt with in Section 2.10, when denaturing gel analysis of RNA was described in detail. For analysis of PCR products, nondenaturing gels are required. Usually, the gels are made with TBE or TAE (Tris-acetate-EDTA) buffer, which have an alkaline pH, so that the phosphate backbones of the DNA molecules are fully deprotonated and negatively charged. Most PCR products are between 500 bp and 5 kb, so the gel density should be from 1.2% to 0.8% w/v agarose in buffer, though this can be manipulated as required. You should be aware that 0.8% w/v agarose gels are fairly fragile, and should be treated carefully. Always support them from underneath.

PCR reactions are usually > 20 µl in volume and if the reaction has been successful, more than 1 µg of DNA should be present. This is more than enough to run on a lane of an agarose gel. Generally, 5 µl is sufficient to give a nice strong band, and this should be made up with 1 µl of 5× sample buffer, which contains a high-density compound such as sucrose and a colored dye, usually bromophenol blue. The sample buffer is designed to help the DNA sample sink to the bottom of the gel wells, and to allow the sample to be seen, both before and during loading the gel, and as a dye front as the gel is subjected to electrophoresis. If the dye front is still visible on the gel, there is no chance that the DNA in each lane has run off the end of the gel. Using only a fraction of the PCR product for analysis by gel electrophoresis means that there is plenty left for downstream applications (e.g. sequencing and cloning). Whilst it is possible to purify DNA fragments after they have been separated on an agarose gel, this is not entirely free from the risk of significant loss of sample, and represents an unnecessary extra step in most cases. Accordingly, it is best to leave some PCR product aside. PCR products are stable for months in the refrigerator. For many downstream applications, and particularly for sequencing, the PCR product will need to be cleaned in some way to remove *taq* polymerase, primers and nucleotides. This is normally achieved using a size exclusion spin column packed with silica beads.

DNA bands in agarose gels are predominantly visualized using an intercalating dye. Ethidium bromide is still the most popular, and is either added to the gel before it cools and solidifies, or alternatively, the gel is soaked in ethidium bromide solution to stain the DNA after bands have been resolved (see Section 2.10). Another popular dye for DNA (but not for RNA, since double-stranded molecules are required) is SYBR green I. This compound can be added to the DNA as part of the sample buffer where it intercalates into the DNA duplex and fluoresces (Section 2.10). The dye exhibits very little fluorescence when not bound to DNA. The advantage of SYBR green I is that no additional staining of DNA in the gel is required, meaning small volumes of dye are used, and toxic waste is not generated in the same amounts as is generated when using ethidium bromide. It would be wrong to think that SYBR green I is entirely safe, however, since any compound that intercalates into DNA is a potential mutagen. Accordingly, you must treat it with caution and always wear protective clothing and gloves when handling any DNA intercalating agent.

4.5 Problems with, and optimization of, PCR

PCR methods have now been developed for all sorts of purposes dependent upon subtle variations in the reagents used. PCR can be used to amplify very long sequences, to amplify sequences very accurately for sequencing, to introduce random and targeted mutations into sequences, to facilitate rapid cloning of sequences, and to quantify DNA and cDNA sequences derived from RNA. It is this last application of PCR that we will discuss in detail in this chapter, but first of all, it is important that the reader is aware of the common pitfalls associated with PCR in general and how these can be prevented.

Nonspecific PCR products

Following a PCR reaction, a small sample of the reaction mixture is loaded onto an agarose gel and subjected to electrophoretic separation. Any DNA present is stained with ethidium bromide or some other intercalating dye and bands visualized using a trans-illuminator. If you are performing basic PCR to amplify a particular DNA sequence, you will expect to see a single bright band having a molecular weight equivalent to that expected. The most common problem at this stage is that you see more than one product. This is indicative of the amplification of at least one nonspecific product. Nonspecific amplification occurs if, by chance, primers anneal to template sequences so that their 3' ends are facing each other and that the distance between the two primer binding sites is not excessive given the processivity of the polymerase enzyme and the extension time being used. Often, it is a single primer sequence that anneals nonspecifically at two sites, and not a heterologous primer pair. This possibility can be checked by running separate PCR reactions, each with only one of the primers. The advantage of finding this out is that you may only have to dispose of and redesign one primer, not both.

Nonspecific primer annealing occurs because of the following problems:

- Annealing is not really nonspecific. It just happens that the exact sequence to which the primer is designed to anneal occurs in the genome multiple times. This is just bad luck, and the only way of stopping this is to design a new primer. This is more likely to occur if primers are very short, or if the primer is designed to target a repetitive region.
- Annealing is nonspecific, due to the primer annealing to template sequences that are not perfectly complementary. This will occur because the correct level of stringency has not been applied to the primer annealing phase of PCR. Low stringency conditions would be: salt or magnesium ion concentrations too high; temperature too low; template:primer concentration too low; the absence of denaturants.

True nonspecific annealing of primers can be minimized by one or possibly a combination of the following:

- Repeat the PCR using a PCR buffer with a potassium chloride concentration <50 mM. If the concentration drops too low, then this will inhibit primer annealing due to charge–charge repulsion between the phosphate backbones of the primer and template DNA molecules, so try small adjustments first of all.
- Optimize the PCR reaction by altering the magnesium chloride concentration from 0.5 mM to 5 mM in 0.5-mM steps. You will find that one

magnesium chloride concentration will be better than the others. The effect will be fairly unpredictable because it depends upon the interaction with nucleotides and primers. As a rule of thumb, however, if magnesium ion concentrations are too low you get poor yields of PCR product, and if the concentration is too high, nonspecific annealing can occur.

- Repeat the PCR reaction using a higher annealing temperature. Try 5°C higher, and then beyond. Ideally for optimizing PCR by temperature, use a gradient block PCR thermal cycler. If you have one of these, it is probably best to run a temperature gradient each time you use a new primer set.

- Non-specific primer annealing is more of a problem during the first few cycles of PCR. Often, the problem actually occurs whilst the reaction is being set up, which is normally done at room temperature, therefore at low stringency (or worse still, on ice – don't do this!). Nonspecific annealing of primer to template is very unlikely to occur at the annealing temperature of the PCR reaction, because the melting temperature of the nonspecific primer/template duplex is likely to be lower than the annealing temperature used. However, if the DNA polymerase is active during the reaction set-up, there is a real possibility that any nonspecific primer annealing will be extended by the polymerase. The result would be a complementary DNA strand with a considerably higher melting temperature than the primer alone. If these nonspecific complementary products are generated during reaction setup, there is a distinct possibility that when the PCR reaction begins, they will anneal to the template, even at the annealing temperature chosen, resulting in the generation of nonspecific PCR products. The obvious way around this is to use a 'hot-start' *taq* polymerase, which is inactive until the first high temperature melting step in the PCR induces a conformational change in the enzyme, and makes it become active.

- Repeat the PCR using serial 10-fold dilutions of the primer(s), but be aware that if a primer becomes limiting, amplification of the correct product may level off earlier in the reaction, reducing yield.

- Add 2% v/v DMSO to the PCR reaction. This reduces hydrogen-bond potential; hopefully, this will affect the nonspecific annealing of primer and template, but be ready to try lower DMSO concentrations if the intensity of the band representing the correct amplification product decreases significantly.

You should be aware that even the presence of a single PCR product of the expected size is not a guarantee of specific amplification. The PCR product should be sequenced, or at the very least should be subjected to restriction enzyme mapping to confirm it represents the true target. It is also worthy of note that very commonly, multiple PCR bands are not due to nonspecific primer annealing at all, but due to contamination of one of the reagents with another primer or primer pair. Hence you get a so-called multiplex PCR, which is where two products are amplified in parallel. It will be almost impossible to modify the PCR protocol to block multiplexing. If the products are of significantly different sizes, then it will be possible to cut out the required band and specifically purify it from the agarose for cloning or sequencing purposes. However, for quantification it is very likely that the presence of a contaminant primer set will affect the kinetics of true target DNA amplification, and this may therefore add significant error to the system. So, on balance, it is far better to dispose of contaminated reagents than continue to use them. You can reduce the potential cost associated with wasting reagents by dividing them into aliquots when they have been purchased so that if a batch is contaminated and needs to be destroyed, it will represent only a small fraction of your total stock. The classic way of checking for template contamination of reagents is to run a so-called no template control, where everything is added to the reaction except template. However, in my experience, contamination with primer is almost as much of a problem as contamination with template. Thus a no primer control, where template is present, is also very important, since primer contamination in the reagents or template will lead to bands in this negative control.

No PCR product at all

This is even more infuriating than having multiple bands, since it may be more of a difficult problem to solve. The simplest explanation for seeing no PCR product on a gel is that the primers have not annealed to the template. In fact, more usually, only one primer is affected, and the other anneals normally, because it absolutely requires two primers to give a PCR product. Primers generally do not anneal to their targets because stringency is too high. The most common problem is that the annealing temperature used is too high or the salt concentration in the buffer is too low. Another major reason for a primer not binding is that it has very strong internal secondary structure, or two primer molecules form a dimer. The potential for this can be reduced by using primer design software that will run checks for internal complementarity and complementarity with the other primer in the pair. If secondary structure cannot be avoided (i.e. the annealing site cannot be moved for some experimental reason) then secondary structure can be limited by adding DMSO up to 2% v/v to the reaction.

It may be that even with DMSO, the band representing the expected amplification product is almost invisible. To help with this, it might be prudent to increase the number of PCR cycles in the reaction, so that more rounds of amplification occur.

A more fundamental problem could be that the polymerase is no longer viable. It is often thought that thermostable polymerases are immortal, but they are not. Frequent freeze–thawing and storage at room temperature for long periods will inevitably lead to depletion of enzyme activity. It is therefore always worth having a control PCR reaction, with primers and a template that you know gives a product every time if the enzyme is competent and the reaction mixture is made correctly. This is run in parallel with any PCR reaction you perform in order to check that the polymerase is still active and that the PCR machine is functioning properly.

Minimizing contamination in PCR reactions

Contamination of PCR reagents with template DNA is a real problem, particularly if the contamination is with previously processed PCR products. There are simple ways of avoiding this. First, set up PCR reactions on a different bench to the one used to run them and/or manipulate PCR products. Second, use filter tips whenever you take aliquots of stock reagents from the tubes, since contamination of stocks with DNA present on the barrels of pipettes is the primary cause of contamination in PCR. Third, perform negative controls; these should include no template and no primer controls to check for template and primer contamination of reagents, respectively. Fourth, aliquot reagents when they are purchased so that if contamination does occur, you only need throw away a small sample of the reagent, with obvious cost benefits.

4.6 Quantitative PCR

As discussed above, PCR was originally designed as a method to amplify the concentration of a particular region of DNA prior to downstream molecular biology or sequencing applications. The mechanics of PCR, however, which double the concentration of DNA during each cycle, mean that it can also be used to quantify how much template DNA is present in a test sample prior to amplification. This is known as quantitative PCR, or qPCR. The way qPCR is undertaken in its most basic sense is that several control PCR reactions are performed, each using a control template DNA at a different, but known concentration. The control template is often a purified PCR product so that its concentration can be accurately determined beforehand, and it can be diluted appropriately for each reaction. All PCR products from the control reactions are run on an agarose gel alongside the test PCR product. The bands are then stained and images of the bands are captured as a digital image. Band analysis software is used to quantify the intensities of all the bands and intensities of the control PCR product bands are used to make a standard curve, which is then used to calibrate the intensities of the test PCR product, and so to extrapolate the concentration of the DNA template in the original test sample.

There is a major problem with this approach for qPCR, however. The technique relies on the assumption that none of the PCR reagents is limiting during all of the control and test PCR reactions. However, PCR reactions using a set, essentially arbitrarily chosen, number of cycles are really end-point assays. Therefore, from the amounts of product generated following each reaction, it is not possible to tell whether one or more reagent has in fact become exhausted during any of the reactions (see *Figure 4.4*). Ideally, the end point chosen would be right in the middle of the exponential phase of PCR amplification, but you cannot guarantee this will be the case. Indeed, the more template DNA there is, the fewer the number of cycles that will be possible before one reagent becomes limiting, which makes the use of end point assays in qPCR all the more problematic.

The way around this problem of using end-point assays in qPCR, is to perform PCR on multiple different dilutions of each test sample template. You might also use different numbers of amplification cycles. Following each reaction, band intensities are checked, and a graph of band intensity versus dilution factor or cycle number is plotted. If the end point is in the exponential phase for all dilutions and cycle numbers, then the plot will show a linear correlation between product intensity and template dilution.

If the curve does level off at lower dilution points, confirming that at higher template concentrations some factor becomes limiting in the reaction, it will still be possible to use data from the more dilute samples for accurate qPCR by making allowance for the dilution factor when comparing samples to controls (*Figure 4.5*).

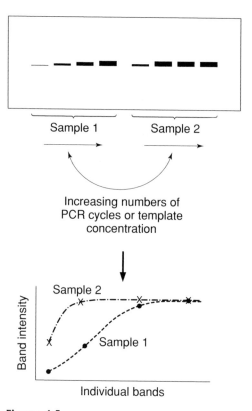

Figure 4.5

Saturation and linearity of PCR reactions. Sample 1 has less of a particular DNA sequence than sample 2. If the PCR reaction is performed multiple times, each with an increasing number of cycles or an increasing amount of starting template, eventually, the end point at which the PCR product is analysed on a gel will be in the (L) phase of the reaction (Figure 4.4). The graph represents the intensities of the PCR product bands to illustrate that the more starting template, the quicker (L) phase is reached. Therefore, if a given number of cycles or starting template dilution were chosen to be used, it would be quite possible to miss real differences in template DNA concentration between two samples. Ideally, one is looking to measure the gradient of the graph, in this case between the first two end points for each sample. Clearly, at this point, the gradient of the sample 2 graph is greater than that for sample 1, confirming that there is more template DNA in sample 2 than in sample 1.

4.7 Real-time PCR: the basics

As an alternative to end-point assays for qPCR, a number of techniques have been developed for so-called real-time PCR, where the concentration of PCR product is monitored at least once every PCR cycle (Higuchi *et al.*, 1992, 1993). In this way, the concentration of the PCR product is seen to build up exponentially, and eventually, if one or more reagents become limiting, the concentration will level off. Since the gradient of the exponential build up of PCR product is directly proportional to the amount of starting template DNA, the use of control templates of known concentration alongside the original test sample, allows calibration of the reaction, and determination of the template concentration in the test sample.

The process of real-time PCR must involve some method of detecting the concentration of double-stranded DNA in the sample chamber, and ideally if very accurate measurements of concentration are required, amplification products of only the target template should be visualized. It is also particularly important that the method used to visualize the DNA should not interfere with the PCR reaction, or the specificity of the primer pair being used. There are a whole host of chemistries now available for real-time PCR quantification of template levels but they fall into two categories. The first involves fluorescent dyes that intercalate into double-stranded DNA, so that the more DNA there is, the more fluorescence there is, which can be measured using a charge-coupled device camera. Originally, the dye used was ethidium bromide, but since this also intercalates into single-stranded DNA, it can affect the extension reaction of DNA polymerase, affecting the kinetics of the process. The second basic real-time PCR method works by monitoring the hybridization of a set concentration of fluorescently labeled probe oligonucleotide to the PCR product; the more PCR product, the more hybridization occurs, and so the more (or less, dependent upon the specific method employed) fluorescence is seen. More and more complex methods based on this second approach are becoming available, but they have a distinct advantage over dye-based methods, since the oligonucleotide probe provides selectivity, and will only monitor the concentration of a PCR product with a particular sequence. Dye methods will monitor double-stranded DNA levels, including those from nonspecific PCR products.

Different real-time PCR strategies

The most common dye-based real-time PCR detection system involves the use of SYBR green I dye (Morrison *et al.*, 1998). This dye has the great advantage over ethidium bromide (and its relative, SYBR green II) of only intercalating into double-stranded molecules, and only fluorescing when it intercalates. Thus, if SYBR green I is present in the PCR reaction, as DNA polymerase extends the primer, fluorescence levels increase (*Figure 4.6*). When the temperature of the reaction is raised and the strands dissociate, the SYBR green I will also dissociate and fluorescence will quench. In the next extension step, however, twice as much template DNA will be available, so the amount of SYBR green I incorporated by the end of extension will also approximately double. Thus if a single reading of fluorescence is taken at the end of each extension step, the value can be plotted on a graph

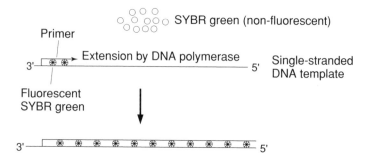

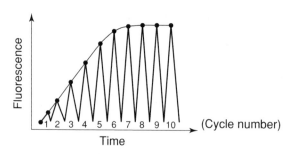

Figure 4.6

Fluorescent intercalating dyes for real-time qPCR. In this experiment, SYBR green I is being used. It does not fluoresce unless it intercalates between the strands of a double-stranded DNA sequence. Therefore, as template DNA is being copied, the level of fluorescence (illustrated as a star) increases to a maximum which is proportional to the amount of DNA in the sample at the end of the extension stage. When the DNA is melted, fluorescence is quenched. The graph shows how this fluorescence level changes with each PCR cycle, with the maximal amount of fluorescence being measured at the end of extension, and the points being joined up to give a DNA amplification curve, which can be used to quantify the DNA using a known concentration of starting template to calibrate the system.

to give the rate of fluorescence increase during the exponential phase of the PCR reaction.

There are a number of problems with the use of SYBR green I. The most obvious is that any amplified DNA becomes fluorescent, so it is important to check the products using agarose gel electrophoresis at the end of the reaction, in order to confirm that there is only a single amplification product. Another way of doing this is to perform melting curves on the fluorescent product. As the temperature increases, fluorescence should decrease (as the double-stranded DNA duplexes dissociate) with a simple curve. If multiple PCR products are present, however, the melting curve will be more complex, with each product having its own melting temperature. The second, less obvious problem with SYBR green I is that the amount of fluorescence in a PCR product is dependent upon the product's length and base composition, since dye molecules intercalate along the DNA molecule.

This means that it is not possible to compare the concentrations of two different templates in the same sample without having control templates of known concentration for each target DNA region.

The 5′ exonuclease assay

This assay utilizes a property of many DNA polymerases, that of the 5′ to 3′ exonuclease (Holland *et al.*, 1991). An oligonucleotide probe is made that is complementary to the DNA template being amplified, and which anneals at a position which is 3′ proximal to the annealing site of the PCR primer. As a DNA polymerase extends the PCR primer in the 5′ to 3′ direction it will come across the oligonucleotide probe, and the 5′ exonuclease activity in the polymerase will degrade the oligonucleotide into its constituent nucleotides, which will dissociate into solution. If the oligonucleotide probe is in large excess, the amount of probe annealing, and so degradation of the probe by the polymerase that will occur during each PCR cycle will depend upon the amount of DNA template present during that cycle *Figure 4.7*).

Visualization of oligonucleotide probe destruction is based on release of a fluorescent nucleotide when the probe is destroyed. The intact probe has a fluorescently labeled nucleotide at one end, and a quencher label at the other end, so the intact probe does not fluoresce. When the probe is degraded, the quencher and fluorescent label become physically separated, and fluorescence occurs. Unlike when using SYBR green I dye, which dissociates from single-stranded DNA during the melting phase of PCR, and so stops fluorescing, the fluorescence of the labeled nucleotide released by the 5′ exonuclease is retained from the point a probe is destroyed. Thus, fluorescence increases as more and more PCR occurs, with a jump in fluorescence occuring during each extension phase (*Figure 4.7*).

Commercially available oligonucleotide probes for this type of assay are marketed by Applied Biosystems, under the TaqMan label. They carry a fluorescein group at their 5′ ends, which emits green fluorescence, and a 3′ quencher, which efficiently accepts resonant energy from the fluorescein group, preventing green fluorescence and causing itself to fluoresce red. This 3′ quencher group blocks the 3′ end of the probe, meaning that it cannot be used as a PCR primer by DNA polymerase. At the start of the reaction, the probes are intact, and the predominant fluorescence in the reaction is red, which is ignored by the detector. As more and more annealed probes are degraded by the 5′ exonuclease activity of DNA polymerase during the extension phase of PCR, the amount of red fluorescence decreases (which does not affect the detector) and the level of green fluorescence increases accordingly (which is detected). Therefore, after a given number of cycles, green fluorescence gets to a point where it can be detected above background noise in the reaction. This point depends upon the sensitivity of the detector for green light. The number of cycles needed to get to this point (known as the 'threshold cycle') in a particular machine with a particular detector is, however, absolutely determined by the starting template concentration. Hence in this assay system, the rate of fluorescence increase is not determined, but instead, the number of PCR cycles required to detect product. The beauty of this assay is that it can be used to compare the quantities of different templates, each with their own specific probe in

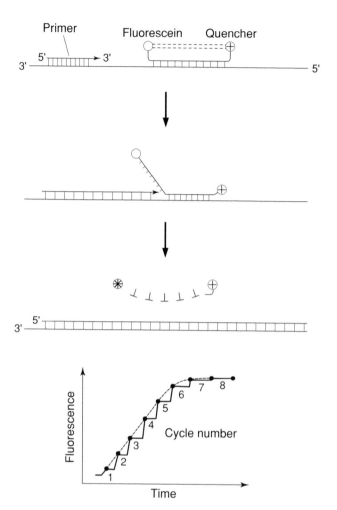

Figure 4.7

Exonuclease dependent real-time qPCR. A fluorescent probe is linked to a quencher via a sequence specific oligonucleotide, and the proximity of the two prevents fluorescence. A separate sequence specific primer is used to initiate PCR, and the polymerase drives off the probe, and its exonuclease activity degrades the probe. Therefore following extension, one unit of probe molecules will have been degraded, resulting in fluorescence. Upon denaturation and annealing, there is no alteration in fluorescence, because the free probe does not fluoresce, and only when the next cycle of extension has been completed does fluorescence rise once more. This can be seen in the graph where the indicative 'saw tooth' pattern is illustrated. If fluorescence is measured at the end of the extension phase, the points can be joined to give a graph of DNA accumulation, which can be used for qPCR as set out in the text.

separate assays performed on the same template mixture. Furthermore, the presence of specific probes means that nonspecific DNA amplification goes unnoticed. The disadvantages of this technique over the use of SYBR green I are that the extension step of PCR must be performed at a lower than normal temperature, since if 72°C is used, the probe may dissociate and not be degraded; secondly, the polymerase used must have a 5′–3′ exonuclease (and not all do!), it is difficult to compare results from experiments using different detection systems because the inherent sensitivity of the detector affects the threshold cycle number; and finally, it is expensive to make the fluorescent probes and new probes must be made for each target sequence.

Dual hybridization probes

This is a relative of the TaqMan system, and follows a similar procedure (Wittwer *et al.*, 1997). However, in this case, two probes are made, which are designed to anneal sequentially onto the target template DNA, 3′ proximally to the PCR primer. The first of the two probes is labeled with a green fluorescein at its 3′ end, and the other probe, which anneals immediately 3′ proximal to the first, is labeled with a quencher fluorophore at its 5′ end. The fluorophore moieties are identical to those used for the TaqMan system. Chemical blocking of the 3′ end of this quencher probe is undertaken to prevent it from being used as a PCR primer. As with the TaqMan system, the quencher fluorophore takes up the resonant energy being emitted by the green fluorescein, and itself fluoresces red. However, in this system, it is the red light that is detected, and the green light which is ignored. Thus, when the probes bind to the target template DNA, red fluorescence occurs and is detected by the monitor. The DNA polymerase used in this case can be devoid of 5′–3′ exonuclease activity, since the probes do not need to be degraded for the system to work, though it does not matter if they are. When even an exonuclease minus DNA polymerase extends the PCR primer along the target template it gets to the bound probes and simply pushes them out of the way, so that they separately become free in solution. This gives the same result as degradation: a blocking of red fluorescence. Furthermore, during the denaturation steps of the PCR reaction, the probes dissociate from the template DNA. This means that they no longer interact, stopping red fluorescence. Thus the probes are a simple method of specifically detecting the concentration of template DNA, since as the template concentration increases during the annealing phase of the PCR reaction, more probe doublets anneal next to each other on the template, and so more red fluorescence occurs (*Figure 4.8*). The interpretation of this reaction in terms of extrapolating it into template concentration is exactly the same as when using SYBR green I. In this case, however, the level of red fluorescence is measured at the end of the annealing phase not at the end of the extension phase of the PCR reaction. The advantage of using these dual probes over the use of SYBR green I, however, is that the probes provide target sequence specificity, only measuring the concentration of a PCR product having a specific sequence. The disadvantage of using this method is that it is costly to make the probes, and new probes need to be made for each target template sequence.

Molecular beacons

This approach is very similar in nature to the dual probe (Tyagi and Kramer, 1996). In this case, however, and more akin to the TaqMan system, a single

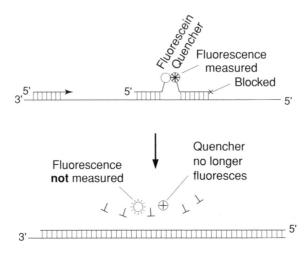

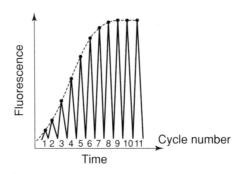

Figure 4.8

Dual-hybrid probes for qPCR. In this case, and unlike that set out in Figure 4.7, the bound probe does fluoresce, therefore as extension proceeds, the amount of fluorescence drops due to the action of exonuclease (or just because the polymerase pushes the probe out of the way). When annealing occurs, though, the amount of probe annealing is dependent upon the concentration of DNA, so if points are taken just before extension, this can be used to measure DNA concentration.

probe is used, which has a 5' fluorescein and a 3' quencher. The fluorescein and quencher groups are quite different to those used for the methods outlined above. Here, resonant energy cannot be passed between the fluorophores at the extremities of the probe, and so the fluorescein does not lose its fluorescence to the quencher. Therefore, when the probe is in place on the template, the result is fluorescence. The probes are made to have very particular properties. When not bound to the template, that is, when pushed off by an exonuclease deficient DNA polymerase (and for this technique, unlike the previous one, exonuclease activity would be disastrous), they are designed to have an inverted repeat at the ends, meaning that they form into a hairpin structure through self-complementarity of the ends of the probe. This hairpin places the fluorescein and quencher immediately adjacent to one another, and so fluorescence is quenched. Due to the fact that only bound probes fluoresce, these probes work as a label for template sequence, and as template sequence levels increase, fluorescence levels at the end of the annealing phase increase (*Figure 4.9*). The advantage of molecular beacons over the dual probe is that only a single probe need be made, with some cost benefit. The major disadvantage, however, is that because the probes must be self-complementary in a very particular way, the sequences at which probes can be designed to bind are very limited, and arriving at the sequence of choice can be very difficult and time consuming. Any secondary structure formation that prevents the stem loop structure from bringing together the fluorescein and the quencher will render useless the probe, since fluorescence will occur whether the probe is bound to its target or not.

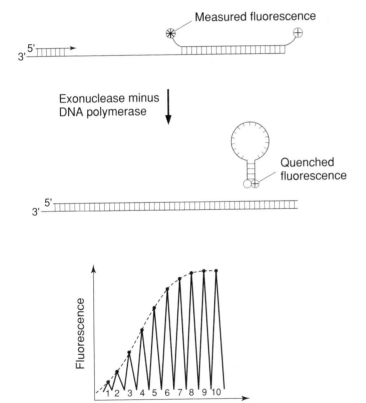

Figure 4.9

Molecular beacons for qPCR. The principle is identical to that in Figure 4.8. The bound probe fluoresces, so the amount of fluorescence seen after annealing is proportional to the amount of target DNA. Here it is important that the DNA polymerase does not have 5' exonuclease activity, because degradation of the probe would stimulate false positive fluorescence.

Scorpions

This is the most complex of the fluorescent probe chemistries (Whitcombe *et al.*, 1999). It involves making an oligonucleotide 'scorpion' that acts as both probe and PCR primer (*Figure 4.10*). The primer portion of the scorpion is complementary to the target sequence and anneals at the

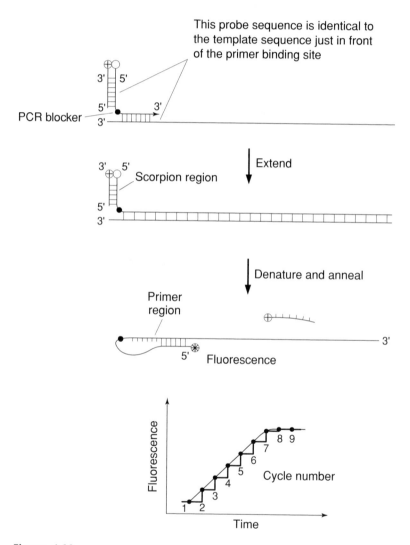

Figure 4.10

Scorpions for qPCR. The principle is the same as with the exonuclease assay in terms of reporting the amount of PCR product. In each round of PCR, a scorpion is incorporated into a product in such a way that it hybridizes with a part of the product and so fluoresces. Therefore, as the concentration of PCR product increases (i.e. at the end of each extension stage), fluorescence increases. A more detailed description of the scorpion is given in the text.

desired point of PCR extension initiation. The primer portion of the scorpion is thus thought of as being in a 5'–3' direction and binds at the 3' end of the single-stranded template DNA molecule. At the 5' end of the primer portion of the scorpion, a linker molecule is attached. This is usually hexaethylene glycol moiety, known as a 'PCR stopper', because if a DNA polymerase were to come across it, it would terminate the polymerization reaction. Immediately following the PCR stopper is the probe sequence portion of the scorpion. This probe sequence is laid out in a 3'–5' direction, with its 3' end and the 5' end of the primer portion flanking the PCR stopper. The probe portion of the scorpion is designed to have the same sequence as the template strand, from a point approximately 3 bp, 3' proxi-mally to the 3' end of the PCR primer. The 5' end of the probe portion of the scorpion (i.e. furthest away from the PCR stopper) is labeled with a fluores-cein group. The detection technique relies on the presence of a second oligonucleotide. This is designed to be exactly complementary to the probe portion of the scorpion, and to have a quencher group on its 3' end. Before the PCR reaction, the two oligonucleotides are mixed and the quencher probe anneals to the probe portion of the scorpion, placing the fluorescein and quencher groups in close proximity, and resulting in no fluorescence. The quencher probe/scorpion duplex is added to the template DNA, and the primer portion of the scorpion (which is single-stranded) anneals to the template. The probe portion of the scorpion remains annealed to the quencher probe. The 3' end of the primer portion of the scorpion is then extended by the DNA polymerase enzyme. When the products are heated during the melting phase of the PCR cycle, the newly formed DNA strand (with its integrated scorpion oligonucleotide) melts away from the template strand, and the quencher probe melts away from the probe portion of scorpion. When the temperature is lowered, and annealing occurs, it is more likely that the probe portion of the scorpion will anneal to its comple-ment on the newly formed DNA strand that it is a part of, than it will anneal to a complementary quencher probe in free solution. This is simply due to the physical linkage between the probe portion of the scorpion and its target DNA sequence. Hence the probe portion of the scorpion folds back on itself and anneals to the newly formed DNA strand, using the PCR stopper as a fulcrum. The fluorescein is no longer proximal to a quencher, so it fluoresces. The scorpion/newly formed DNA product can be used as a template for PCR, since the PCR stopper prevents anything beyond the primer portion of the scorpion being copied and the fluorescein blocks the probe portion of the scorpion so this is not used as a point of extension by DNA polymerase. Of course, for further amplification to occur, a second oligonucleotide PCR primer will be needed, that anneals to the template at the 3' end of the newly formed strand, and defines the 3' end of the template sequence being amplified, but this is the same as for all PCR reactions. The 5' end of the template to be amplified in each round of PCR is defined by the primer portions of free scorpions in solution, which are provided in excess. Hence during each PCR cycle, half of the PCR products become associated with a scorpion, and so fluorescently labeled. The remaining free scorpions in solution will be mopped up by annealing to quencher probes, which are provided in stoicheometric amounts with the scorpions.

All this means that in an apparently complex way, the scorpion simply measures the concentration of template sequence during each annealing/extension phase of the PCR reaction as a readout of increased fluorescence over time. The design of scorpions is less tricky than the design of molecular beacon probes, but they are not straightforward since secondary structure in the scorpion will play havoc with the experiment. However, the probes become incorporated into the PCR products, so the extension temperature used does not have to be as low as is the case when using 5′ exonuclease assays. The main problem with scorpions is that because of their chemical modifications, they are quite expensive.

4.8 Reverse transcription-PCR measurement of RNA levels (qRT-PCR)

There has been much discussion of PCR, the mechanics of the process, and particularly its use in quantifying the absolute concentration of a specific DNA sequence, or at least the relative concentration of that sequence in two samples of DNA. But how does this help us in our quest to measure gene expression? Well, simply, if one wants to measure the relative concentrations of an RNA molecule having a particular sequence in two preparations of total RNA, all that needs to be done is that the RNA is converted into cDNA, and then the cDNA can be quantified using qPCR as defined above. This process in total is referred to here as qRT-PCR. I do this to differentiate this technique from RT-PCR, which is a more generic term for the conversion into cDNA of an RNA molecule and amplification of that cDNA. In general, straightforward RT-PCR is used to clone or sequence a cDNA, though most books use the name RT-PCR to refer to these types of experiment, and to measurements of gene expression as well.

There are many different methods for performing qRT-PCR, and many commercially available kits and enzymes to do so. This can be confusing, but it is important to choose a method for cDNA production and qPCR that is appropriate for the specific goal of your experiments. It goes without saying that in addition to being optimal for your purposes, the method must behave in a robust and reproducible manner, and this generally means that the reverse transcription reaction chosen converts almost exactly the same proportion of RNA into cDNA (ideally 100%, but this is not essential) irrespective of the type of RNA sample and the concentration of the particular RNA molecule being copied. Set out below are a number of examples of how this might be tackled.

4.9 qRT-PCR methodologies

Two-tube qRT-PCR

The simplest way of making cDNA is to use a specialist reverse transcriptase enzyme. These are described in detail in Section 3.12. It should be noted here, though, that like all DNA polymerases, reverse transcriptases require an oligonucleotide primer to be annealed to the template RNA, which provides a 3′ hydroxyl group from which point deoxynucleotides are added by the enzyme. It is best that the reverse transcriptase enzyme chosen will have significant RNase H activity (so there is no amplification of RNA to DNA – i.e. RNA molecules are not reverse transcribed more than once). It is also very important that enzymes with the highest temperature stabilities are used, since raised temperatures in the reverse transcription reaction reduces secondary structure in the template and primer, and limits nonspecific primer annealing. Because of these two requirements, the most commonly used specialist reverse transcriptase enzyme for cDNA production is the AMV reverse transcriptase.

Because qRT-PCR is almost always used to quantify the cDNA product of one RNA molecule having a specific sequence, it is common to use a single, sequence-specific reverse transcription primer that is designed to anneal only to the RNA molecule whose cDNA product is the target for the qPCR stage. This is the basis of the classical two-tube qRT-PCR reaction (*Figure 4.11*). In the first tube, an aliquot of total RNA is placed and a specific primer is used to convert RNA molecules of a given sequence into cDNA via AMV reverse transcriptase. The reaction is then stopped by heating at 70°C for a few minutes, and an aliquot is taken to serve as the template for a standard qPCR reaction in the second tube. The fact that there is substantive carry over of non-target RNA into the qPCR reaction is not an issue provided the DNA polymerase used for qPCR does not have significant reverse transcriptase activity. The fact that a reverse primer is carried over from the reverse transcription reaction is not a problem either, since this same primer is used as one of the pair of primers for qPCR. Once the second tube qPCR reactions have been set up using the reverse transcription reaction product as template, the reactions proceed, and the cDNA concentration is calculated exactly as described in Sections 4.6 and 4.7 when discussing qPCR and real-time qPCR.

More often than not in qRT-PCR, one wishes to use an RNA preparation to measure the concentration of one RNA molecule, then another with a different sequence, and then another, and so on. If you do this, be aware of potential experimental variability, which is a common complaint amongst those routinely carrying out qRT-PCR. The point at which most experimental variability is introduced into a qRT-PCR experiment is during the first tube reverse transcription step. This manifests itself as the observation that each time you perform a reverse transcription reaction, a different proportion of the RNA is converted into cDNA. This is due to two problems the first of which is RNA instability. Every time you remove the RNA preparation from the freezer, thaw it and withdraw an aliquot, the amount of RNA will have reduced since the last time you thawed it. Eventually, and it will take a surprisingly short time, there will be so little RNA left undegraded, that the RT-PCR signal will be too low to measure accurately, if at all. The second source of experimental variability in reverse transcription reactions is due to the fact that reverse transcriptase enzymes are notorious for their thermal instability, meaning that they quickly become inactive after they have been taken out of the freezer a few times. In order to mitigate against these major sources of error, I would advise two courses of action. First, I implore you to learn from my own mistakes and convert all of the RNA in your sample into cDNA and store aliquots of the cDNA for sequence specific qPCR at a later date. The cDNA will be more stable than the RNA, its storage in ready, qPCR-sized aliquots will reduce the problems associated with

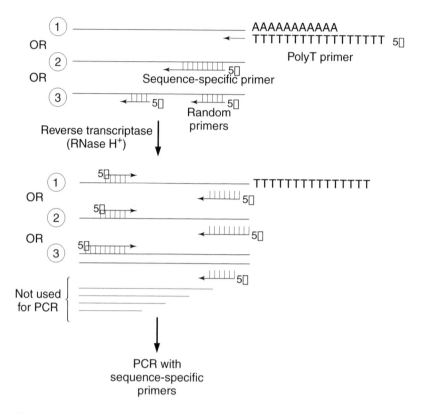

Figure 4.11

Two tube RT-PCR. Firstly, total RNA is reverse transcribed to produce first-strand cDNA using either (1) a polythymine primer (if the RNA is polyA⁺) (2) a sequence specific primer, or (3) a pool of random primers (which produces a pool of differently sized products, only some of which will be suitable for PCR amplification because only some will have binding sites for the reverse PCR primer). Following cDNA production and RNase H action to degrade the template RNA, PCR is performed using two sequence-specific primers. So, in this way, total cDNA can be produced first, and this can be divided into aliquots to allow quantification of a different cDNA products using qPCR.

freeze–thaw cycles, and the expression levels of multiple genes can be assessed using a single RNA preparation without loss of signal and with minimal introduction of error. Secondly, purchase only sufficient reverse transcriptase to perform the reverse transcription reactions you have planned to undertake imminently, and not large amounts with a hope that it will be used in the future. When you go back to it in a couple of months, it will probably have gone off.

The way to produce total cDNA from total RNA, and so have a single first tube reverse transcription reaction followed by multiple sequence specific qPCR reactions, rather than having multiple two-tube qRT-PCR reactions each targeting a different sequence, is to use reverse transcription primers that anneal to all RNA molecules. As discussed when talking about total cDNA production for array hybridization (Section 3.12), it is possible to use polythymine primers to reverse transcribe polyA$^+$ mRNA molecules. The problem is that not every mRNA is polyA$^+$ in a eukaryotic cell, and that prokaryotic mRNAs are not polyA$^+$ at all. Hence I would suggest the use of random primer mixtures. Ideally, these will be random decamers, since they result in, on average, longer cDNAs than the more commonly used random hexamers. All the specificity in the qPCR stage comes from the choice of PCR primer pair so it is essential that a large majority of cDNAs are long enough to encompass a region including binding sited for both PCR primers; hence, the longer the better. For the same reason, the production of total cDNA using random primers will involve longer reverse transcription reaction times than specific cDNA production. It is also important to consider the concentration of deoxynucleotides added, since it is important that they do not run out.

One-tube qRT-PCR

Whilst double-tube RT-PCR reactions are the most common RT-PCR reactions, they have a number of problems. It takes more time to run two reactions rather than one, and more to the point, it takes more time to set up two reactions rather than one. Also, setting up multiple reactions provides greater opportunity for a technical mistake to be made. Because of this, one-tube qRT-PCR reactions are becoming increasingly common (*Figure 4.12*). They either involve the use of thermostable DNA polymerases

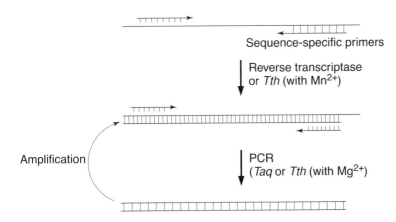

Figure 4.12

One tube RT-PCR. This is the method used mainly if RT-PCR for only one sequence of RNA is to be performed. Sequence-specific primers are used to reverse transcribe the starting RNA and PCR amplify the resultant cDNA. Normally, either a single enzyme with both reverse transcriptase and DNA polymerase activity is used (in the case of *Tth*, Mn^{2+} ions must be removed and replaced by Mg^{2+} ions following the reverse transcription step) or a mixture of AMV reverse transcriptase and a thermostable DNA polymerase (in which case the RNA template must be completely free of salt contamination because the buffer solution must be very finely balanced to allow activity of both enzymes).

with significant reverse transcriptase activity to catalyze both the reverse transcription and PCR steps in one tube, or a mixture of AMV reverse transcriptase and a thermostable DNA polymerase. In the first case, a highly thermostable enzyme is being used, so even the reverse transcriptase portion of the reaction can be processed at even higher temperatures than when using AMV reverse transcriptase, and the enzyme will be more robust and tolerant of freeze–thaw cycles, reducing experimental variability due to differential degradation of the enzyme, which is a significant problem when using reverse transcriptase enzymes. However, I would add the same note of caution when considering using RNA preparations multiple times here that I did when discussing double-tube reactions, above. If you are planning to use an RNA preparation more than three times (the standard experimental replicate for qRT-PCR expression analysis for one gene) then I would strongly advise you to routinely reverse transcribe your purified RNA into cDNA and store the cDNA aliquotted in such a way that single-tube qPCR reactions can be run on the aliquots. Single tube qRT-PCR reactions are, therefore, only to be recommended for RNA preparations that are being used a very limited number of times.

The first thermostable DNA polymerase to be used for single tube RT-PCR was the *Tth* enzyme. This has significant reverse transcriptase only in the presence of manganese ions, so the reverse transcription reaction involves a reaction buffer containing manganese. Carry-over of manganese significantly impairs the function of *Tth* as a DNA polymerase for PCR, however, so the user has to add an EGTA-containing chelator buffer to the completed reverse transcription reaction, and then a separate magnesium chloride solution to provide the *Tth* enzyme with its requirement for Mg^{2+} ions for efficient PCR. Hence, whilst this is, strictly speaking, a single-tube RT-PCR experiment, the user cannot run the reaction overnight, for example, since user intervention is required. You can get away without this chelation step for qRT-PCR in some cases. If you use a buffer containing manganese and magnesium ions, and accept that the PCR products will have significant errors (because the presence of manganese causes the polymerase to introduce incorrect bases). As well as meaning that the RT-PCR product should never be cloned and sequenced (if cloning the product or producing the exact sequence is important to you) the presence of errors can cause PCR termination. The result of this is that the RT-PCR product bands can appear quite smeary on an agarose gel due to the presence of significant amounts of shortened products. PCR errors can also affect the kinetics of PCR, since PCR products that are truncated cannot be used as templates in further rounds of PCR. This will reduce the accuracy of the qRT-PCR technique, to a point where the method is only semi-quantitative.

Thermostable reverse transcriptase/DNA polymerases that do not require manganese do exist. For example the *Ctherm* enzyme. This makes true single tube qRT-PCR possible, though it should be remembered that *Ctherm* carries a proofreading activity, which slows down the extension rate of the enzyme, making reaction times long, but products ideal for cloning and sequencing. It appears, however, that this enzyme is no longer manufactured, and the reason may be that manufacturers are moving away from the use of DNA polymerases with reverse transcriptase activity for one-tube

RT-PCR and towards the use of reverse transcriptase/DNA polymerase mixtures. These are usually mixtures of AMV reverse transcriptase and thermostable polymerases such as *Taq* or *Tth*. The difficulty in producing these mixtures is in manufacturing a buffer within which both enzymes work efficiently. Accordingly, these reagents are perhaps less forgiving of contamination of templates with salts etc. The result is usually poor reverse transcription, meaning product concentrations are under-represented. In a qRT-PCR approach, this is particularly problematic, since it is quite possible for different comparator RNA preparations to be differentially contaminated with different salts.

So, in conclusion, one-tube RT-PCR has its uses, but it is not really appropriate for accurate qRT-PCR. Personally, for gene expression analysis, I would always advocate total cDNA synthesis followed by qPCR using the cDNA as template.

Quantification of qRT-PCR products

qRT-PCR based gene expression experiments usually involve comparing cDNA concentrations using qPCR after reverse transcribing two RNA preparations, one using RNA isolated from control cells, and one using RNA isolated from experimental cells. The qPCR approach can either be an endpoint assay using serial dilutions of cDNA template, or ideally can be a more sophisticated real-time qPCR approach, but in either case, the approaches are exactly as described for standard qPCR in Sections 4.6 and 4.7. Thus qRT-PCR is ideal for measuring the expression of a particular gene in a particular cell type under particular physiological conditions and is an essential accessory to check apparent gene expression changes suggested from large-scale array-based hybridization experiments. When used in conjunction with real-time qPCR approaches (Section 4.7) it is highly accurate with an excellent dynamic range.

All the experimental controls (e.g. the use of housekeeping controls) and experimental design pointers described for northern blotting analysis of the concentration of individual RNA molecules, as described in Section 3.8, apply to qRT-PCR experiments.

4.10　SIP-PCR and the virtual northern blot

One of the real benefits of performing qRT-PCR rather than, say, northern hybridization is that due to the amplification phase of the process, very small amounts of RNA can be quantified using qRT-PCR where much larger amounts of RNA are required for northern hybridization. However, the dynamic range of a northern hybridization reaction is considerably better than qRT-PCR when using agarose gels stained with ethidium bromide to visualize qRT-PCR amplicons. This means that quantification is easier using northern hybridization because it does not require multiple experiments, each with a different dilution of starting RNA to get accurate results.

One way of overcoming the problems associated with the requirement for large amounts of RNA to run a successful northern blot experiment (without simply preparing more RNA, which might not even be possible when dealing with precious tissue samples) is to amplify the concentration of total RNA, or a cDNA copy of that RNA, and so the signal strength, prior to northern blotting and hybridizing the immobilized RNA/cDNA target to a labeled DNA probe washed over the surface of the blot (Eberwine *et al*, 1992). This is outlined in Section 3.12, where the use of reverse transcriptase and RNA polymerase to amplify RNA levels is explained. However, another way of doing this would be to use reverse transcriptase and DNA polymerase, i.e. RT-PCR.

The conversion of total RNA into total cDNA using reverse transcriptase has been discussed previously, as has the amplification of a cDNA with a particular sequence by PCR. However, the amplification of total cDNA by PCR has not. Indeed, how might this be achieved? As described, PCR requires that two oppositely oriented primers anneal to a DNA molecule before it can be amplified. Whilst random primers can be used to make cDNA, which only requires a single anchor point, they cannot successfully be used to amplify a complement of total cDNAs. This is because the multiple binding sites for random primers within each cDNA molecule would result in a very large number of PCR amplicons, many of which would be very small and would cause problems during northern blot hybridization, because their small size would mean they would hybridize nonspecifically to the labeled probe being used to give specificity. The only alternative is to have a single primer or pair of primers that anneal at both ends of every cDNA to be amplified. For this, you need to have a known sequence at each end of each cDNA molecule, which can be used as a primer binding site. This process is called sequence independent primed PCR (SIP-PCR) because the primers bind outside the variable sequence of the cDNA molecules being amplified (*Figure 4.13*).

Primer binding sites can be added to each end of a double-stranded cDNA in a number of ways. First, it is straightforward to generate a population of first strand cDNAs with a polythymine tract at their 5′ ends. The use of a polythymine primer, which anneals to the polyA tail at the 3′ end of mRNA molecules and primes cDNA production using reverse transcriptase will do this. Of course, the usual caveat applies, that there are some eukaryotic mRNAs which are not polyA$^+$. Furthermore, no prokaryotes have polyA$^+$ RNA (as far as we know), though problems associated with low RNA levels, and so the need for amplification in the first place, are generally limited to eukaryotes.

Attaching a known sequence onto the 3′ ends of cDNA molecules is more problematic. The first approach to be considered is where one ligates a single-stranded oligonucleotide onto the 5′ ends of all RNA molecules prior to their conversion into cDNA. The enzyme that does this is called an RNA ligase, and that from the T4 bacteriophage is commercially available for this purpose. The oligonucleotide is then copied by reverse transcriptase as part of the RNA molecule, producing a single known sequence at the 3′ ends of each first strand cDNA produced. However, RNA ligations are very tricky and inefficient processes and are not ideal for use with small amounts of substrate. Furthermore, if capped eukaryotic mRNAs are being used as substrates, the cap must be removed first using tobacco acid pyrophosphatase to reveal a 5′ phosphate, which is absolutely required for oligonucleotide ligation.

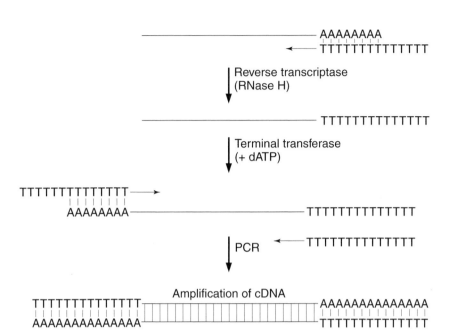

Figure 4.13

PolyA SIP-PCR. Reverse transcription using a polythymine primer produces a first-strand cDNA which is extended at its 3' end using terminal transferase to put on a 3' polyadenine tract. This is used for second-strand cDNA amplification using the same polythymine primer used for reverse transcription, and the PCR reaction proceeds from this point doubling the template cDNA concentration at each round.

An alternative to RNA ligases for attaching a known sequence onto the 3' ends of first-strand cDNA molecules is to use an enzyme called a terminal transferase. M-MuLV reverse transcriptase has a terminal transferase activity, meaning that it adds a short string of cytosines at the 3' end of each first strand cDNA molecule it generates (Peliska and Benkovic, 1992). These cytosines can then be used as PCR primer binding sites using a polyguanine primer, which is longer than the string of cytosines added, so that following the first round of PCR, primer annealing is more efficient, and so that nonspecific annealing onto runs of cytosines within the cDNA is not a problem.

A more efficient terminal transferase-dependent approach, which allows the use of thermostable reverse transcriptases that do not have their own inherent terminal transferase activity is so-called polyA-PCR (Brady *et al.*, 1990). Here a polythymine primer is used to synthesize first-strand cDNA, adding a polyT sequence onto the 5' end of the first strand cDNA, as set out above. Next, bovine terminal deoxynucleotide transferase is added, together with dATP. It is essential that all dNTPs from the reverse transcriptase reaction are removed prior to addition of terminal transferase, since it will add any available deoxynucleotide to the 3' end of the first strand cDNA molecule and the whole point of this technique is you want to produce a known sequence tag. The best way of doing that is to only use one deoxynucleotide (Schaefer, 1995) and the use of dATP results in 3' polyadenylation of the first-strand cDNA, meaning that the same polythymine primer can be used for second-strand cDNA production, and so PCR, as was used for reverse transcription (*Figure 4.13*). Following PCR, the copy number of each cDNA will be elevated, and because of the fact that the process does not depend on the sequence of the cDNA, it is highly likely that the kinetics of amplification of cDNAs with different sequences will be very similar.

Once the amplified double-stranded cDNA population has been produced, it can be used to perform a virtual northern blot (Franz *et al.*, 1999). The cDNA molecules are separated according to size using nondenaturing agarose electrophoresis (see above for a discussion of DNA agarose gels). The separated cDNAs are then transferred to a nylon+ membrane, where they are immobilized. The procedure is almost exactly as that set out for northern blotting in Section 3.4. However, this is, strictly speaking, a Southern blot which is being performed, since DNA is being transferred and immobilized. Accordingly, therefore, the agarose gel containing separated cDNA molecules must be soaked in alkaline buffer prior to transfer in order that the double-stranded DNA molecules are denatured, and can be used as targets for the specific, labeled DNA probe being used in the Southern hybridization reaction.

4.11   *In situ* qRT-PCR

Because qRT-PCR allows the quantification of very small amounts of RNA, it is quite possible to take it to a single-cell scale. The aim would be to take a section across a tissue, and measure the expression of a gene in all the individual cells that make up the tissue, allowing you to conclude whether there is any difference in expression in different parts of the tissue sample. I'm not trying to say this is easy, nor am I able to give a very detailed analysis of sample preparation. However, if you have histology facilities, and are used to preparing tissue sections on glass slides for visualization down a microscope, then with a little extra work, you can start to do *in situ* qRT-PCR, for example, see Stein *et al.*, 1997.

The tissue sample is normally fixed with formaldehyde or glutaraldehyde and then embedded in paraffin before being cut into thin sections for mounting onto the slide. To prepare for *in situ* RT-PCR, the paraffin must be removed by heating the slide at 95°C for 5 min, then re-fixed by immersing it in xylene. When this has been done, cells within the sample must be permeabilized to allow RT-PCR reagents to enter them. For this, low concentrations of proteinase K are used, which literally digests parts of the outer envelope. Needless to say, this takes some optimization, and is the most common point at which scientists give up on *in situ* RT-PCR. But persevere, for when it works, you are overcome with a wonderful sense of achievement! Once the cells have been permeabilized, they must be treated with DNase I to remove genomic DNA, which will ruin the RT-PCR experiment. To check that the DNase I has done its job, *in situ* PCR-only controls must be run in parallel with the RT-PCR reactions to prove there is no DNA contamination of the sample. At the same time as DNase I treatment is taking place, RNase inhibitors are washed over the sample. Whilst the use of tissues that have been treated with RNA stabilizing reagents before fixing will help to keep the RNA intact, and the fixing process itself with protect intracellular RNA, as soon as the cells are permeabilized, external RNases can get to work and should be kept at bay as much as possible with the use of RNase inhibitors (see Table 2.1).

The RT-PCR reaction is identical for *in situ* approaches as it is for *in vitro* approaches. There are special heating blocks onto which glass slides can be placed, thus allowing the reaction temperature to be controlled, and in the PCR phase, cycled accordingly. It is best to use a two-stage process for *in situ* RT-PCR, with cDNA production being followed by the addition of PCR reagents and enzymes, allowing for a denaturation step to destroy residual reverse transcriptase activity before embarking on PCR. It is only going to be feasible to perform sequence specific RT-PCR. On each occasion, reagents are simply washed onto the tissue *in situ*.

One of the benefits of *in situ* RT-PCR is that because the amplified cDNA molecules are fixed in space, visualization of their concentrations is a simple process, which can be achieved in two main ways. The first is to incorporate radioactive or DIG labels into the RT-PCR products *in situ* using labeled nucleotides. Once unincorporated nucleotides have been washed away, the labeled products can be visualized, either with autoradiography, or using chemiluminescence, or even using silver precipitate deposition, as set out in Section 3.12. Furthermore, if more sophisticated fluorescent labels are incorporated into the RT-PCR products these can be visualized directly using laser excitation and a CCD device to capture emitted fluorescence.

Alternatively, if a label is not incorporated into the RT-PCR product *in situ*, the product can be visualized using *in situ* hybridization with a radioactive, DIG- or fluorescent-labeled sequence-specific probe, with the target:probe interaction being visualized as for direct incorporation of label.

Finally, it should be mentioned that the very real problems associated with sample preparation and optimization of *in situ* RT-PCR reactions may be a thing of the past if sophisticated laser dissection equipment were used to specifically cut out parts of tissue slices immobilized onto slides in order that qRT-PCR might be performed. The tissue fragments would be used to purify RNA on a minute scale, which would then be used for real-time qRT-PCR according to one of the chemistries outlined above. Whilst this would represent a slightly blunter instrument than *in situ* RT-PCR for determining the exact cellular location of RNA expression, it would be far more suitable for quantification of expression levels, which is something that *in situ* RT-PCR is not best used for, because the signal has to be integrated over a wide area, and differential background effects across the sample can interfere with this process.

Further reading

Lorkawski, S and Cullen, P (Eds) (2003) *Analyzing Gene Expression, A Handbook of Methods, Possibilities and Pitfalls*. Wiley VCH, Weinheim, Germany, pp. 267–373.
McPherson, MJ and Moller, SG (2000) *PCR: The Basics*. Bios Scientific Publishers, Abingdon, UK.

References

Barnes, WM (1994) PCR amplification of up to 35 kb DNA with high fidelity and high yield from lambda bacteriophage templates. *Proc Natl Acad Sci USA* 91: 2216–2220.

Brady, G, Barbara, M and Iscove, NN (1990) Representative in vitro cDNA amplification from individual hematopoietic cells and colonies. *Methods Mol Cell Biol* 2: 17–25.

Cariello, NF, Swenberg, JA and Skopek, TR (1991) Fidelity of *Thermococcus litoralis* DNA polymerase (Vent) in PCR determined by denaturing gradient gel electrophoresis. *Nucleic Acids Res* 19: 4193–4198.

Chien, A, Edgar, DB and Trela, JM (1976) Deoxyribonucleic acid polymerase from the extreme thermophile *Thermus aquaticus. J Bacteriol* 127: 1550–1557.

Eberwine, J, Yeh, H, Miyashiro, K, Cao, Y, Nair, S, Finnell, R, Zettel, M and Coleman, P (1992) Analysis of gene expression in single live neurons. *Proc Nat Acad Sci USA* 89: 3010–3014.

Franz, O, Bruchhaus, I and Roeder, T (1999) Verification of differential gene transcription using virtual northern blotting. *Nucleic Acids Res* 27: e3.

Higuchi, R, Dollinger, G, Walsh, PS and Griffith, R (1992) Simultaneous amplification and detection of specific DNA sequences. *Biotechnology* 10: 413–417.

Higuchi, R, Fockler, C, Dollinger, G and Watson, R (1993) Kinetic PCR analysis: real-time monitoring of DNA amplification reactions. *Biotechnology* 11: 1026–1030.

Holland, PM, Abramson, RD, Watson, R and Gelfand, DH (1991) Detection of specific polymerase chain reaction products by utilizing the 5′-3′ exonuclease activity of *Thermus aquaticus* DNA polymerase. *Proc Natl Acad Sci USA* 88: 7276–7280.

Kermekchiev, MB, Tzekov, A and Barnes, WM (2003) Cold-sensitive mutants of *Taq* DNA polymerase provide a hot start for PCR. *Nucleic Acids Res* 31: 6139–6147.

Lundberg, KS, Shoemaker, DD, Adams, MW, Short, JM, Sorge, JA and Mathur, EJ (1991) High-fidelity amplification using a thermostable DNA polymerase isolated from *Pyrococcus furiosus. Gene* 108: 1–6.

Marchuk, D, Drumm, M, Saulino, A and Collins, FS (1991) Contruction of T-vectors, a rapid and general system for direct cloning of unmodified PCR products. *Nucleic Acids Res* 19: 1154.

Merkens, LS, Bryan, SK and Moses, RE (1995) Inactivation of the 5′-3′ exonuclease of *Thermus aquaticus* DNA polymerase. *Biochim Biophys Acta* 1264: 243–248.

Morrison, TB, Weis, JJ and Wittwer, CT (1998) Quantification of low-copy transcripts by continuous SYBR Green I monitoring during amplification. *Biotechniques* 24: 954–962.

Myers, TW and Gelfand, DH (1991) Reverse transcription and DNA amplification by a *Thermus thermophilus* DNA polymerase. *Biochemistry* 30: 7661–7666.

Peliska, JA and Benkovic, SJ (1992) Mechanism of DNA strand transfer reactions catalyzed by HIV reverse transcriptase. *Science* 258: 1112–1118.

Saiki, RK, Scharf, S, Falcona, F, Mullis, KB, Horn, GT, Erlich, HA and Arnheim, N (1985) Enzymatic amplification of beta-globin genomic sequences and restriction site analysis for diagnosis of sickle cell anemia. *Science* 230: 1350–1354.

Saiki, RK, Gelfand, DH, Stoffel, S, Scharf, S, Higuchi, R, Horn, CT, Mullis, KB and Erlich, HA (1988) Primer-directed enzymatic amplification of DNA with a thermostable DNA polymerase. *Science* 239: 487–491.

Schaefer, BC (1995) Revolutions in rapid amplification of cDNA ends: new strategies for polymerase chain reaction cloning of full-length cDNA ends. *Anal Biochem* 227: 255–273.

Stein, U, Walther, W, Wendt, J and Schild, TA (1997) In situ RT-PCR using fluorescence-labeled primers. *BioTechniques* 23: 194–195.

Suggs, SV, Wallace, RB, Hirose, T, Kawashima, EH and Itakura, K (1981) Use of synthetic oligonucleotides as hybridization probes. 3. Isolation of cloned cDNA sequences for human β-2-microglobulin. *Proc Natl Acad Sci USA* 78: 6613–6617.

Tyagi, S and Kramer, FR (1996) Molecular beacons: probes that fluoresce upon hybridization. *Nature Biotech* 14: 303–308.

Whitcombe, D, Theaker, J, Guy, SP, Brown, T and Little, S (1999) Detection of PCR products using self-probing amplicons and fluorescence. *Nature Biotech* 17: 804–809.

Wittwer, CT, Herrmann, MG, Moss, AA and Rasmussen, RP (1997) Continuous fluorescence monitoring of rapid cycle DNA amplification. *BioTechniques* 22: 130–138.

Wu, DY, Ugozzoli, L, Pal, BK, Qian, J and Wallace, RB (1991) The effect of temperature and oligonucleotide primer length on the specificity and efficiency of amplification by the polymerase chain reaction. *DNA Cell Biol* 10: 233–238.

Protocol 4.1 PCR amplification of a specific cDNA sequence

EQUIPMENT

PCR tubes
Bench-top centrifuge
Thermal cycler

MATERIALS AND METHODS

1. For standard *Taq* polymerase-driven PCR, use 50 mM KCl, 10 mM Tris-HCl (pH 9.0) containing 1.5 mM $MgCl_2$, 0.2 mM of each dNTP and 0.1% v/v Triton X-100 and 1% v/v DMSO if secondary structure in the primer is likely to be a problem.
2. First, add 100 pmol of each sequence specific oligonucleotide primer.
3. Second, add 20% v/v of first-strand cDNA reaction product (Protocol 3.3) which has first been treated for 5 min at 70°C.
4. Finally, add 5 U of *Taq* DNA polymerase.
5. A total volume of 50 μl is best (i.e. including 10 μl of cDNA product).
6. Flick the tube and briefly centrifuge.
7. Do not forget to add mineral oil if required.
8. Use the following PCR protocol as a starting point: 5 min at 95°C; 30 cycles of 1 min 95°C, 1 min at 55°C (or lowest T_m for primers, minus 5°C, optimally), 1 min at 72°C; one cycle of 5 min at 72°C. For low abundance cDNAs, up to 35 cycles may be needed.

Protocol 4.2 Single-tube RT-PCR

EQUIPMENT

PCR tubes
Thermal cycler

MATERIALS AND METHODS

1. The reaction buffer for one-tube RT-PCR is really tricky to get right. Have a try with 50 mM Tris-HCl, pH 8.3 containing 50 mM KCl, 5 mM MgCl$_2$, 0.25 mM each dNTPs and 0.1% v/v Tween-20. If you get poor results, I suggest you buy a one-tube RT-PCR kit which will come with a proprietary optimized buffer, for which you will not be able to obtain the recipe.
2. Use 500 ng of total RNA in a 20 μl reaction
3. Add 100 pmol of forward primer and 120 pmol of reverse primer.
4. Add 20 U of AMV reverse transcriptase and 5 U of *Taq* DNA polymerase.
5. Incubate for 30 min at 47°C, then 5 min at 95°C, then 30 cycles of 95°C for 45 s, 55°C (or 5°C less than the T_m for the lowest primer) for 45 s and 72°C for 1 min, followed by a single 5 min extension at 72°C.

Protocol 4.3 PolyA plus SIP-PCR

EQUIPMENT

Eppendorf tubes
Thermostatically controlled heating block
Petri dishes

MATERIALS AND METHODS

1. Produce first-strand cDNA as set out in Protocol 3.3, using polythymine (18–25) primer.
2. Heat inactivate at 80°C for 5 min.
3. Dialyze the cDNA reaction through a 0.22 μm filter disc floating on nuclease-free water in a Petri dish for 10 min.
4. Make up the cDNA in buffer to a final concentration of 100 mM sodium cacodylate, pH 6.8, containing 1 mM $CoCl_2$, 50 μM DTT and 0.1 μM dATP to a total volume of 50 μl.
5. Add 30 U of bovine terminal deoxynucleotide transferase and incubate for 60 min at 37°C.
6. Heat inactivate and dialyze as above, and use 10 μl as the template for a PCR reaction, as in Protocol 3.1.

Protocol 4.4 Virtual northern blot

EQUIPMENT

Nylon+ membranes
Oven
Hybridization oven

MATERIALS AND METHODS

1. Run the ladder of SIP-PCR products on a nondenaturing agarose gel (see Section 2.9).
2. Rinse the agarose gel in distilled water.
3. Wash the gel in 1.5 M NaCl, 0.5 M NaOH for 30 min with gentle rocking.
4. Rinse in distilled water.
5. Wash the gel in 1.5 M NaCl, 1 M Tris base, pH 8.0, twice, 15 min each time.
6. Transfer the DNA within the gel onto a nylon+ membrane using capillary action (see Section 3.2) using 3 M NaCl, 0.3 M sodium citrate (20× SSC) as transfer buffer overnight.
7. Fix cDNA onto the membrane by baking at 70°C for 1 h.
8. Prehybridize in 20× SSC containing 25% v/v formamide and 1× Dehardt's solution (0.4 g l^{-1} Ficol, 0.4 g l^{-1} polyvinyl pyrrolidine, 0.4 g l^{-1} bovine serum albumin, fraction V, plus 0.1% (w/v) SDS and 0.1 mg ml^{-1} heterologous DNA final concentration) for 30 min at 42°C.
9. Add labeled probe (see Section 3.5) and hybridize for 1 h at 42°C.
10. Wash twice at 50–68°C for 10 min depending upon experience with your probe in 0.2× SSC (30 mM NaCl, 3 mM sodium citrate), pH 7.0, containing 0.1% w/v SDS.

Differential display, subtractive hybridization, amplification suppression and SAGE techniques for measuring gene expression

5

5.1 Introduction

I have discussed at length methods for determining relative amounts of a specific and global RNA species in different preparations of total RNA from cells or tissues treated in different ways. However, all the techniques described so far require varying degrees of prior knowledge. In the case of northern hybridization or RT-PCR, where the abundance of an RNA with a specific sequence is being measured, you need to know at least part of the sequence of the gene encoding the RNA in the organism of study, or of a highly homologous gene in a closely related organism. This is because sequence specific probes require specific oligonucleotide primers in order to effect their production by DNA polymerase I *in vitro*, so at the very least, the sequence of the primer binding site (perhaps only within a homologous gene) must be known beforehand. In the case of RT-PCR, two, sequence-specific primers are needed, each to define one end of the sequence to be amplified and so quantified. In the other technique described, that of array-based hybridization, the sequences of a number of genes, and in many cases, of the entire genome, must be known in order to synthesize DNA probes for immobilization onto the array. But what happens if you want to find out which genes are differentially expressed in cells growing in one comparator physiological state compared with those growing in another without knowing which genes to look for, and without having the entire genome sequence, or indeed, any sequence information? The techniques described in this chapter all have the common aim of finding RNA molecules that are present in one total RNA preparation, but absent in another. The total RNA preparations might each come from a different cell type, or from the same cell type, but growing in a different physiological state. These approaches therefore do not really allow quantification of the amounts of an RNA species in the two comparator total RNA preparations (though they can help in this regard – e.g. SAGE, see Section 5.6) and are not really devised for measuring gene expression *per se*. Instead, they might be considered as methods which ask a binary question; is the RNA there, or is it not? Of course, a truly binary gene expression result is very hard to find because there is not really any such thing as a promoter that is absolutely turned off. However, in many cases, the difference in expression of a given gene between two comparator physiological growth conditions is large enough that any quantification of RNA produced from the gene might as well be binary. Hence these techniques are routinely and successfully used to characterize large gene expression differences between comparator physiological conditions. Absolute differences in RNA level can be determined using RT-PCR or northern hybridization in order to learn more about the actual expression differences once a significant difference has been highlighted.

5.2 Determining differences between genomic complements

Methods for determining differences in RNA complement between two comparator total RNA preparations have been derived from techniques devised initially for comparative genomics, where differences in genomic complement between two comparator genomes are determined (Avison, 2004). The question here is: which genes are only found in one of the two comparator genomes (i.e. a truly binary situation); and the result can inform the scientist of possible reasons why two comparator organisms express different phenotypes. In fact, when used for gene expression studies, these comparative genomics techniques simply substitute looking for differences in genomic DNA complement for looking for differences in total cDNA complement between two comparator cDNA pools. Each cDNA pool is produced from a different total RNA pool using reverse transcriptase and either a polythymine primer (i.e. to reverse transcribe polyA+ mRNAs) or a mixture of random hexamers as primer (see Section 4.9). The added advantage of these techniques is that they do not just inform you that some cDNAs or genomic DNAs are present in one comparator and not in the other, they highlight the actual DNA molecules that differ in their abundance so that they can be cloned and sequenced. Hence, they have the potential to generate considerable amounts of information.

Comparative genomic techniques for determining the differences in genomic complement between the unsequenced genomes of two comparator organisms are broadly divided into three types: differential display, subtractive hybridization and amplification suppression.

Differential display

Effectively, all the DNA in a genome is digested by a single restriction enzyme into a pool of differently sized fragments. The fragments are then separated by gel electrophoresis and the bands are visualized (displayed) by staining the DNA within them with ethidium bromide or some other DNA intercalating dye (Malloff *et al.*, 2001). The pattern of bands depends on the complement of genes in the genome, thus if two digested genomes from two comparator organisms having subtly different genomic complements are separated side by side on a gel (*Figure 5.1*), the banding patterns may appear different, and bands appearing only in one profile, but not in the other can be excised from the gel, the DNA within them eluted and cleaned, ligated into a cloning vector and, ultimately, sequenced. There are significant technical problems with this approach. First, there are so many fragments produced that it is very difficult to separate them. For example, the *E. coli* genome would produce approximately 1100 fragments using a restriction enzyme that recognizes a specific six-base pair sequence. Thus finding differences in fragment banding patterns on a gel is like finding a needle in a haystack. One way around this is to use two-dimensional separation of the fragments, so that they are more likely to form discrete bands/spots, though this is far from ideal. The second problem is that point mutations affecting restriction enzyme recognition sites result in so-called restriction fragment length polymorphisms (RFLPs) each of which manifests itself as two apparently missing bands and one apparently extra band in one of the pools of comparator genomic DNA fragments. Thus differential display has a very high false positives rate, which leads to a significant amount of follow-up work.

Subtractive hybridization

This technique, pioneered by Britten and Kohne (1968) is a more robust technique for comparative genomics than differential display. The two

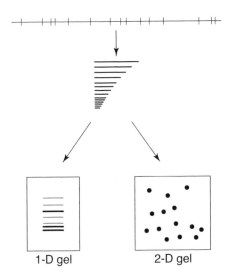

Figure 5.1

Differential display. Genomic DNA is digested with a restriction enzyme and the fragments separated by 1-dimensional or, more usually, 2-dimensional gel electrophoresis and the bands/spots stained with a DNA intercalating dye. This gives a fingerprint for the particular genome, and different genomic fingerprints can be compared.

comparator genomes are digested with the same, blunt-end cutting, restriction enzyme to create two pools of fragments. All the restriction fragments within the pool from one comparator genome are end labeled with biotin (*Figure 5.2*). These are referred to as the 'driver' fragments; the fragments from the other comparator genome are referred to as the 'tester' fragments; these are not labeled in any way. The tester and driver fragments are then mixed, with the driver fragments being provided in large excess. The mixture is heated to melt the double-stranded fragments, and then cooled. Since the driver fragments are in excess, it is far more likely that hybridization between an individual, single-stranded driver fragment and a complementary single-stranded target fragment will occur than that two strands of a target fragment will re-anneal. Thus, the only re-annealing of tester fragments that will occur is with fragments that are unique to the comparator genome from which the tester fragments were generated. All other fragments will either represent hybrids between a driver and a tester, or will represent driver fragments. All biotin-labeled fragments (i.e. all those that

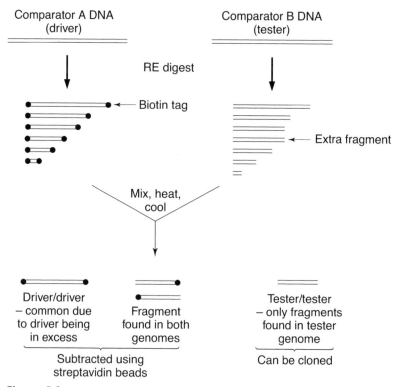

Figure 5.2

Subtractive hybridization. Restriction fragments from the driver genome are biotinylated and mixed with nonbiotinylated, but otherwise equivalent, restriction fragments from a tester genome. The driver is provided in large excess, so following hybridization, any over-abundant driver DNA, together with hybrids of fragments found in both comparator genomes can be subtracted using streptavidin-coated Sepharose beads. The only fragments remaining will be those representing DNA sequences unique to the tester genome.

have at least one driver strand within them) are then removed (subtracted) using streptavidin-linked beads. The only fragments remaining in solution will be fragments that represent genomic sequences unique to the tester comparator genome. These restriction fragments can be ligated into a cloning vector, and the insert can be sequenced. This approach is far better than differential display, since there is positive selection of sequences present in only one of the comparator genomes, rather than negative selection (i.e. looking for gaps in an overcrowded gel lane). Furthermore, RFLPs will not create false positives using subtractive hybridization, since hybridization between individual strands of driver and tester fragments does not require that the lengths of the fragments are the same.

Amplification suppression

Genomic DNA from both comparators is digested with the same restriction enzyme, which is designed to produce a single-stranded overhang at each end of each fragment (a sticky end). Both ends of each fragment from one comparator – again called the driver – are ligated to a single-stranded linker sequence containing a PCR primer binding site (*Figure 5.3*). The fragments from the other comparator (the tester) are also ligated to a linker, but this carries a binding site for a PCR primer with an entirely different sequence. The tester and driver fragments are then heated and mixed, with the driver being provided in large excess. As with subtractive hybridization, the excess driver fragments mean that upon cooling the mixture, hybrid products will be formed if a fragment with a particular sequence is present in both driver and tester samples. A few sequence differences between the tester and driver strands (i.e. point mutations) will not prevent hybridization. The hybrid fragments will therefore have a different PCR primer binding site at each end. The remaining, re-annealed driver fragments will have the driver-specific PCR primer binding site at both ends. Fragments that are unique to the tester pool will be the only fragments to re-anneal in such a way that the tester-specific PCR primer binding sites are found at both ends. Therefore, when PCR is performed using the tester/driver mixture as template and a single tester-specific PCR primer, none of the hybrid fragments, or the driver fragments will be amplified (i.e. amplification will be suppressed by the absence of appropriate primer binding sites at one or both ends). On the other hand, fragments that are unique to the tester pool, representing DNA sequences that are unique to the tester comparator genome, will be amplified successfully and can be cloned and/or sequenced. The amplification step makes cloning more straightforward, but the use of linkers makes the technique more difficult than subtractive hybridization.

5.3 Differential display techniques for measuring gene expression

The aim of differential display for RNA sequences is to convert every RNA into a cDNA, and then to display these cDNAs on a gel. The procedure is repeated with a different preparation of RNA, and the two cDNA lanes are compared and bands unique to one comparator are cut out of the gel and cloned. During cDNA production, labels can be incorporated, making

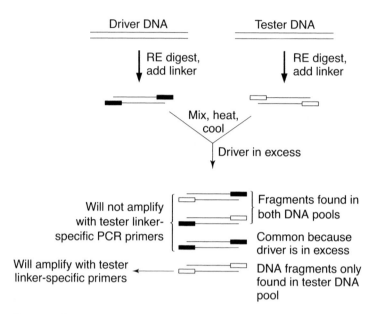

Figure 5.3

Amplification suppression. Driver and tester genomic DNAs are digested with a restriction enzyme (RE) and the ends ligated to single-stranded linker sequences. The two pools are mixed, with the driver DNA being in excess, and hybridization results in a pattern of products where only those tester-specific DNA sequences will have tester-specific PCR primer binding sites at both ends. Hence, using PCR and a primer that targets the tester-specific linker, only these tester/tester hybrids will amplify. Amplification of the other hybrids will be suppressed.

visualization easier. The real problem with this technique is that large numbers of similarly sized cDNAs will be produced, meaning that the displayed bands will overlap, making differences almost impossible to locate, even when using two-dimensional gel electrophoresis. Hence the variations on this theme described below are all designed to reduce the number of cDNAs produced and displayed, so differences in banding pattern are more likely to be seen.

RNA arbitrarily primed PCR (RAP-PCR)

The aim here (McClelland *et al.*, 1994) is to dramatically reduce the number of bands displayed in each, but to perform a large number of display experiments for a pair of comparator total RNA pools, so covering all possible cDNAs overall (*Figure 5.4*).

The procedure works by having a bank of random primers. Each experiment uses a different primer from the bank. The primer is used to prime cDNA production using reverse transcriptase, at low temperatures to maximize the number of cDNAs produced (this temperature can be changed in order to get the required number of products). The cDNA pool is then used as template for a PCR reaction using two random primers (one

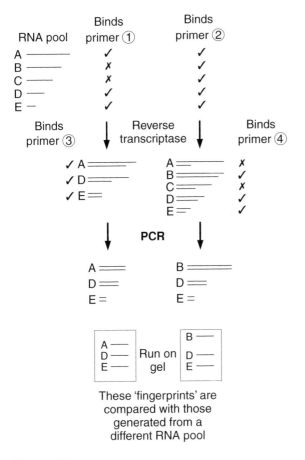

Figure 5.4

RAP-PCR. Several aliquots of an RNA pool are converted into cDNA, each using a different random primer. Not all RNAs will be converted to cDNA using a given primer (e.g. primer 1 in this example). These cDNA aliquots are then further divided into aliquots, and each is subjected to PCR using a pair of random primers (one of which will be the one used for reverse transcription). The end result is a number of different random cDNA fingerprints for the RNA sample, which can be visualized on a gel. The collective fingerprints from different RNA samples are compared in order to locate differences in RNA complement.

of which is the same as that used as the reverse transcription primer). Thus only a small number of cDNAs will have, by chance, a binding site for the second PCR primer in the opposite orientation to the first, resulting in a PCR product, and the sizes of the PCR products generated will span a wide range. Accordingly, when separated using an agarose gel, only a small number of bands will be visible for each preparation of total RNA and each pair of random primers. Thus it is possible to look for differences in banding pattern (indeed, bands don't have to be entirely absent in one comparator, just significantly reduced in intensity). The differently displayed cDNA bands can then be sequenced to reveal the identity of the differentially expressed genes in the two organisms.

To get a full screen of the entire RNA pool, a number of different pairs of primers will be required, each used in a separate display experiment. An 8 base-pair sequence appears approximately every 65 000 base pairs in DNA (assuming the composition is entirely 25% of each nucleotide). Since an RNA on average is approximately 1000 base pairs in a bacterial cell, the use of a single, random, 8 base-pair reverse primer will convert about 1.5% of the RNAs into cDNAs having an average length of 500 base pairs; PCR using, in addition, a second random 8 base-pair primer will amplify around 0.75% of these cDNAs. The maximum possible number of RNAs in a bacterial cell (i.e. the number of genes in the genome) is around 1000–6000 depending upon the organism. Thus the use of a pair of 8 base-pair random primers for RAP-PCR will result in the amplification of less than one product per total RNA pool. However, the use of low stringency conditions for reverse transcription and PCR will promote nonspecific primer annealing and will result in the same number of products as when using a 6 base-pair random sequence, (30–200 at high stringency, which is probably ideal). Thus, when using random 8 base-pair primer pairs and low-stringency conditions, around 30 pairs of primers will be required, each in a separate display experiment, in order to display cDNAs from all possible RNAs. To reduce the number of primer pairs required still further, it is possible to determine the base composition of coding DNA in a given genome prior to differential display, and so facilitate the preferential use of random primers that will anneal to sequences that are over-abundant in coding DNA, and therefore RNA. This approach is called 'targeted display'.

Differential display RT-PCR (DD RT-PCR)

This approach is in many ways similar to RAP-PCR. The major difference is that the reverse transcription primer is a polythymine primer, meaning that is only suitable for conversion of polyA$^+$ mRNA molecules into cDNAs (Liang and Pardee, 1992). In order to reduce the number of cDNAs produced, the polythymine primer has at its very 3' end either an adenine, guanine or cytosine. This means that the oligonucleotide will only prime reverse transcription of mRNAs in which immediately 5' proximal to the start of the polyadenine tail is found a thymine, cytosine or guanine base, respectively (*Figure 5.5*). Thus using each primer, approximately one third of all the mRNAs will be reverse transcribed. From this point, the procedure is exactly as for RAP-PCR, with each cDNA preparation being divided into aliquots, and with PCR being performed using the appropriate polythymine primer, and a different random primer for each aliquot. Thus for each aliquot, a different pattern of RT-PCR products will be seen, and comparison of the patterns obtained using the same pair of primers for two different polyA$^+$ mRNA preparations is used to locate bands unique to one, or having a significantly enhanced intensity in one.

Restriction display PCR (RD-PCR)

In this method polyA$^+$ mRNA is used as the starting product and is reverse transcribed using polythymine primer. From this first-strand cDNA

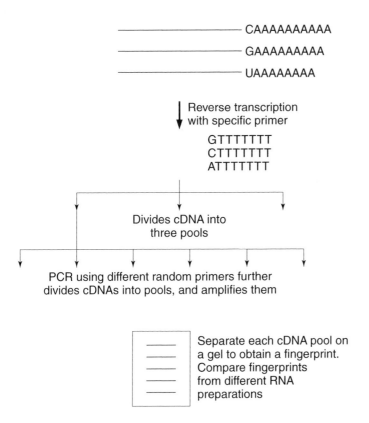

CAAAAAAAAAA
GAAAAAAAAA
UAAAAAAAA

Reverse transcription
with specific primer

GTTTTTTT
CTTTTTTT
ATTTTTTT

Divides cDNA into
three pools

PCR using different random primers further
divides cDNAs into pools, and amplifies them

Separate each cDNA pool on
a gel to obtain a fingerprint.
Compare fingerprints
from different RNA
preparations

Figure 5.5

DD RT-PCR. The principle is the same as with RAP-PCR, in Figure 5.4. However, in this case, the first aliquotting step divides mRNAs based on the nucleotide immediately before their polyadenine tails. Hence, following cDNA production all mRNAs will be reverse transcribed, and only three aliquots will be generated (cf. rather more with RAP-PCR. The second stage of breaking up the cDNA pool uses PCR with a group of different random forward primers, each tried on a different aliquot of cDNA. The reverse primer used for PCR is the reverse transcription primer. Hence as with RAP-PCR, the end result is a collection of different fingerprints for each RNA sample which are consistent from one experiment to the next if the same primers are used. The fingerprints generated from different RNA preparations can be compared.

product, a second strand is synthesized using RNase H and DNA polymerase I, as set out in Section 3.12.

Next the double-stranded cDNA pool is incubated with a restriction enzyme, which will cut at some position within some of the cDNAs leaving an overhang, or 'sticky end' at the 5' end of the cDNA with respect to the coding sequence. By no means all cDNAs will be cut, so a significant fragment of the cDNAs will remain blunt-ended at their 5' ends. Different restriction enzymes can be used, each cutting a different fraction of the cDNAs, creating a number of aliquots of the cDNA, each containing a group of sticky-ended cDNAs. It is possible that some of the cDNAs will be cut twice by a restriction enzyme, producing two or more products, one with a sticky end at the 5′ end and another (or possibly more than one) with sticky 3′ and 5′ ends. Each cDNA preparation that has been cut with each restriction enzyme is mixed with a specific linker sequence that has a sticky end which is complementary to the sticky end produced by the restriction enzyme. Hence different linkers will be used when different restriction enzymes have been used. The linker sequence carries a PCR primer binding site. Addition of DNA ligase will ligate the linker to both ends of the cDNA fragments (i.e. to both ends if the fragment results from two or more cuts within a cDNA). Finally, PCR is performed with primers complementary to the linker sequence. Thus cDNAs with a linker and an adapter, and those fragments with a linker sequence at both ends will all be amplified and form discrete bands that can be displayed (*Figure 5.6*). This method is much less popular than DD RT-PCR or RAP-PCR. This is due to the large amount of work required, the poor reproducibility of the technique (due to the variable efficiency of DNA ligase), and the greater number of products produced in each experiment, meaning overcrowded gels and results that are difficult to interpret.

5.4   Subtractive hybridization techniques for measuring gene expression

The aim of subtractive hybridization is to compare two pools of DNA (or cDNA if made from RNA by reverse transcriptase) referred to as the 'tester' and 'driver' pools. Both strands of each driver DNA molecule are labeled such that they can be recognized and removed (subtracted). The tester DNA molecules are not labeled. The procedure works by mixing the tester and driver pools, with the driver being in large excess. The mixture is heated to melt all the strands and then cooled to allow re-annealing/hybridization. Hybridization between complementary tester and driver strands occurs if the same cDNA is present in both pools; that is, the gene is expressed at similar levels in the two growth conditions used to isolate the two total RNA samples, and so generate the two cDNA pools. The presence of a single driver-derived strand in such hybrids should be sufficient so that the label targets this hybrid for subtraction. Of course, since the driver cDNA is in excess, there will be a lot of re-annealed driver cDNAs in the mixture, and there may be some driver cDNAs that do not have complementary cDNAs in the tester pool. All these re-annealed cDNAs will be labeled at both ends, so will be subtracted along with the hybrids. Therefore, this method cannot locate genes that are expressed in the driver growth condition but not in the tester growth condition. It is only capable of locating genes expressed solely in the tester condition, since the resultant unique tester cDNAs will re-anneal and will consequently not be labeled for subtraction, and will remain in solution following subtraction. In order to find reciprocally expressed genes, therefore, the experiment must be repeated a second time, with the driver and tester RNA pools being treated oppositely. The methods outlined below refer to different methods of subtraction, and how the appropriate tag is applied to the driver cDNA pool, so that it is targeted for the subtraction process.

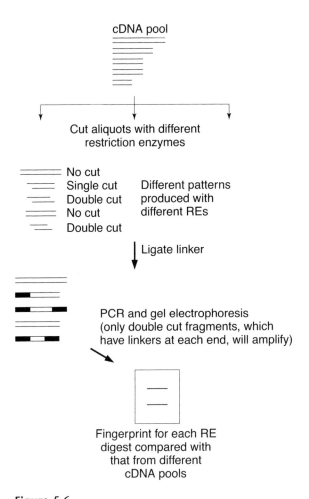

Figure 5.6

RD-PCR. As with RAP-PCR and DD RT-PCR, the aim is to divide the cDNA pool into various aliquots, each with a particular fingerprint. The aim here is to use a number of different restriction enzymes to digest aliquots of a cDNA pool. The display portion locates only those restriction fragments that have been cut at both ends, since only these will ligate to a linker sequence at both ends, and only these will be amplified by PCR using linker-specific primers. The end result is a fingerprint for each restriction enzyme. The collective fingerprints from different cDNA preparations can be compared.

Standard subtractive hybridization

Standard subtractive hybridization for comparing gene expression profiles is usually performed in the same way as subtractive hybridization for comparative genomics. The obvious additional step is that the two total RNA preparations need to be converted into cDNAs. cDNA production usually involves a polythymine primer, if polyA$^+$ mRNA is being used, or a random hexameric primer if not. The first-strand cDNA is converted into second-strand cDNA using RNase H and DNA polymerase I (Section 3.12). For efficient subtraction, at least one end of each strand of each driver cDNA must be tagged with an appropriate label. In this case, the label is a biotin tag. Tagging is achieved by converting the cDNAs into sticky-ended fragments using a restriction enzyme, and then each end is ligated to the same short, double-stranded linker sequence that carries a biotin tag at its 5' end. The tester is similarly treated, but ligated to the same linker sequence, but one that is not tagged with biotin (*Figure 5.7*).

When the tester and biotinylated driver cDNA fragments are mixed (with the driver being in excess) and the mixture heated and cooled to stimulate hybridization between cDNA strands common to both tester and driver pools, all hybrid cDNAs will have a biotin tag on the 5' end of at least one strand. All re-annealed driver cDNAs will be biotin tagged at both 5' ends. The only totally unbiotinylated cDNA fragments will be those exclusively from the tester cDNA pool – that is, representing those genes expressed only in the tester growth condition – and the unwanted cDNAs can be subtracted using streptavidin-coated beads, since streptavidin and biotin interact very strongly. The presence of the linker sequence at each end of the remaining

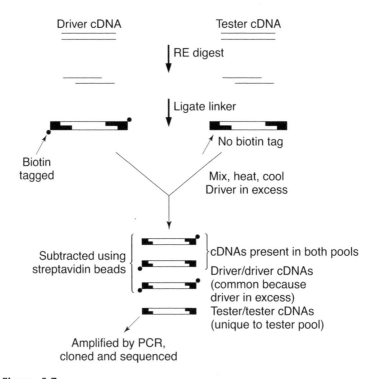

Figure 5.7

Reverse transcription subtractive hybridization. This follows the same principle to that set out in Figure 5.2, though cDNAs are used as starting material and not genomic DNA. The driver cDNA restriction fragments are ligated to a double stranded, biotinylated linker; the tester cDNA fragments are ligated to the same linker, but in this case, it is not biotinylated. Following subtractive hybridization using streptavidin beads, truly tester-specific cDNAs can be amplified using linker-specific PCR primers.

tester strands means that they can be amplified by PCR, cloned and sequenced.

Problems with this technique include inefficient ligation of the biotinylated linker to driver cDNA fragments, which makes for false positives, because unbiotinylated driver strands that hybridize to tester strands will not be subtracted and can be cloned. A second problem is that the streptavidin beads can become saturated with biotinylated fragments, meaning some driver/tester hybrids, or driver/driver homo-hybrids will not be subtracted, adding to the tally of false positives. To overcome the first of these potential pitfalls, it is advisable to perform PCR using primers designed to anneal to the tester-specific linker sequence following subtraction, but before cloning. The only amplification that will occur will be with re-annealed, linked tester cDNA fragments, none of the artifactually unsubtracted cDNA fragments will amplify. To overcome the second potential pitfall, it will be necessary to perform at least two incubations with streptavidin beads, and possibly more.

Enzymatic degrading subtraction

This is a complex procedure (*Figure 5.8*) that actually removes, by degradation, cDNAs that are found in both tester and driver samples, or are unique to the driver (Zeng *et al.*, 1994). First, cDNA is produced from both tester and driver mRNA samples using polythymine primers. If amplification of the cDNA signals is required, it is possible to use polyA SIP-PCR at this stage (Section 4.10). This is not always necessary, however. If this is not to be done, the cDNA must be made double-stranded as set out in Section 3.12).

Next, the Klenov DNA polymerase from *E. coli* is added exclusively to the tester double-stranded fragment. The 3′ to 5′ exonuclease activity of this enzyme means that, when incubated in the absence of nucleotides, both 3′ ends of the double-stranded tester cDNA fragments will be chewed back, making the 5′ ends overhang. After allowing this to happen, [α^{35}S]dNTPs are added, and are used by the 5′ to 3′ polymerase activity of the Klenov DNA polymerase to fill in the recessed 3′ ends, returning the cDNAs to a double-stranded state, and meaning that the 3′ ends become both radioactively labeled and contain thionucleotides. The driver cDNA pool is left unmodified, so the ends of its constituent fragments contain no radioactivity and no thionucleotides.

The modified tester and unmodified driver cDNAs are then mixed, with the driver being in large excess, and the mixture is heated to melt the strands and cooled to allow re-annealing/hybridization. If a cDNA is found in both tester and driver pools, then the end result will almost always be hybridization, and will result in a product where only one 3′ end is radioactive and contains thionucleotides. Alternatively, if a cDNA is unique to the driver pool, its strands will re-anneal, resulting in a cDNA where neither 3′ end is modified. If, on the other hand, a cDNA is unique to the tester pool, its fragments will re-anneal, and these will be the only products with thionucleotides contained within both 3′ ends.

All but these doubly thionucleotide-containing fragments are enriched by nuclease digestion in two ways. The mixture is incubated with DNA exonuclease III, which digests from the 3′ ends of any individual DNA strands

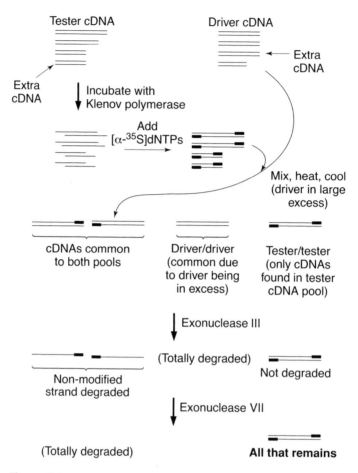

Figure 5.8

Enzymatic degrading subtraction. Rather than using biotin tags to subtract driver cDNAs (Figure 5.7) here, both 3′ ends of the tester cDNA fragments are converted into thionucleotides. This modification protects from exonuclease III attack, meaning that following hybridization with excess, unmodified driver DNA, only truly tester-specific cDNAs, which will contain thionucleotides at both ends, will be immune to the actions of exonuclease III. The exonuclease VII degradation step is solely to completely remove any single-stranded, originally tester cDNA-derived DNAs, which are left following exonuclease III attack. If they are allowed to remain, they could hybridize and lead to false positives. Hence the two exonucleases are added simultaneously, not sequentially as the figure implies.

that do not contain thionucleotides. Thus both strands of re-annealed driver cDNAs will be degraded. Furthermore, the driver-derived strand of hybrid driver/tester cDNA will also be degraded, leaving the tester strands from these hybrids undamaged, but single-stranded. Double stranded cDNAs resulting from re-annealing of strands found exclusively in the tester cDNA pool will contain thionucleotides at both 3′ ends, will not be degraded by DNA exonuclease III, and will therefore remain double stranded and intact. At the same time, the mixture is incubated with DNA exonuclease VII, which cleaves from the 5′ ends of all single-stranded DNA molecules. Hence all the thionucleotide containing tester-derived single-stranded cDNA fragments released when their partner non-thionucleotide containing driver-derived hybrid partner strands are degraded with DNA exonuclease III will be degraded by DNA exonuclease VII. The only fragments that will remain will be those exclusively found in the tester cDNA pool, which will be double-stranded, capped with thionucleotides and radioactively labeled. Hence they can be displayed on a gel, being visualized using autoradiography, and can be cloned and sequenced.

One major advantage of this method is that it can be used for sequential subtraction using different driver pools in order to answer the question, 'Which cDNAs are only found in the tester pool, and are not found in any of the driver pools?' If the cDNAs come from mRNA preparations, then the question really amounts to 'Which genes are expressed in this condition, but not in any of these other conditions?' The method is rarely used, because it looks daunting, yet it is surprisingly straightforward, and in many ways is more efficient than using biotin/streptavidin interactions to pull down hybrid products and so subtract them from the pool. The main reason for this is that the unwanted cDNAs are not just moved into solid phase, onto beads that can get saturated, leading to incomplete subtraction, but are completely degraded.

5.5 Amplification suppression techniques for measuring gene expression

Amplification suppression techniques are commonly used in the laboratory to get an idea concerning cDNA abundance differences between two pools, each derived from RNA isolated from cells or tissues growing in a different physiological state. The methods are quick, cheap and generally simple to interpret. The basis is the same in each: only amplify cDNAs that are unique to one of two pools. How the amplification of other cDNAs is suppressed depends on the particular method chosen.

Suppressive subtractive hybridization

The basic principle of this technique (Diatchenko *et al.*, 1996) is to locate cDNA sequences present only in the 'tester' cDNA pool, and absent in the 'driver' cDNA pool (*Figure 5.9*). The two cDNA pools are generated from RNA preparations using reverse transcriptase and either a polythymine primer or pool of random primers. The RNA is digested with RNase H, and a second-strand cDNA is synthesized using DNA polymerase I (Section 3.12) creating a blunt-ended double-stranded cDNA representative of each RNA in the preparation. These blunt-ended cDNAs can be used themselves, but because some of them are long they may be poor substrates for PCR, so commonly, the cDNA pool is incubated with a blunt-end cutting restriction enzyme, such as *Rsa*I, which cuts with a 4 base-pair recognition sequence, meaning that the fragments are on average about 250 base pairs long.

The pool of tester cDNA fragments is incubated with DNA ligase and a single-stranded linker sequence. The linker is unphosphorylated at its 5′

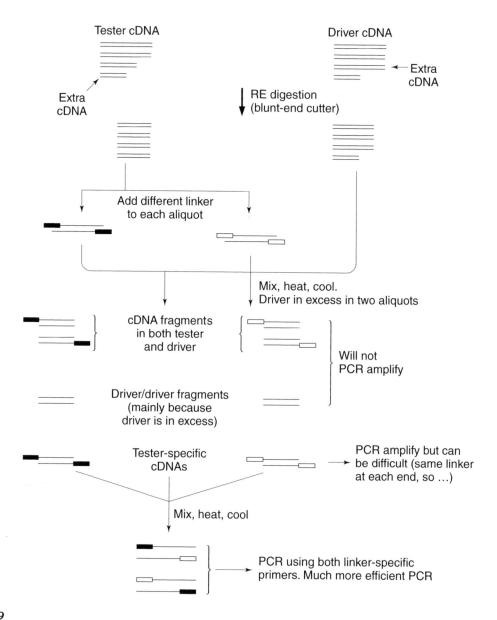

Figure 5.9

Suppressive subtractive hybridization. This amplification suppression technique involves the use of excess, untagged driver cDNA and a minority of linker-ligated tester cDNA. Because the driver is in excess, only truly tester-specific cDNAs will have a linker (and so a PCR primer binding site) at both ends. Therefore, only these will amplify using linker-specific PCR primers. However, unlike classical amplification suppression (Figure 5.3), the use of cDNAs in this approach has the potential to generate small linker-tagged fragments. These small fragments can be very poor PCR substrates because they can fold up due to sequence complementarity between the two identical linkers, one at each end. To get over this potential problem, the tester cDNA is divided into two aliquots, each is provided with a different linker. Therefore, following individual hybridization reactions, the two reactions are mixed and re-hybridized, producing a proportion of hybrids that are tester cDNA specific and contain a different linker at each end. PCR using two linker-specific primers would amplify these hybrids very efficiently.

end, so can only be ligated onto the 5′ end of each strand of the cDNA fragments, creating an overhang, and, most importantly, a PCR primer binding site. However, it should be remembered that the presence of an identical primer binding site at each end of a DNA molecule does not always make for very efficient PCR, particularly when short fragments of cDNA are being used, such as in this technique. This is because during PCR there is the potential for the single strands to fold up due to complementary base pairing between the PCR primer binding sites. To get over this, the tester cDNA pool is often divided into two portions, A and B, with each being ligated with a different linker sequence. In this way, PCR would be very efficient if a different PCR primer binding site were present at each end of the template cDNA hybrid (*Figure 5.9*).

The driver cDNAs are not ligated to linkers, and are used in large excess to create hybrid mixtures with each of the tester aliquots in exactly the same way that hybrid mixtures are made when performing subtractive hybridization. The double-stranded hybrids created when mixing each aliquot of tester with driver independently would be those carrying a portion-specific linker at one 5′ end and no linker at the other 5′ end. These hybrids represent those cDNAs present in both tester and driver cDNA pools; that is, genes expressed in equal amounts in both tester and driver physiological states. There would also be a large amount of driver/driver homo-hybrids formed, because the driver cDNA is in excess. However, because neither tester/driver nor driver/driver hybrids have PCR primer binding site containing linkers at both ends, they cannot be amplified by PCR with linker-specific primers. cDNA fragments found only in the tester pool will re-anneal carrying the same linker sequence at each 5′ end. Because of the possibility that such templates would not be efficiently amplified by PCR, the two separate portions of tester cDNA fragments, which have already been hybridized to driver cDNA, are themselves mixed. The mixture is heated and cooled, and now cDNAs unique to the tester pool can hybridize in such away that a significant proportion of cases, a different linker sequence will be present at each 5′ end. These hybrid fragments will amplify very efficiently when PCR, using both linker-specific primers, is performed. If tester/tester hybrids exist that have the same PCR primer binding site at each end then they might amplify, but if they did not then this would not be a disaster. However, amplification of all the other possible combinations of hybrid will be suppressed.

There is the potential for significant background amplification using this technique. This is because very low abundance cDNA fragments, even if present in both driver and tester pools, may not hybridize, making them possible candidates for PCR amplification, and so generating false leads. Because of this, it is always safest to use the PCR amplification product as template for a second PCR reaction using primers that bind to regions within the linkers that are 3′ proximal to the original PCR primer binding sites (i.e. a nested primer pair). This will significantly reduce the proportion of background amplification, since the chances that nonspecifically amplified sequences will carry binding sites for two independent PCR primers is very small indeed, yet all of the cDNA fragments unique to the tester pool, and so carrying linker sequences, will be amplified efficiently (*Figure 5.9*).

To further reduce the false positive rate, it is now commonplace to use subtractive hybridization to remove cDNAs common to both tester and biotin-tagged driver pools (with streptavidin beads) and then use the resulting subtracted mixture to perform amplification suppression using the same driver. Of course, even if this extra layer of complexity is employed, it is still imperative to check all putative differentially expressed genes really are differentially expressed using a truly quantitative measure of RNA levels such as northern hybridization or qRT-PCR.

Differential subtraction chain

One major problem when using amplification suppression or subtractive hybridization techniques is that the driver cDNA pool needs to be in large excess to the tester cDNA pool. This is often ignored as a problem in the literature, since these were essentially derived as comparative genomics tools, and genomic DNA is readily available. However, since all the cDNA must be prepared from RNA, there is an implicit requirement for large amounts of RNA to be purified from cells or tissues growing in a particular physiological state. This may not be simple. Differential subtraction chain (DSC; Luo *et al.*, 1999) allows for a 1:1 driver to tester cDNA ratio (*Figure 5.10*).

Tester and driver double-stranded cDNA pools are treated with the same restriction enzyme. The enzyme must produce sticky-ended fragments. The tester pool is mixed with a double-stranded, sticky-ended adapter sequence, whose overhang is complementary to the overhang on the sticky end of the cDNA fragment. Thus the same adapter is located at each end of the tester cDNA fragments which are thus blunt ended. The driver cDNA fragments are kept adapter-free.

The tester and driver cDNA pools are mixed, heated and allowed to hybridize. Because the concentrations of cDNAs in both pools are the same, there will be no overabundance of tester/driver hybrids representative of cDNAs present in both cDNA pools, as there is with amplification suppression and subtractive hybridization techniques where driver cDNAs are in large excess. Thus there will be almost equal amounts of tester/tester, driver/tester and driver/driver cDNAs for these equally abundant cDNAs. However, for cDNAs only found in the driver pool, only driver/driver hybrids will be seen, and for cDNAs found only in the tester pool, only tester/tester hybrids will be seen.

The first complication of DSC is the use of mung bean exonuclease. This enzyme degrades single-stranded DNA molecules, including overhangs at the ends of double-stranded molecules. Since driver/tester cDNA hybrids will have an overhang at each end on the tester-derived strand (because this tester cDNA will have been ligated to an adapter, which is not present on the complementary driver strand), the addition of mung bean nuclease will remove the adapter, making the driver/tester hybrid cDNA essentially indistinguishable from the driver/driver hybrid. Hence, effectively, more driver cDNA has been generated and tester cDNA has been removed, increasing the apparent driver/tester ratio. However, for tester-specific cDNAs (which is what you will want to locate) this step has no effect, because tester/tester hybrids are blunt ended, so are not affected by the nuclease.

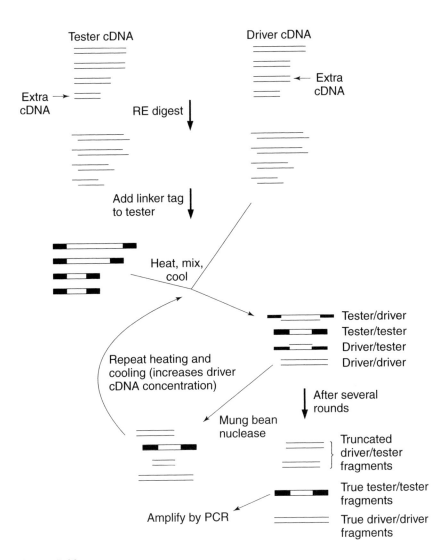

Figure 5.10

Differential subtraction chain. The aim of this method is to provide a way of suppressing amplification of tester/tester hybrids that happen to form in the presence of a 1:1 ratio of driver cDNA, but that are not true tester-specific cDNAs. In achieving this aim, the need for a dramatic excess of driver cDNA is negated. The principle of the approach is that on each round of the reaction, tester cDNAs which hybridize with driver cDNAs (i.e. representing those cDNAs that are found in both tester and driver cDNA pools) are effectively converted into driver cDNAs, therefore raising the apparent concentration of driver cDNA for another round of hybridization. In the end, only true tester-specific cDNAs come through the screen and can be amplified by linker-specific PCR. The linker which is added to tester cDNA differentiates the tester and driver cDNAs. Therefore, in hybrids, there will be a single-stranded linker sequence overhanging at both ends. These ends are degraded using mung bean nuclease, converting a tester cDNA into a driver cDNA for another round of hybridization.

The mixture is then heated and cooled to cause another round of hybridization (the heating step inactivates the nuclease, so it will not degrade single-stranded molecules before hybridization occurs). Now, upon hybridization, the proportion of spurious tester/tester hybrid formation (due to cDNAs present in both tester and driver pools) will reduce, because the driver:tester ratios for those cDNAs will have increased. Tester/tester hybrids representing true tester-specific cDNAs will be equally abundant as in the previous round. More tester/driver hybrids can be converted into driver/driver hybrids using mung bean nuclease, and the whole process can be repeated again and again until the amount of tester cDNA representative of cDNAs originally present in both tester and driver cDNA pools falls to a proportion that is so low that false positives will occur below a reasonable threshold. Three rounds of DSC are sufficient to mimic a 100-fold driver/tester cDNA ratio, which is sufficient for most purposes, yet it is very difficult to purify enough RNA to achieve this ratio without DSC.

The final stage of DSC is simply PCR, since amplification of driver/tester or driver/driver hybrids will be suppressed if adapter-specific PCR primers are used. Because there are often significant problems when using a single PCR primer, which binds at both ends of the sequence to be amplified, DSC is often run as two separate aliquots, each using a different adapter sequence. Prior to PCR amplification, the two aliquots are mixed, heated and cooled to allow hybridization of tester strands, with some of the products having a different adapter sequence at each end, and so amplifying efficiently using PCR with two different adapter-specific primers, as is the case for amplification suppression techniques (above).

Amplified differential gene expression

This is a procedure that combines elements of amplification suppression and differential display (*Figure 5.11*). It is also useful because it does not require anything more than a 1:1 driver/tester ratio (Chen *et al.*, 2001).

Both tester and driver cDNA pools are incubated with a frequent cutting, sticky-end producing restriction enzyme such as *Taq*I. Each cDNA could be cut into two fragments, one with the sticky end at the 3' end and a blunt end at the 5' end (type A fragments), the other with the sticky end at the 5' end and a blunt-ended 3' end (type B fragments). Alternatively, a cDNA could be cut more than once, creating type A and B fragments plus, in addition, any number of fragments with a sticky end at both ends (type C fragments), depending upon the exact number of times the cDNA is cut with the enzyme. There may also be some cDNAs without cut-sites for the restriction enzyme. These will remain blunt-ended at both ends and are called type D fragments. Because the enzyme used is a frequent cutting enzyme, type C fragments will be more common than type A and B fragments.

The pools of tester and driver cDNA fragments are incubated with DNA ligase and a double-stranded, sticky-ended adapter sequence, which has an overhang complementary to the sticky end generated by the restriction enzyme. A different adapter sequence is used to ligate to the driver fragments than is used to ligate to tester fragments. Remember that blunt-ended, type D fragments will not be linked to an adapter, and so they will

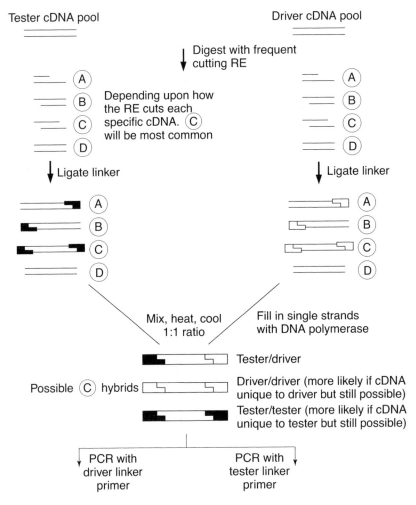

Figure 5.11

Amplified differential gene expression. This technique is not designed to completely suppress amplification of all cDNAs that are present in both tester and driver cDNA pools. It merely suppresses amplification of those cDNAs that hybridize together by chance when the tester and driver pools are mixed in a 1:1 ratio, heated and cooled. Clearly, simply by chance some non-tester-specific cDNAs will re-anneal. However, the approach simply amplifies any difference in the levels of specific cDNAs between the tester and driver pools. This is because the more tester or driver specific a cDNA is, the more likely it will be to re-anneal, and so be amplified using PCR and primers targeted to a tester or driver specific linker sequence. This amplification of gene expression differences can be visualized using array-based hybridization approaches, and whilst it will no longer be quantitative, this approach certainly enhances qualitative differences in gene expression.

not be visible using this technique. Furthermore, single blunt-ended fragments of types A and B will also not be visible, since adapters are required at both ends. Thus cDNAs without any, or with only one, cut site for the restriction enzyme used cannot be compared in tester and driver pools using this method. It is sensible, therefore, to repeat the whole process with at least one different restriction enzyme, since the chances that a cDNA will not have at least two cut sites for two frequent-cutting enzyme will be very slim.

The tester and driver pools are next mixed, heated and cooled to allow hybridization. The tester and driver cDNA total amounts must be almost identical. There will be three classes of type C hybrid fragments that might be formed. Type A, B and D hybrids will form in the same three ways, but will be ignored, since no measure of their abundance can be made. As with DSC, the hybrids formed will be either tester/tester, driver/tester or driver/ driver types. It is clear that if a cDNA is present in far greater amounts in one cDNA pool than the other, then homo-hybrids will be more likely to form than tester/driver hetero-hybrids. Thus if a cDNA is only present in the tester pool, then tester/tester hybrids will predominate, and if a cDNA is only present in the driver pool, then driver/driver hybrids will predominate.

To read the profile of predominant hybrids in each cDNA pool, the mixture is divided into two aliquots, and each is used as template for PCR with primers designed to amplify only one type of homo-hybrid. This can be achieved because a single primer is used, which binds to one adapter sequence. Hence, amplification of hetero-hybrid type C fragments, and of any type A, B and D fragments is suppressed, and all homo-hybrids will be amplified. The amplification products from each aliquot are run on a gel side by side, and the banding pattern used to find cDNAs that are abundant in the driver or in the tester pool, but rare in the other, using a classical differential display approach. Alternatively, the two mixtures of PCR amplicons can be used to probe arrays, allowing both quantification of differences, and identification of the cDNAs differentially expressed.

As a differential display technique, this method suffers from many of those drawbacks previously described. However, there are two main advantages. The use of a restriction enzyme increases the relative range of sizes of the fragments whose abundances are being compared, increasing the possible space between bands on the gel. The major advantage is that the use of amplification suppression as part of this technique actively amplifies the differences in cDNA complement, and dramatically reduces the signal intensity of cDNA fragments found at very similar levels in both cDNA pools. Indeed, when coupled with micro-array analysis, this is a very powerful technique for measuring gene expression levels, and overcomes one fundamental problem of micro-array analysis: that of dynamic range.

5.6 Serial analysis of gene expression

This is a powerful technique to profile the gene expression pattern of a cell or tissue. It does not require the same investment of resources as setting up a cDNA array for gene expression profiling, but does not work very well without knowing the genome sequence of the organism of interest, or at least a large fraction thereof. It is generally used as a high throughput screen for gene expression studies, and can be used to profile gene expression in both absolute (i.e. which genes are expressed at all and which are not) and semi-quantitative terms (Valculescu *et al.*, 1995). The aim is to create a SAGE profile, which is essentially a picture of which cDNAs are present in a pool derived from a total mRNA preparation, and how common each of them is. Thus, different SAGE profiles can be produced for mRNA preparations produced from cells or tissues growing in different physiological conditions, and can be compared.

The principle is simple (*Figure 5.12*). PolyA$^+$ mRNAs from a particular cell or tissue type growing in a particular physiological state are converted into cDNAs using a polythymine primer that is biotinylated on its 5' end. The cDNAs will be, in the vast majority of situations, significantly longer than 250 bp, so will be cut at least once with a 4-bp cutting restriction enzyme such as the sticky-end-producing enzyme *Nla*III (which will cut, on average every 256 bp). The cDNAs are digested with this so-called 'anchoring restriction enzyme', named because it anchors the sequence at the 5' end of each cDNA fragment that is biotinylated at its 3' end. Of course, the enzyme may well cut more than once in each cDNA, producing an array of fragments that have been removed from the biotinylated end. These are ignored, however, and the biotinylated fragments are purified using streptavidin-coated beads. The resulting purified fragments will be biotinylated at their 3' ends and will have a sticky end, with half an *Nla*III recognition site at their 5' ends. To this 5' end, a double-stranded linker sequence is ligated, with ligation efficiency being increased by using a 3' overhang on the linker, which is complementary to the 5' overhang on the *Nla*III-generated fragment. Addition of the linker does three things. First, it completes the *Nla*III recognition site. Secondly, it provides a recognition site for the so-called 'tagging restriction enzyme', *Bsm*FI. Third, it provides a PCR primer annealing site for amplification of fragments.

Like many type III restriction enzymes, the tagging enzyme, *Bsm*FI does not cleave at its recognition site, and in this specific case, cleavage produces a blunt end approximately 14 base pairs 3' proximal to the recognition sequence. Thus, the addition of the tagging enzyme to the linker-ligated, biotinylated cDNAs will result in the generation of a 'tag' from each cDNA. The tag is a short, blunt-ended, double-stranded DNA molecule which includes the complete linker sequence at its 5' end, and about nine bases of cDNA from the point at which the linker is attached at its 3' end. The resultant pool of tags is treated with DNA ligase. The 5' end of each linker sequence is chemically modified before its use in SAGE, in order to prevent ligation of the linker ends. Hence, in the presence of DNA ligase, there is no possibility of tags ligating head to head (i.e. linker end to linker end) or of them forming long chains; the only possible ligation event would be tail to tail, producing a so-called 'ditag' having a linker sequence at each end, and pointing inwards. The point of this, is that the pool of ditags can be amplified, en mass, using PCR with a primer that anneals to the linker sequence. This process makes it possible to use small amounts of cDNA, and so small amounts of mRNA to generate a SAGE profile for a particular growth condition.

The final stage of SAGE is to remove the linker sequence from each ditag using the anchoring restriction enzyme, *Nla*III. This produces ditags that have 5' and 3' complementary sticky ends. These sticky-ended ditags are

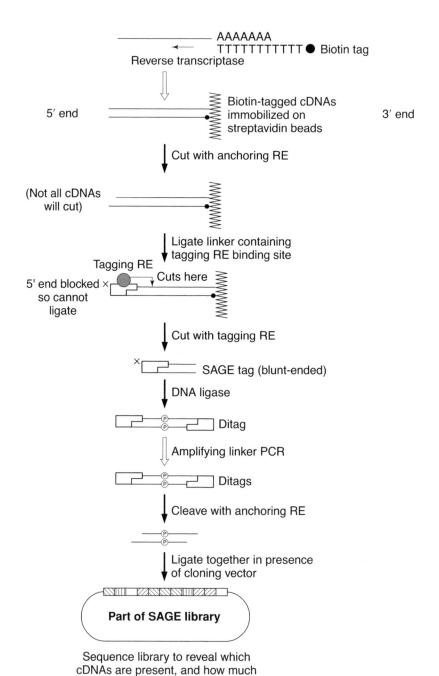

Figure 5.12

Serial analysis of gene expression. The text gives a detailed description of this technique. cDNAs are biotinylated during the reverse transcription step, anchoring them to beads making downstream reactions more easy to control, and washing steps more straightforward. Most cDNAs will cut with the anchoring restriction enzyme, but not all. Linkers are added to those that do separated from the cleaved linker sequences on the basis of size using polyacrylamide gel electrophoresis, and the ditags are gel purified. The addition of an aliquot of *Nla*III digested cloning vector and DNA ligase to the purified sticky-ended ditags causes them to ligate together into strings of ditags, and for some of these concatamers to be ligated into the cloning vector. The end result is a mixture of cloning vector molecules, each with a range of different ditags in different orders. The mixture is transformed into a competent *E. coli* strain, and resultant colonies are picked, plasmid DNA purified and the inserts sequenced. The screening and insert sequencing of large numbers of recombinant vectors will give an overview of the abundance of ditags, and from simple sequence analysis, using the *Nla*III sequence as a punctuation mark, the individual 9 base-pair tag sequences can be revealed. Through relating the tag sequence back to the genome sequence, you can learn which genes were represented in the cDNA pool, and so which were being expressed during the physiological conditions of the study. Furthermore, it can be calculated how often each tag is present in the SAGE profile, providing a reasonable estimate of its expression profile in different physiological conditions where SAGE profiles have been determined.

There are a number of possible problems with SAGE. First, very AT-rich tags are commonly lost during the PCR amplification phase, meaning that expression of genes where sequences 3' proximal to the 3' terminal *Nla*III cut site happen to be AT-rich (this is an entirely random event) is often underestimated (Margulies *et al.*, 2001). Type III restriction enzymes are notoriously tricky to work with, and *Bsm*FI is no exception. If incubation is not controlled, then it can cut at any position between 12 and 16 bases downstream of the recognition sequence.

A major problem with SAGE is that it is difficult to interpret when genome sequences, or at least EST cDNA library sequences are not available for an organism. It is possible to use a SAGE tag as a probe to screen a cDNA library in the hope that the entire cDNA can be cloned and sequenced, in order to determine what gene it represents. However, this approach is fraught with danger, since short probe sequences are notorious for binding non-specifically to target sequences, resulting in the generation of many false positive

---

cut, and the linkers attract a tagging restriction enzyme that cuts downstream of the anchoring restriction enzyme site. The result is a single blunt-ended tag representing the cDNA. Increasing the amount of a certain cDNA present in a cDNA preparation (if it is there at all, even) increases the amount of a particular SAGE tag that will be released using the process. The tags are ligated together to make ditags, amplified by PCR using primers that target the linker sequence (if necessary to amplify the signal) and then are freed as sticky-ended ditags using the anchoring restriction enzyme. These ditags are ligated into long strings which are ligated into a cloning vector. In the entire reaction, a lot of very different strings of ditags will be formed, but the proportion that a given tag makes up of the overall SAGE library will be proportional to the abundance of a given cDNA represented by the tag in the original cDNA pool. Sequencing large numbers of cloned strings of tags will eventually reveal not only which cDNAs are represented in the cDNA pool, but also their relative abundance. To make this work accurately, >50 000 tags must be sequenced.

clones. Even if a genome sequence is present, it is sometimes difficult to identify the gene that a SAGE tag relates to. A 9 base-pair sequence occurs, on average every 262 144 bases. This means that if a coding sequence is on average 1000 bases, then 1 in 262 coding sequences will have an identical 9 base-pair sequence. If there are 40 000 coding sequences in a genome, as there may well be in the human genome, then by chance, about 153 of these would share an identical 9 base-pair sequence at some point. However, things are slightly better than this, since the analysis can assume that the 9 base-pair sequence will include the *Nla*III cut site that is closest to the 3′ end of the coding sequence. On average, the 3′ terminal *Nla*III cut fragment will be 256 base pairs long, meaning that a 9 base-pair sequence will occur within the 3′ terminal fragment once in 1024 cases. Accordingly, if there are 40 000 coding sequences in the genome, 39 of them could share an identical 9 base-pair sequence within their 3′ terminal cut fragment. Because of this fundamental problem, the search for tagging restriction enzymes that give longer and so more discriminatory SAGE tags is ongoing.

SAGE is a technique that requires approximately 5 µg of polyA⁺ mRNA, which means around 500 µg of total RNA. This is a large amount, and is only really possible to achieve using large volumes of tissue culture cells or large pieces of tissue, and then the real problem of heterogeneity amongst cell types within the tissue emerges. Micro-SAGE is a variation on the SAGE theme (Datson *et al.*, 1999), which requires about 100-fold less RNA. The principle is the same as SAGE, but the biotinylated polythymine primer is anchored to the side of a PCR tube, meaning that the resultant cDNAs remain anchored. This reduces the potential for sample loss. The process also benefits from an additional ditag PCR amplification step. Other variants, for example SAGE-Lite (Peters *et al.*, 1999), actually amplify the cDNA copy number prior to digestion. This is through the use of M-MuLV reverse transcriptase for first-strand cDNA production, which carries terminal transferase activity, and adds a number of cytosines to the end of the cDNA. These are used as a binding site for a polyguanine primer, which, together with the polythymine reverse transcription primer, can be used to PCR amplify the cDNA in a form of SIP-PCR (Section 4.10). The risk of doing this, as with any cDNA amplification procedure, is that not every cDNA will amplify to the same extent. If there were sequence-dependent reasons for differential amplification, then this would affect the apparent stoicheometry of different tags in the resulting SAGE library, with potentially catastrophic results.

Further reading

Avison, MB (2004) Comparative genomics, digging for data. *Methods Mol Biol* 266: 46–69.

Lorkawski, S and Cullen, P (Eds) (2003) *Analyzing Gene Expression, A Handbook of Methods, Possibilities and Pitfalls.* Wiley VCH, Weinheim, Germany, pp. 478–622.

References

Britten, RJ and Kohne, DE (1968) Repeat sequences in DNA: hundreds of thousands of copies of DNA sequences have been incorporated into the genomes of higher organisms. *Science* 161: 529–540.

Chen, ZJ, Shen, H and Tew, KD (2001) Gene expression profiling using a novel method: amplified differential gene expression (ADGE). *Nucleic Acids Res* 29: e46.

Datson, NA, van der Perk-de Jong, J, van den Berg, MP, de Kloet, ER and Vreugdenhil, E (1999) MicroSAGE: a modified procedure for serial analysis of gene expression in limited amounts of tissue. *Nucleic Acids Res* 27: 1300–1307.

Diatchenko, L, Lau, YF, Campbell, AP, Chenchik, A, Moqadam, F, Huang, B, Lukyanov, S, Lukyanov, K, Gurskaya, N, Sverdlov, ED and Siebert, PD (1996) Suppression subtractive hybridization: a method for generating differentially regulated or tissue specific cDNA probes and libraries. *Proc Natl Acad Sci USA* 93: 6025–6030.

Liang, P and Pardee, AB (1992) Differential display of eukaryotic mRNA by means of the polymerase chain reaction. *Science* 257: 967–971.

Luo, JH, Puc, JA, Slosberg, ED, Yao, Y, Bruce, JN, Wright, TC, Becich, MJ and Parsons, R (1999) Differential subtraction chain, a method for identifying differences in genomic DNA and mRNA. *Nucleic Acids Res* 27: e24.

Malloff, CA, Fernandez, RC and Lam, WL (2001) Bacterial comparative genomic hybridization: a method for directly identifying lateral gene transfer, *J Mol Biol* 312: 1–5.

Margulies, EH, Kardina, SL and Innis, JW (2001) Identification and prevention of a GC content bias in SAGE libraries. *Nucleic Acids Res* 29: e60.

McClelland, M, Ralph, D, Cheng, R and Welsh, J (1994) Interactions among regulators of RNA abundance characterized using RNA fingerprinting using arbitrarily primed PCR. *Nucleic Acids Res* 22: 4419–4431.

Peters, DG, Kassam, AB, Yonas, H, O'Hare, EH, Farrell, RE and Brufsky, AM (1999) Comprehensive transcript analysis in small quantities of mRNA by SAGE-lite. *Nucleic Acids Res* 27: e39.

Valculescu, VE, Zhang, L, Vogelstein, B and Kinzler, KW (1995) Serial analysis of gene expression. *Science* 270: 484–487.

Zeng, J, Gorski, RA and Hamer, D (1994) Differential cDNA cloning by enzymatic degrading subtraction (EDS). *Nucleic Acids Res* 22: 4381–4385.

Measuring gene expression using reporter gene assays

6

6.1 Introduction

Up to now, the methods described in this book have been concerned with estimating gene transcription rates by measuring RNA levels, either absolutely or in one cell relative to another. You will now appreciate that isolating RNA and measuring RNA levels are tricky techniques, which are prone to variability and potential error. If you simply want to find out whether a single gene has its expression switched on or off under certain growth conditions, or in certain cells, then direct measurement of transcript levels can seem rather daunting. Is there an easier way? Well, the answer is 'yes', provided you are prepared to accept that the measurements you make might give a false impression concerning the absolute up- or down-regulation of transcription, but will give confirmation that up- or down-regulation occurs. The answer is to use reporter genes, which encode either easily assayable enzymes, or protein epitopes against which commercially available antibodies are available for western blot analysis. Hence, the ouput of gene expression measured is not the ephemeral, difficult-to-locate transcript, but the stable, easy-to-quantify reporter protein.

Promoter-probe reporters

When using reporter genes in the most simplistic sense, all that needs to be done is to splice the promoter for the gene of interest, whose level of expression is to be measured, upstream of a promoterless reporter gene in what is generally referred to as a promoter-probe vector and put the recombinant vector into the cell of interest. When cells are growing in the physiological conditions of interest, measuring production of the reporter protein will give a general feeling of the activity of the test gene's promoter, and so by inference, a measure of the expression of the test gene. Of course, since transcription initiation rate, which is dictated by promoter activity, is only one aspect of control of gene expression, and because transcriptional elongation rate, post-transcriptional and translational control events are very important, there is the potential for the use of promoter-probe reporter gene vectors to miss significant gene-specific transcriptional and post-transcriptional control events. General effects on transcription and translation rates, common to all genes in a cell at a given time are likely to be reported fairly accurately by promoter-probe reporter vectors, which means that their use, whilst having the disadvantage of only providing a semi-quantitative measure of gene expression, does allow the integration of specific control of promoter activity with a number of general gene expression control parameters that cannot be taken into consideration when looking at transcript levels alone. One aim of this chapter is to discuss some of the many and varied commercially available promoter-probe reporter vectors, and by discussing them, to help you understand how other such vectors you may come across, but that are not mentioned directly here, can be used.

End-product gene expression reporters

As just mentioned, transcription is only the first part of gene expression. Therefore, to find that the transcript of a gene is differentially abundant in one cell relative to another does not mean that the protein encoded by that transcript will show the same difference in abundance. Chapter 1 sets out a number of translational control points, any one of which could block the translation of an RNA, or indeed the translation of all RNAs in a cell. So anyone who finds transcriptional up-regulation of a gene should seriously consider checking whether there is a commensurate increase in protein levels. I will refer to this as assaying the end product of gene expression. Promoter-probe reporter vectors are unlikely to work for this purpose, because the reporter protein product is usually alien to the cell, and may well be uncoupled from post-transcriptional control of gene expression events. In some cases, it will be easy to quantify the end product of gene expression. For example, if the gene encodes a protein that has an easily assayable enzyme activity. Here, measurement of that activity allows a specific enzyme activity per unit of cell extract to be calculated for each growth condition, or cell type, and so a relative change in protein production can be determined. Specific protocols will not be given here for such assays, but you should check for the existence of specific assay methods in the literature prior to concerning yourselves with more generic, and possibly indirect methods described in this chapter. One method for quantifying a protein that does not have an assayable enzyme activity is the use of western blot analysis. Here, the aim is to use a protein-specific antibody to pick out and quantify the presence of a specific protein in the background of a cell extract containing many different proteins. The principles and practice of the western blot approach will be discussed in Section 6.3.

If there is no specific way of quantifying a protein encoded by your gene of interest, either by activity measurement, or by western blotting, then the best hope you have of being able to calculate a gene-to-protein expression ratio is to modify the gene such that it encodes a hybrid protein product, with the C-terminal portion being a reporter protein. This allows quantification of the hybrid protein product through measurement of the enzyme activity of the added on portion, or the reaction of this portion with a specific antibody in western blot analysis. This approach differs from the use of promoter-probe vectors, because the presence of the entire native promoter, untranslated region and most of the coding sequence for the gene, in addition to the reporter gene, means that specific post-transcriptional and translational control signals embedded within the sequence of the native test gene will exert their effects on hybrid protein production. So, hybrid protein level should mirror the level of the native test protein in the cell far more closely than can be expected when using promoter-probe reporters. It is not possible to be certain that the amounts of hybrid protein and native test protein are exactly the same, however, since alteration of a gene in this way may well change the kinetics of transcription and translation somewhat, but as a relative measure of protein production in two different cells, or the same cell growing in two different conditions, it is usually adequate. Therefore, the second aim of this chapter is to explain how reporter genes can be used to measure gene expression in a way that is as close to measuring native protein product levels (i.e. end-product gene expression) as is possible without actually doing so. The measurement of total native protein levels, known as 'proteomics' will be discussed in Chapter 7.

6.2   A guide to measuring enzyme activity

Enzymes are proteins that catalyze the conversion of a substrate or substrates into a product or products. They work by making the conversion more likely to occur (i.e. more energetically favorable) than it would spontaneously in solution. The principal way this is achieved is that the substrate-to-product conversion is broken up into energetically manageable chunks, occurring through an intermediate or series of intermediates, each being stabilized by intermolecular interactions with the enzyme. The overall rate of an enzyme-driven reaction is calculated ultimately by the rate at which substrate is destroyed or product accumulated, depending upon what is being measured. Whilst there may be very many steps to the reaction, each with different kinetics, we will only concern ourselves with basic, steady state enzyme assays, where the property 'V', the rate of substrate destruction or product accumulation (in moles per unit time) observed in the assay, is king.

Basic principles of steady-state kinetics

The rate at which an enzyme works can be limited by the concentration of substrate ([S]). This is a complex relationship, and depends upon how strong the interaction is between the substrate and the enzyme (i.e. this interaction can be rate limiting for the entire reaction). So in a steady-state enzyme assay, when [S] is low, V is likely to be dependent upon [S]. Above a certain value of [S], however, V will no longer be dependent upon [S], because some other kinetic component of the enzymatic reaction will become rate limiting. This means that in an assay that contains a particular, fixed concentration of enzyme ([E]), simply by increasing [S], V cannot become faster than the maximal attainable reaction rate, V_{max}. In the simplest case, a graph of [S] against V gives an inverse exponential curve which levels off at V_{max}.

To take this point one step further, in an assay of enzyme activity where [S] is not rate limiting, the rate of the reaction in the assay, V, will effectively be reporting V_{max}. If the assay were repeated with more enzyme, V would increase. This shows that V_{max} is not only dependent upon the affinity of enzyme for substrate, but on the absolute amount of enzyme in the reaction. Each mole of enzyme catalyzes the conversion of a certain number of moles of substrate into product per unit time, a property reflected by the catalytic rate constant (turnover number), k_{cat}, of an enzyme for a given substrate, which will not be dealt with further here. When performing reporter gene expression studies, we want to quantify and compare the amount of reporter enzyme in two extracts from cells grown in different physiological conditions. This will only work if we are measuring a V value which is as close to V_{max} as possible for the enzyme preparations in both assays (i.e. we use a situation where [S] is not rate limiting). Therefore the difference in V between the two assays will tell us what the difference in [E] in the two assays is, and so what the difference in reporter gene expression in the two growth conditions is. If we use a value of [S] that is limiting in either reaction, we risk incorrectly representing the difference in reporter gene expression between the two growth conditions. Worse still, if we are careless and use different [S] values for different reactions, then we may find that we get entirely irreproducible gene expression data.

Calculating the rate of an enzyme-driven reaction in a spectrophotometric assay

The units of V in a reaction are moles per minute (mol min^{-1}), or some derivative thereof (often nmol s^{-1}). But this only tells you how much catalysis is going on in the reaction. It does not tell you how much enzyme there is in the cell extract which was used for the assay. As discussed above, if we make sure [S] is not limiting in any assay, so we can assume we are calculating V_{max} in all assays, we can be confident that the difference in calculated V values between assays is dictated solely by a difference in [E].

V can be calculated using data from all sorts of different enzyme activity assays. Most commonly, one might measure the rate at which product accumulates. An excellent way to do this directly would be to have a product that absorbs light of a particular wavelength that is not absorbed by any other reagent in the enzyme reaction. Thus the observed rate of the enzyme reaction, V_{obs} would have units of change in absorbance (ΔAU) at a given wavelength per unit time. A linear correlation exists between the amount of light of a given wavelength absorbed by a solution of any compound and the compound's concentration. This is known as the extinction coefficient (E_0) and can be simply calculated. First, make a stock solution of the chemical by weighing out an appropriate amount of pure solid and dissolving it in an appropriate solvent. Next, make a number of doubling dilutions of the stock solution and determine the amount of light of a set wavelength absorbed by each relative to solvent, whose absorbance is set to zero. From this data, a graph can be drawn, whose gradient equals E_0 and has units of absorbance units per molar concentration (AU M^{-1}). The other factor that determines the amount of light absorbed by a solution is how far the light source has to travel through the solution. Absorbance readings are taken using a spectrophotometer, and these almost always take cuvettes with a 1 cm light path. So if you use this type of apparatus to determine E_0, the units will be AU M^{-1} cm^{-1}. E_0 values for many different compounds can be looked up, and will be provided with these units. Remember, therefore that if the light path of your apparatus is not 1 cm, you need to make an appropriate correction.

In many cases, it might not be possible to directly quantify the product of an enzyme using a spectrophotometer, perhaps because its absorbance spectrum is not significantly different from substrate. To get over this problem and determine V_{obs}, some enzyme reactions are performed in the presence of all the reagents and enzyme(s) needed for a coupling reaction to take place, which produces a product that does have a unique absorbance spectrum. For example, to measure the rate of pyruvate production by the enzyme pyruvate kinase, one might perform the reaction in the presence of the enzyme lactate dehydrogenase and the co-factor NAD^+, because lactate dehydrogenase will convert any pyruvate produced by pyruvate kinase into lactate plus NADH (1 mole of each per mole of pyruvate and NAD^+). Accumulation of NADH can be visualized because it absorbs light at 340 nm. Therefore the rate at which NADH accumulates is equal to the rate at which pyruvate is being produced by pyruvate kinase. Therefore, to convert the V_{obs} for NADH accumulation into V for pyruvate kinase, the E_0 of NADH at 340 nm should be used.

When using a 1 cm pathlength cuvette in a spectrophotometer, to convert V_{obs} (in ΔAU min^{-1}) into V (in mol min^{-1}) in all cases, use E_0 (in AU M^{-1}) to calculate the change in molar concentration per minute, $\Delta M/t$ (in M min^{-1}) ($= V_{obs}/E_0$) and then use the known reaction volume in the cuvette, v_r (in liters) to calculate V (in mol min^{-1}) ($= (\Delta M/t) \times v_r$, where t is the time, in minutes). So, in total:

$$V = \frac{V_{obs}}{E_0} \times v_r.$$

Non-spectrophotometric assays of enzyme activity

There are other ways of determining V_{obs} for an enzyme apart from using a spectrophotometer to measure absorbance of light by products or substrates. There might be more direct readouts. For example, the product of the reaction, or of a coupled reaction, may directly produce photons of light, which can be quantified (as photons per second) using a luminometer. Another way is to have a radioactive substrate that, when converted into product, has the radioactive atoms removed, meaning they can be separated from the remaining substrate and that remaining substrate can be quantified using a scintillation counter, or autoradiography to count the rate of radioactive particle emissions (as emissions per second).

The disadvantage of this type of measurement of enzyme activity is that it cannot be monitored in real time, which can be done with a spectrophotometer. Instead these are end-point assays. The enzymatic reaction is allowed to proceed for a set amount of time, something is done to the reaction to 'develop' it (in the case of the light production) or to remove the degraded radioactive portions of substrate, and the resultant mixture is subjected to quantification. The result of an end-point assay is the amount of product produced (or substrate degraded) in a given unit of time in the enzymatic reaction. But the problem is that if one cannot see the enzyme working, how does one know that the V_{obs} value obtained from a single end-point assay is accurate. Even if [S] is very high at the start of the reaction, so $V_{obs} = V_{obs(max)}$ for the enzyme, [S] will decrease during the reaction, meaning that it could well become rate limiting at some point, making $V_{obs} < V_{obs(max)}$, and even to the extent that [S] = 0, where $V_{obs} = 0$. Looking at the output of a spectrophotometric enzyme assay, one can see the rate of the enzyme start to decrease, and eventually to level off as [S] is reduced to a point where it is rate limiting, to a greater and greater extent, and then the reaction stops. It is then very simple to only use the gradient of the reaction before leveling off begins in order to calculate V_{obs} because at this point in the reaction, $V_{obs} = V_{obs(max)}$. For an end-point assay, it is a little bit of a stab in the dark, and it is very likely that V_{obs} will result in an underestimation of $V_{obs(max)}$ for the enzyme in the reaction, and worse, the proportion underestimated will be higher as the relative amount of enzyme in the extract increases (because the more enzyme, the faster [S] will decrease, and so will become rate limiting earlier in the reaction). To mitigate this potential pitfall, it is important to perform at least two parallel end-point assays with each preparation of enzyme, each proceeding over a different period of time. If the determined V_{obs} is identical for both assays, then it is clear that both have captured points in the linear range of the reaction, where [S] is not limiting. However, if the two V_{obs} values are different (with the V_{obs} for the longer assay being significantly lower than the other) then the experiment must be repeated using assays over shorter times until you are convinced that the reaction is linear. If you are having problems, then increasing [S] might help. Though be careful, because some enzymes are actually inhibited in the presence of very high substrate concentrations, so don't increase [S] too much.

The other major problem with these indirect end-point assays is that it is more difficult to calculate V from V_{obs} values obtained (i.e. there is not a simple extinction coefficient that is used). If one is using a luminescent product or radioactive substrate to measure the enzyme activity, then it is possible for you to calibrate the luminometer or scintillation counter yourself simply by providing the equipment with a number of solutions having known amounts of substrate/product in them, and using the same capture protocol that will be used in the enzyme assay to quantify the emissions from these standards. Hence a graph of substrate/product amount versus photons or radioactive counts per minute can be drawn, with the gradient being 'counts (or photons) per minute per mole of substrate (or product)', which is a constant that can be used to calculate the number of moles of substrate or product in the end-point assays and so (given it is known how long each assay was run for prior to the end point, and the starting amount of substrate and presumably product, which would be zero) one can extrapolate to the rate of enzyme activity, V, which is the number of moles of substrate or product converted per unit time in the assay.

Going from *V* to reporter enzyme expression level

Your enzyme assay (however it reports product accumulation or substrate depletion) will give you a *V* value with units of mol min^{-1} or some derivative thereof. You can directly compare *V* for two assays, A and B, each using a different cell extract in order to calculate the relative difference in reporter enzyme production in the two cell populations used to make the extracts. Simply divide V_A by V_B; if the result is 1, then there is no difference in gene reporter gene production in the two cell populations; if the result is $\neq 1$, then there is differential reporter enzyme production in the two cell populations. However, this simple comparison of *V* values can only be used if the amount of cellular protein added to each assay is the same. It is not good enough to use the same volume of cell extract, because it may be that one extract is a more dilute solution of cellular proteins than the other. So, if you want to come to any conclusion about the level of reporter enzyme production in the two populations of cells it is not sufficient to compare the enzyme activity (*V*) calculated from assays using cell extracts from the two populations, you must compare the specific enzyme activities in the two cell extracts themselves. Specific activity has units of moles substrate/product depleted/generated per minute per gram of total protein, or some derivative thereof (e.g. nmol min^{-1} mg^{-1}). In order to do this, you need to be able to measure the concentration of protein in each cell extract, (or at least to be able to roughly quantify it) and so calculate the amount of protein present in each enzyme assay. There are two basic ways of doing this: the direct measurement of protein concentration (see below) and the indirect measure of protein concentration. The indirect measure involves the assumption that each cell used to make the cell extract contains the same amount of protein. Therefore, if the density of cells used to make each of the cell extracts is known, a rough estimate can be made of the relative concentration of protein in each cell extract. To determine the density of cells used to prepare each cell extract involves counting the cells immediately prior to cell lysis. For cells in suspension, cell density can be estimated spectrophotometrically, the greater the cell density, the greater the turbidity reading obtained when passing light through the sample. Many different researchers will use different wavelengths of light for this measurement of 'optical density' (OD). Popular values are 420 nm or 600 nm, but it doesn't really matter, so long as you are consistent and always use the same wavelength for your experiments. When using a cell type for the first time, run a wavelength scan to find the wavelengths with the highest turbidity readings, and pick one of them.

The concentration of protein [P] in a cell extract (in mg l^{-1}) can be determined using a whole variety of protein assay methods including various commercially available protein assay reagents (see below). For calculation of specific enzyme activity (moles substrate/product converted per minute per gram of protein) in a cell extract, you need to know the *V* value (in moles converted per minute) calculated from an enzyme assay using a known volume of cell extract, v_e (in liters) having an accurately calculated protein concentration [P] (in grams per liter). Overall, the following formula is used:

$$\text{specific enzyme activity} = \frac{V}{[P] \times v_e}.$$

Factoring in the method for calculating V from raw assay data (above), the following formula is used to work from the change in absorbance versus time data generated from a steady state spectrophotometric assay of enzyme activity into specific activity of the enzyme in the cell extract used in the assay:

$$\text{specific enzyme activity} = \frac{\left(V_{obs} \middle/ E_0 \times v_r \right)}{[P] \times v_e}.$$

Assay of protein concentration

The Bradford dye assay (Bradford, 1976) is the most commonly performed assay of protein concentration. It is based on the equilibrium between three forms of Coomassie blue G dye. Under strongly acid conditions, the dye is most stable as a doubly protonated red form. Upon binding to protein, however, it is most stable as an unprotonated, blue form. Bradford assay reagents of various sorts are available commercially, and come as an acid-treated solution (so should be treated with caution). When mixed with protein, the solution turns blue, and the protein concentration in the sample is estimated by measuring the absorbance of the solution at 595 nm [P_{obs}], and can be accurately determined (i.e. in mg ml^{-1}) using a standard curve of known protein concentration (usually of a commercially available powdered source such as bovine serum albumin) versus absorbance. Other commercial protein assay reagents are available, which use different dyes.

Do we really need to do all these maths?

The answer is most definitely 'no'. For gene expression experiments, all we care about are relative amounts of reporter enzyme levels in two or more different cell extracts from cells growing in different physiological states. We don't care about the absolute number of units of enzyme in each cell extract, just how one relates to the other. As such, we can happily use V_{obs} values for enzyme assays, so long as we take into consideration the amount of cell extract in each assay, and we can just as well take [P_{obs}] values as well to calculate crude, but no less accurate relative specific enzyme activity values for each cell extract, which can be used to perform calculations that tell us all we need to know about relative levels of reporter enzyme production in each physiological state.

The equation to use in this case would be:

$$\text{relative specific enzyme activity} = \frac{V_{obs} \middle/ v_{e,1}}{[P_{obs}] \middle/ v_{e,2}}$$

where $v_{e,1}$ is the volume of cell extract used in the enzyme assay and $v_{e,2}$ is the volume of cell extract used in the protein assay. If $v_{e,1} = v_{e,2}$, then they can both be omitted from the calculation.

Alternatively, if pre-lysis cell density values are being used to estimate protein concentration in each cell extract, the following, more simple, equation can be used:

$$\text{relative specific enzyme activity} = \frac{V_{obs}\Big/v_{e,1}}{OD}.$$

This will only work if both the volume of suspended cells used to make each extract and the volume of buffer used to extract each population of cells are identical, otherwise corrections will have to be introduced into the calculation.

So if it can be broken down to something so simple, why did I spin it out? Well, it is important that you understand the basics of enzyme activity assays so that you can appreciate where things might go wrong. If you can be confident that V_{obs} is directly proportional to V_{max} and is therefore only affected by [E], and [S] is not limiting in any experiment, and further, that [P_{obs}] is in the linear range, then you will be able to produce accurate relative specific activity data that can be directly compared across a number of experiments run at different times. If any of the above does not apply, however, beware, because the data you get may be entirely irreproducible. The section on statistics (Chapter 8) may help you to devise methods for estimating how significant the differences in relative reporter enzyme production values you calculate may actually be.

6.3 Western blotting: a beginner's guide

Northern blotting has been dealt with in Chapter 3. Western blotting entails the same principle, but instead of visualizing a specific RNA sequence using a probe, the western blot is designed to visualize the presence of a particular amino acid sequence, and if done correctly, can be used to quantify how much of the amino acid sequence is present in an immobilized sample. The probe in a western blot experiment is an antibody raised against a particular amino acid sequence, which provides specificity to the reaction. In common with all blots, western blots start with the immobilization of the protein or mixture of proteins that is to be probed for the presence of a particular amino acid sequence onto a membrane. The most commonly used membrane material for western blotting is polyvinylidine difluoride (PVDF), which is a hydrophobic membrane, and associates with proteins through hydrophobic interactions. Nitrocellulose, which is also hydrophobic at high salt concentrations, is sometimes used for western blotting, but PVDF is far superior, since the salt concentration can be kept within physiological range, allowing efficient and strong target:probe interactions. As with northern blots, there are several ways in which proteins, or mixtures thereof, can be applied to the surface of a PVDF or nitrocellulose membrane. As a quick and dirty approach, dot blotting is adequate, where a drop of the solution to be tested is spotted onto the surface of the membrane. This can be used in a qualitative analysis aimed at determining whether a particular amino acid sequence is present within the sample or not, but it is not really suitable for quantification, since it is difficult to ensure that all the protein in the sample becomes attached to the membrane, without saturating its binding capacity. Furthermore, the presence of lots of different proteins, sandwiched together in close proximity, is likely to severely impair the ability to probe for the presence of the target amino acid sequence. Therefore, the prior separation of individual proteins in a mixed sample using electrophoresis (typically sodium dodecyl sulfate polyacrylamide gel electrophoresis (SDS-PAGE)), followed by transferring the ladder of protein bands onto a membrane using electro-transfer is the most common western-blotting technique. This approach reduces the chances both of saturation of the membrane and of interference between target and nontarget sequences for the probe. The aim of this section is to divide the western blot into its constituent parts and explain how each works, what can go wrong with it, and what can be done about it.

SDS-PAGE

The principles of gel electrophoresis were described in detail in Section 2.9. To recap briefly, molecules are passed through a matrix of strands, effectively working like a sieve, and the larger the molecule, the more friction there is as it passes through the matrix, causing its progress to be retarded. Therefore, after a given amount of time, small molecules will have passed further through the gel than large ones, resulting in physical separation. In order to force the molecules through the gel matrix, an electric current is used. Therefore, the molecules to be separated must be made charged, and all must have the same sign of charge. They become attracted to the electrode with the opposite charge, and since this is placed deliberately at the far end of the gel from the point at which they are loaded, this charge–charge attraction provides the force to drive the molecules through the gel. Another factor, which must be taken into consideration when separating molecules according to size, is that the molecules must all be denatured (i.e. all secondary structure within them must be broken up). This is necessary, because tightly folded molecules would be subjected to considerably less friction when passing though the gel matrix than would relatively unstructured molecules of the same, or even smaller size. Thus to separate predictably according to molecular mass, and not volume, denaturation is essential.

There are 20 (unmodified) amino acids used as building blocks for proteins, which are broadly grouped based on their physical properties. Proteins have a particular composition of amino acids that means they have a particular make-up of physical properties. The most important one of these when considering gel electrophoresis is the net charge of the protein at the pH at which the protein will be separated from others by electrophoresis. Some proteins will be negatively charged, some will be positively charged and some will be entirely neutral in charge. For electrophoresis to be successful, all the proteins must be made to have the same charge, and to this end, sodium dodecyl sulfate (SDS) is added to the mixture of proteins to be separated as part of a concentrated 'sample buffer'. It is a highly negatively charged molecule, and so it coats proteins with this negative charge. It also helps to cancel out charge–charge and hydrogen-bonding interactions that hold together tertiary structure within a protein, and also are important in forming protein complexes. Hence, as well as equalizing charge, SDS also acts as a denaturant. The final pieces of secondary structure are disrupted by the addition of a reducing agent within the sample buffer, usually beta-mercaptoethanol, which breaks sulfhydryl links between certain amino acids. The last constituent of the sample buffer is a dye, usually bromophenol blue, which allows the sample to be visualized when loading it onto the gel, and also acts as an indicator of how far the sample has progressed through the gel. The final act before loading the sample onto the gel is to heat it using a 95°C heating block or boiling water bath, since this heat results in complete denaturation of the proteins, and the SDS then gets to all parts of the denatured protein and prevents secondary structure formation once the protein mixture cools.

The basic constituents of the polyacrylamide gel are described in Section 2.10. The gel matrix associated with an SDS-PAGE gel does not materially differ from that used to separate small DNA and RNA molecules, though there are some subtle differences to the electrophoresis conditions. First, because proteins are smaller than nucleic acids, protein PAGE gels generally have a higher percentage of acrylamide (i.e. smaller mesh sizes in the matrix) than nucleic acid PAGE gels. The obvious difference is the presence of SDS, both within the gel and the running buffer, so that denaturation of proteins remains throughout the electrophoresis run. Other than that, the buffer is a simple Tris buffer at very slightly alkaline pH (see *Table 6.1* for gel and buffer

Table 6.1 Recipes for SDS-PAGE gels

Use of gel	Ratio of acrylamide (%) to bis-acrylamide (%)	Constituents
Separation of 40–160 kDa proteins	10:0.05	2.5 ml of 40% w/v acrylamide:0.2% bis-acrylamide stock 2.5 ml of 1.5 M Tris base, pH 8.8 containing 0.4% w/v SDS 4.89 ml of water
		To set, add 100 µl of 10% w/v APS (made fresh) and 10 µl of TEMED
Separation of 10–50 kDa proteins	15:0.4	5 ml of 30% w/v acrylamide 2 ml of 2% w/v bis-acrylamide 2.5 ml of 1.5 M Tris base, pH 8.8 containing 0.4% w/v SDS 0.38 ml of water
		To set, add 100 µl of 10% w/v APS (made fresh) and 20 µl of TEMED
Stacking gel	5:0.1	1.66 ml 30% w/v acrylamide 0.05 ml 2% w/v bis-acrylamide 2.5 ml 0.5 M Tris-HCl, pH 6.8 containing 0.4% SDS 5.73 ml water
		To set, add 50 µl of 10% w/v APS (made fresh) and 10 µl of TEMED

recipes). Another peculiarity of the protein PAGE gel is the use of a 'stacking' gel. In nucleic acid PAGE, wells are formed in the top of the gel matrix (the gel being run vertically) by placing a 'comb' into the gel before it sets. These wells are filled with sample. However, the sample therefore takes up a significant depth of gel (around 5 mm) prior to separation commencing, and this simple fact reduces the resolving power of the gel; molecules that happen to be 5 mm above other, identical molecules because of this well filling effect, will still be 5 mm above after the gel has been run. This minor reduction in resolution is not so much of a problem for nucleic acid gel electrophoresis, because the differences between molecules are large. However, for protein gels, extra resolving power is required, and is supplied by the use of the stacking gel. A stacking gel is identical to the main gel except that the mesh size is larger (i.e. the percent acrylamide and bis-acrylamide is lower). The stacking gel is poured above the main gel, once the former has set, and the wells are formed in the stacking gel. The samples are placed in the wells, and electrophoresis commences. The difference in position between identical protein molecules in the sample does not change as it passes through the stacking gel, but as the sample hits the main gel, there is a sudden reduction in progression rate for the front edge of the sample, allowing the top edge of the sample to catch up. Hence this sudden change in density reduces the apparent sample depth from >5 mm to around 1 mm, with a consequential increase in resolving power (*Figure 6.1*).

Pouring an SDS-PAGE gel

Pouring SDS-PAGE gels can be tricky unless you invest in dedicated gel casting and running apparatus, which I would definitely recommend. Perhaps the best known example is the Bio-Rad mini-protean system. It should be used as follows.

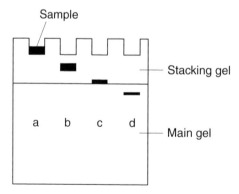

Figure 6.1

The use of stacking gels in SDS-PAGE. The sample is loaded into the well, and takes up more than 0.5 cm, when running through the gel, this would mean that every protein band would be spread out (lane b). However, when the sample running though the stacking gel hits the edge of the main gel matrix (lane c), it concertinas up, meaning that the thickness of each band running through the main gel is very much less than that running through the stacking gel (lane d), increasing the resolution.

Washing the glass plates

First, thoroughly wash the two glass plates that will contain the gel, using standard washing-up liquid in water. A foam rubber bung is excellent to scrub debris and old acrylamide from the surface of the plates. Rinse the plates well using distilled water then finally with 70% (v/v) ethanol in water. Dry them by propping them up (the gel casting block is an excellent support), with the bottom of each plate resting on a piece of paper towel. Finally, dry the plates thoroughly but gently with a piece of soft tissue, and inspect them carefully for chips or cracks, which may result in an imperfect seal and leakage of gel contents during casting. Also check for any bits of debris stuck to the surface of the plates, since these will inhibit polymerization of the gel. If the plates still appear dirty, go back and wash them again. It is not necessary to use gel coating reagents for dry SDS-PAGE gels, such as those used when working with more fragile gels such as denaturing sequencing gels. If looked after, gel plates can last for a long time, and can be re-used over and over again. Make sure you have a supply of new plates in stock just in case, however, which should be cleaned as above prior to their first use. Plates in service can be stored immersed in water with a little washing-up liquid. It is advisable to keep the small and large plates separate, so that you will notice if one size is running out before the other (the smaller plates are most likely to break). Once a gel has been run and discarded, rinse the gel plates under the tap, and place back into the washing-up liquid solution. There is no point thoroughly cleaning them at this point because you will need to do so again anyway when you come to re-use them.

The gel casting apparatus

The gel casting apparatus includes a plastic clamp (also used to run the gel) and a casting block. Two glass gel plates are placed into the clamp, with the shorter plate at the front, separated by a thin plastic spacer (usually, 0.5–1 mm). Make sure the gel plates are clamped in place absolutely level. Use the bench top to do this if it is level; check with a spirit level first. Put the clamp on the surface of the bench and slide the glass plates and spacers all the way to the bottom until each is flush with the bench top. Any slight gap at the bottom can result in leakage of the gel during pouring. The clamp is then tightened, though there is no need to over-tighten, which is likely to break the plates; a comfortable finger tightness is enough to prevent leakage of gel through the spacers at the side during pouring. Once the plates and spacers are positioned within the clamp, the whole lot is clipped into the casting block, forcing the bottom of each gel plate into a soft rubber block, therefore preventing leakage at the bottom of the gel during pouring.

Making and pouring the solutions

Whilst the gel plates are drying, the main gel solution (*Table 6.1*) can be made up at the acrylamide percentage, and acrylamide to bis-acrylamide ratio required, although the setting reagents, TEMED and APS solution (which must be made up fresh – that is, within a couple of hours of use –

each time it is used) should not be added until immediately prior to pouring. Make up all gel solutions in a clear plastic container and invert the solutions gently to mix. The presence of SDS will cause frothing if you mix too vigorously, which will result in bubbles in the gel. After adding the setting reagents, pipette the main gel solution into the gap between the two glass plates, 1 ml at a time gently and steadily, leaving enough room for around 0.5–1 cm of stacking gel and the gel comb above it. Pour the main gel whilst the casting block remains vertical and level (check the bench with a spirit level first). It is not necessary to tip the casting block to one side, or to pipette down the edge of the spacers, since the gel is fairly dense, and will sink well. The most likely problem here would be air bubbles being caught in the gel, though this can be avoided if the gel is poured gently. Once the main gel solution has been poured, watch the gel for a while to make sure it is not leaking from the bottom. If it is, then it may be possible to seal the leak with Vaseline, but this is often not satisfactory. Sadly, the best way of dealing with this is to start again. To make sure that the surface of the main gel is flat, the best way is to pipette a layer of water-saturated butanol, which will sit on the surface of the gel, leaving it perfectly flat once set. Pipette this very carefully, since if it mixes too much with the gel solution it can inhibit setting, or even dilute out the reagents. You will have made more gel solution than you will need. Leave the remainder in the plastic tube. When it has set, you will know that the gel between the plates has also set, since setting is a chemical reaction, and has nothing to do with cooling, or the environment in which the gel solution finds itself. Setting will take around 15–20 min.

The stacking gel

Whilst the main gel is setting, make up the stacking gel to the strength required. When the main gel has set, pour off the water saturated butanol, and wash the surface of the gel with distilled water. Squirting it in using a water bottle with a fine nozzle is the best approach: the surface of the gel will be quite tough. The surface of the main gel must be dried reasonably well before pouring the stacking gel. This is done by rotating the entire casting apparatus 90° and inserting the corner of a folded piece of tissue paper into the gap between the gel plates, which will suck out by capillary action, assisted by gravity, the water. Next, return the casting apparatus to the level bench top and pour the stacking gel gently using a pipette. Insert the comb slowly, and initially at an angle of 45°, gradually leveling off as you move it into position. This will prevent air bubbles forming between the teeth of the comb. Make sure there is 0.5–1 cm of stacking gel above the main gel. The remaining stacking gel will set at the same time as the stacking gel between the glass plates.

Loading and running an SDS-PAGE gel

Once the gel has been poured and has set, it is ready to be run. It is possible to store the gel overnight in the refrigerator. To do this, wrap the gel (and clamp, if being used to run the gel) in Clingfilm. It is not advisable to store home-made gels for longer than this, but you can purchase ready-made, pre-cast gels, and these come bathed in storage buffer, enclosed in a plastic pouch to stop them drying out. When you are ready to use it, place the gel into the running apparatus. Most casting clamps are not removed, but are simply unclipped from the casting block and clipped into the running block. Alternatively, in some systems (and this would include the use of pre-cast gels) the gel is slid into a slit in the running block. In this latter case, the spacers and glass plates must not be damaged or pulled away from the gel when inserting it into the running block. Once the gel has been placed within the running block, remove the comb from the gel. This should be done in one steady and not too rapid movement. It is possible that the wells may be damaged, or even that the stacking gel may come away from the main gel, due to the vacuum pressure generated when pulling out the comb. Pulling one end slightly more forcefully than the other (but not to the extent that the walls separating the wells are bent) is likely to help. Once the comb has been removed, use a water bottle to rinse out any debris in the wells. Again, be fairly gentle, so that the walls separating the wells are not damaged.

The aim of a running block is to separate two reservoirs of buffer, one in contact with the top of the gel, the other in contact with the bottom. In SDS-PAGE, the running buffer in each reservoir is identical, but there must be no leakage of buffer from one to the other, since this will cause the level of the top reservoir to reduce until it no longer comes into contact with the gel, stopping electrophoretic separation, and/or cause a short circuit which bypasses the gel. Thus, when the gel (with or without casting clamp) is fitted into the running block, you must be certain there are no leaks. Fill the reservoirs up with buffer (don't just fill the top, or the bottom of the gel might dry out), and watch to see if the level of buffer in the top reservoir falls. If it does, it may be possible to seal the edges of the gel with Vaseline, though the most common problem is that the rubber seals around the edge of the running block have become worn or perished. They should be replaced at regular intervals, and can normally be bought separately from the rest of the running block. The standard SDS-PAGE running buffer is shown in *Table 6.2*.

Once you are happy that there is no leakage of buffer, you can load and run the gel. Add sample buffer (*Table 6.2*) to the protein solution and heat it at 95°C for a few minutes immediately prior to loading it. Give the sample a brief centrifugation to pellet down any crystals or other debris, and load

Table 6.2 Buffers for SDS-PAGE and western blotting

Buffer	Constituents
Sample buffer	250 mM Tris-HCl pH 6.8, 5% w/v SDS, 0.25% w/v bromophenol blue, 25% v/v glycerol, 200 mM DTT (added just before use)
	Mix this 50:50 with sample
Electrophoresis running buffer	14.4 g l⁻¹ glycine, 3 g l⁻¹ Tris base, 1 g l⁻¹ SDS
Western transfer buffer	14.4 g l⁻¹ glycine, 3 g l⁻¹ Tris base

the sample steadily with a Hamilton syringe, allowing the sample to settle into the base of the well as you go. Each well will take 10–20 µl of sample, and don't run more than 50 µg of protein in total in each lane of a mini-gel. Make sure that the sample does not spill from one well to the next. Air bubbles in the syringe can cause a sudden burst of air in a well, causing sample to be lost. This is not only bad because it can cause contamination of one well with the sample from an adjoining well, but if quantification is the aim of running the gel, then indeterminate loss of sample is a very bad thing. The gel can be run at different voltages depending upon what is required. Beware using too high a voltage, since this can cause the gel to heat up and even melt. Low voltages can result in diffusion of proteins. Between 50 and 150 V should be suitable for most apparatus. The length of time needed to resolve a gel will depend upon the molecular weight(s) of the proteins you want to resolve. It is important that the gel is run long enough to separate the proteins, but not so long that proteins are either lost from the end of the gel, or form very diffuse bands that are difficult to see, which occurs close to the bottom of the gel. The size of the protein you want to separate from others will also inform the amount of acrylamide in the gel, and the acrylamide:bis-acrylamide ratio used. It may be that some trial and error will be required when dealing with very large, or more likely, when separating very small proteins. The use of protein size markers will help greatly in this regard; as each gel is run it can be stained to check the positions and intensities of the bands of known molecular weight. However, a more interactive way of doing this is to use protein size markers that are pre-stained. It is possible to buy ready prepared stained marker proteins that run at a particular molecular weight, but are also a particular color. In some cases, each protein is a different color. These pre-stained markers are very useful when running gels since, provided you can see the marker lane, you can instantly know where on the vertical plane of each gel a protein having a particular molecular weight will be. Therefore, you will be in total control of separation, and will be sure that the protein you are interested in has not run off the bottom of the gel, or become too diffuse a band.

Visualizing proteins in an SDS-PAGE gel

There are a whole host of different ways of staining proteins, from simple stains such as Coomassie brilliant blue R, which chemically links to certain amino acids, to very sensitive staining techniques such as silver staining, which is far less discriminatory between the amino acid compositions of different proteins, and so is better for quantification. Such staining procedures will be dealt with in Chapter 7, where protein quantification will be discussed in more detail. As far as western blotting goes, however, the only real use of staining is to check that proteins of the required molecular weight have been transferred onto the PVDF membrane. Here, a simple Coomassie stain of the post-transfer gel will tell you. The use of pre-stained molecular weight markers when running the gel will also confirm that a particular molecular weight of protein has been transferred onto PVDF, but with the added advantage that there is no need for staining the proteins remaining within the gel. The standard SDS-PAGE gel stain is Coomassie brilliant blue –R250 (2.5 g l^{-1}) made up in water containing methanol (45% v/v) and glacial acetic acid (10% v/v). Staining would be for 1 h, or more if protein samples are low concentration, with rocking. Destain the background with 20% v/v ethanol, 10% v/v glacial acetic acid in water. Change the destain solution several times, each for 30 min. Gels can be stored in water when destained sufficiently, and should be dried between two sheets of wet cellophane after being stretched over a piece of glass and clamped in place with bulldog clips. Alternatively, for a more reliable crack-free gel drying experience, invest in a vacuum gel drier.

Western transfer of SDS-PAGE gels

It is quite possible to perform a dot western blot, where a sample of cell extract is spotted onto a piece of PVDF membrane, and can then be probed for the presence of a particular protein. This approach is often used when purifying proteins using liquid chromatography, where the different samples to be tested come from individual fractions of eluent from the chromatography column. However, for quantitative approaches, such as measuring gene expression, the potential for saturation of a membrane with sample concentrated onto a small spot is high, and the close proximity of proteins is highly likely to cause interference between target protein and probe. Therefore the separation of proteins using SDS-PAGE followed by transfer onto a PVDF membrane is the best approach to western blotting.

PVDF membranes need to be treated carefully. Clearly, they can easily become contaminated with protein from all sorts of places, which may affect the western blot results, particularly if the protein you want to probe is present on your skin, or on bench surfaces. Wear gloves whilst handling membranes and keep the backing paper on at all times, but don't assume that the backing paper will prevent contamination, since it is only loosely attached. Only cut the membrane from the roll at the last minute, and do not put the roll down on the bench. It is quite possible to cut out an appropriately sized piece without putting the roll down. If necessary, get a colleague to hold the roll. Once separated from the roll, it is best to immediately place the membrane into equilibration and then transfer buffer. If you do have to put the piece of membrane down, keep the backing paper on, and place it onto a piece of paper towel, not directly onto the bench top.

PVDF is a very hydrophobic substance, and if membranes are placed directly into aqueous solutions, for example transfer buffer (for transferring SDS-PAGE gels, the transfer buffer is the same as the running buffer, but without SDS – see *Table 6.2*), then it is highly likely that patches of the membrane will remain dehydrated. This can be a real problem, because these patches will not take up transferred proteins properly. If the membrane you are using looks blotchy (it is normally very obvious since the blotches are white, with the hydrated membrane being gray), then throw it away and start again. The best way to ease the transfer of PVDF into a hydrated state is to initially soak the membrane in methanol. Gently float the membrane on the surface so that pockets of air do not form between the fibers, and gradually submerge it. Once it has been soaked in methanol for a few minutes, pour off the excess liquid and apply transfer buffer. From this point (until you have read the result), never let the membrane dry out. This is extremely important, and these membranes dry out very quickly. Due to surface tension, a layer of liquid will always coat the membrane when excess liquid is poured away, but it will not take long for evaporation to reduce this reservoir, and almost as soon as this happens, the membrane, or part thereof will revert to its dehydrated state. If this happens after proteins have been transferred then all is lost, because methanol assisted re-hydration is likely to result in stripping of many of the proteins from the membrane.

Once the membrane has been hydrated and equilibrated in transfer buffer it can be used for western transfer. Transfer of nucleic acids from gels onto membranes has been dealt with in Chapter 3, where capillary transfer is the most common method. Capillary action could theoretically be used to drive proteins out of SDS-PAGE gels, and so onto the PVDF membrane, but the very small mesh size in polyacrylamide gels means that this process would be very slow compared with its application to nucleic acid containing agarose gels. Hence a bit more force is required to push proteins out of polyacrylamide gels, and passing an electric current across the gel is an excellent way of providing such force. The proteins will retain their coating of SDS following SDS-PAGE, and this is important, because it means that proteins will be negatively charged and will move towards the anode in an electro-transfer approach. Simply, the gel is placed on top of a sheet of PVDF, and the two are sandwiched with a few sheets of filter paper soaked in transfer buffer (which is essential to allow flow of electric current. The sandwich is placed on top of a metal sheet, which will form the positive electrode, and the negative electrode sheet is squashed on top. The thickness of the gel, PVDF membrane and filter papers keep the plates from touching, so the only way electric current can flow is through the filter papers and gel, driving the proteins from the gel and onto the PVDF membrane, where they are trapped due to the very small pore size of PVDF, through which the vast majority of polypeptides cannot pass.

As discussed above, it is possible to use pre-stained molecular weight markers in the gel to inform you as to whether transfer has occurred, and if so, what molecular weight of protein has been passed onto the membrane. Because the size of the protein dictates how fast it will pass from the gel and onto the membrane, the longer you allow transfer to take place, the larger the maximal size of protein transferred will be. Clearly, the gel sieve size, which is dependent upon the concentration of acrylamide in the gel, and the acrylamide:bis-acrylamide ratio, will also affect the rate of transfer for all proteins. Therefore, when transferring using a particular percentage gel for the first time, it is best to err on the side of long transfer times, reducing the time on the second occasion, dependent upon what the size markers tell you about how far above the size of your target protein had been transferred. It is better to go over time rather than under time, because to look at even pre-stained molecular weight size markers on the membrane will involve removing the gel, and there is no real way the gel can be put back in place without significant distortion of the bands, if transfer has not run far enough and needs to be restarted.

Once transfer has been completed, the filter papers and gel are removed from the membrane, which must be gently peeled away from the bottom layer of filter papers and placed in transfer buffer. Gently brush the surface of the membrane with a gloved finger to remove any debris and pieces of gel. During western transfer, the SDS comes away from the proteins (transfer buffer is usually the same as gel running buffer, but without the SDS). Therefore, proteins return to their native charge, but have little or no secondary structure, and so their hydrophobic cores are exposed. This is important, because immobilization of proteins with PVDF is facilitated by hydrophobic interactions. These interactions will persist throughout the probing process so no fixing of proteins to PVDF is required in the vast majority of cases (unlike the situation of immobilization of nucleic acids to nylon or nitrocellulose). There are a few exceptions to this rule, however. A number of proteins have little secondary structure normally, which means their complement of hydrophobic amino acids will be very low. This can make them bind very weakly to PVDF, and simply float away as soon as the membrane is placed into an aqueous solution. To guard against this eventuality, it is always best to run two western transfers in parallel on the first occasion you come to probe for the presence of a specific protein; one membrane should be treated to fix proteins onto it, and the other should not. The results will inform you whether the fixation was necessary, and the fact that one was fixed will validate any negative result obtained. From then on, fix or don't fix as appropriate. Fixation is best performed by glutaraldehyde cross-linking. Simply submerge the membrane into the appropriate buffer for the probing procedure (usually PBS–TWEEN, see below) and add 0.05% v/v glutaraldehyde for a few minutes. Beware that glutaraldehyde solutions go off very quickly, so don't be tempted to by a large bottle if you will only be doing experiments infrequently; buy a fresh bottle every time you have planned to use it.

The finishing touch before the membrane is ready to probe is to block the membrane. To prevent the antibody probe from binding to the membrane, thus giving a positive background, which will ruin your experiment, the exposed surface of the PVDF (i.e. where protein bands have not been transferred) needs to be covered up. The best way of doing this is to use 10% w/v milk powder, which contains a number of proteins that will bind avidly to the membrane and efficiently block it. Unless you work on milk proteins, in which case, use a solution of 10% w/v bovine serum albumin, but otherwise, baby milk powder is far less expensive!

Probing proteins on western blots

In western blotting, the probe used to detect the presence of a particular amino acid sequence is an antibody. Antibodies recognize so-called 'epitopes'. These are particular three-dimensional structures (and not just containing amino acids) that fit into the variable region of a particular immunoglobulin (IgG) molecule. The use of primary antibodies, specific against a particular epitope does not itself allow visualization and quantification of the immobilized protein, however. For this, a labeled secondary antibody, raised against the common region of the IgG antibody of the animal in which the primary antibody was raised. The label on the secondary IgG is usually an enzyme, which can be used to visualize the presence of the primary antibody, and so target protein due to the generation of chromogenic products by the enzymes (*Figure 6.2*). These substrates are discussed in detail in Section 3.5.

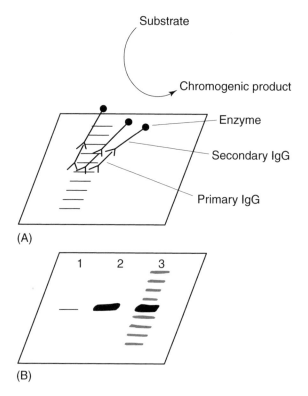

(A)

(B)

Figure 6.2

The western blot. (A) Proteins are separated using SDS-PAGE and transferred onto a PVDF membrane as a ladder of bands. A primary antibody, specific to one epitope present on one of the proteins will (if that protein is present) interact with its target, and effectively become immobilized onto the blot. If a secondary antibody is required to develop the reaction, it is raised against the common region of the primary antibody, and conjugated with an enzyme such as horseradish peroxidase, which will convert a substrate into a colored or fluorescent product. (B) As more total protein is loaded onto the gel and transferred onto the membrane, the signal associated with the amount of antibody binding increases. However, if too much protein is loaded, this can lead to saturation of the signal, and even non-specific interaction between antibodies and proteins (see lane 3).

Antibodies are washed over the blots in PBS–TWEEN buffer (0.3 g l⁻¹ KCl, 10 g l⁻¹ NaCl, 1.4 g l⁻¹ Na₂HPO₄, 0.3 g l⁻¹ KH₂PO₄ containing 0.2% v/v TWEEN-20) containing 0.5% w/v milk powder. If the antibody is a serum total IgG, use a dilution of around 1:1000 to 1:10 000 serum:buffer. This can be increased if the proteins are not visible, and can be reduced if there is nonspecific protein staining. If the primary IgG is a monoclonal, then more is required, so a ratio of 1:100 is not unusual. An incubation time of about 1 h with primary antibody will be enough, with the membrane being rocked constantly. Then pour off the primary antibody, wash the blot a couple of times with PBS–TWEEN, and then incubate with secondary antibody according to the manufacturer's instructions. Because secondary IgG's are purified, they are used at very much lower concentrations than are primary IgG's.

Visualization of the enzyme on the secondary antibody is usually through chemiluminescence, with the bands being exposed to high-speed X-ray film, as described in Section 3.7.

6.4 Promoter-probe reporter enzymes and assay of their activity

There are a number of commercially available promoter-probe vectors for prokaryotic and/or eukaryotic use, which encode easily assayable enzyme reporters. Furthermore, some enzymes described in Section 6.5 as being end-product gene expression reporters, are also available as promoter-probe reporters, but are not predominantly used as such, so are discussed there, rather than in this section.

The use of any enzyme as a reporter requires that its activity is easily assayable, and for that property read 'colorful'. This could be that the products or substrate of their reactions have unique absorbance spectra in a spectrophotometer, so V_{obs} can easily be determined. Alternatively, the product or substrate can be quantified in different ways, such as by the generation of luminescent products after the enzyme has worked on a substrate, or the breakdown of radioactive substrates by enzymes.

Beta-galactosidase in bacteria

This is a favorite promoter-probe reporter enzyme of bacteriologists. The enzyme was originally characterized from *E. coli*, where it is called LacZ. In this specific case, the enzyme has evolved to hydrolyze the beta-glycosidic bond that links the monosaccharides glucose and galactose, which together form the disaccharide sugar, lactose, but beta-galactosidase activity is simply the cleavage of a beta glycosidic bond between galactose and any other molecule that can be attached to galactose using this type of linkage. So-called 'chromogenic' beta-galactosidase substrates have been developed which release a colored dye when their internal glycosidic bond has been cleaved. LacZ can be used as a promoter-probe or end-product gene expression reporter in any bacterium that does not produce its own beta-galactosidase enzyme (and beware, because many do).

Like other *E. coli* genes, the guanine–cytosine content of *lacZ* is approximately 50%. This means that if the gene is used as a reporter in bacteria with extremes of guanine–cytosine content, where the codon usage profile will inevitably be different from that in *E. coli*, it is quite possible that translation of the protein will be slow simply because of a scarcity of appropriate tRNAs to recognize the actual codons used. Sadly, this is an uncertain, and pretty much unquantifiable problem, so there is nothing much to be done about it. Given that LacZ is a large protein, relative to other bacterial proteins, this potential problem is probably greater for LacZ, than for some other bacterial reporters.

Measurement of LacZ activity in bacteria

The classic chromogenic LacZ substrate used to measure beta-galactosidase activity in bacterial cell extracts is *ortho*-nitrophenyl galactoside (ONPG), which has a beta-glycosidic bond joining a galactose sugar and a dye moiety. When the beta-glycosidic bond within ONPG is hydrolyzed by LacZ, the *ortho*-nitrophenyl dye group is released from the sugar and, as a result, becomes bright yellow. So, as the enzyme works on the colorless ONPG substrate, the solution becomes more and more yellow. The rate at which yellow color develops is directly proportional to the amount of enzyme present. The generation of the yellow color is monitored using a spectrophotometer set to determine absorbance of 420 nm wavelength light. Thus, extracts of cells carrying the reporter gene/hybrid are added to buffer containing ONPG and mixed in a bijou bottle and an aliquot is taken at an arbitrarily chosen zero time point, transferred to a plastic cuvette where sodium carbonate is added to stop the reaction. The absorbance reading at 420 nm of the sample is then taken, and is used to zero the spectrophotometer. The remainder of the reaction mixture is then incubated for a set period of time before another aliquot is removed to a fresh spectrophotometer cuvette, spiked with sodium carbonate and used to take a second reading. If some of the ONPG has been hydrolyzed between the times when the reactions in the first and second aliquots were stopped with sodium carbonate, then the second spectrophotometer reading will be more than the first (i.e. more than zero). This change in absorbance is used to determine the amount of LacZ enzyme activity in the reaction by dividing the change of absorbance by the number of minutes separating the times at which the reactions in the two aliquots were stopped with sodium carbonate. Section 6.2 gives general information about performing enzyme activity measurements, and Section 6.3 describes the pitfalls associated with a end-point assays, which the above essentially represents.

If you do not have any clue as to how active the promoter cloned upstream of a *lacZ* promoter-probe reporter gene is, then it is worth setting up a LacZ activity assay with double the normal volume, allowing for twice as many activity measurements to be taken. Thus, if there is no significant increase in absorbance at 420 nm after the first aliquot is taken, the remaining reaction can be left to proceed for longer before another aliquot is taken. The last aliquot can be left overnight if necessary. It is very important that the reactions are left to proceed at a set temperature, so that activity data can be compared from one experiment to another. Therefore, set up the reaction in the bijou and place it in a water bath (air incubators do not allow sufficiently rapid equilibration of temperature). It is not necessary to agitate the reaction, but it is worth briefly inverting the bijou prior to taking an aliquot.

The standard method for converting raw LacZ activity (V_{obs}) data in bacteria (change in absorbance per unit time) to relative specific enzyme activity data (i.e. taking into consideration the amount of protein in each cell extract) simply involves measuring the density of cells (i.e. turbidity of cells, see Section 6.2) used to produce each extract and assuming that each cell carries approximately the same amount of protein. You might be tempted to try and get more accurate data by measuring protein concentrations, and/or looking up the extinction coefficient for the released ONP product of ONPG hydrolysis by LacZ. There is nothing to stop you from doing this, but the benefit of using a standard assay method to generate LacZ reporter activity data with given units (in this case, the Miller unit, in honor of Jonathan Miller, who developed the assay) is that studies concerning promoter strength can be compared from one research group to another, even across the decades. For example, it is possible to not just compare the strengths of one promoter under two different physiological conditions, but the strengths of two different promoters in the same physiological condition as well. The number of Miller units (Miller, 1972) present in a bacterial cell extract is calculated using the formula:

$$\frac{1000 \times [A_{420} - (1.75 \times A_{550})]}{A_{600} \times t \times v}$$

where A_{420} is the absorbance of the yellow ONP product; A_{550} is the scatter from cell debris, which, when multiplied by 1.75 approximates the scatter observed at 420 nm; t is the reaction time in minutes (i.e. how long it took to achieve the observed A_{420}); v is the volume of culture assayed (in milliliters); and A_{600} reflects cell density prior to the assay.

Luciferase in eukaryotic tissue culture cells

Luciferase is a general name for enzymes that catalyze the conversion of a substrate into a product with the concomitant generation of light, which can very easily be measured. The rate at which light is generated is proportional to the rate at which the enzyme works, so luciferase reporter enzymes have been very popular for many years. The majority of eukaryotic promoter-probe reporter vectors use luciferase genes. There is no real reason why luciferase reporters cannot be used in bacterial cells, though commercial promoter-probe reporters are optimized for use in mammalian cells, and there may be implications for their use in bacterial cells, so by all means try, but proceed with caution.

There are two main luciferases used as eukaryotic reporter enzymes, and other variants which have more specialized purposes. The first is firefly luciferase from the organism *Photinus pyralis*. It catalyses the conversion of the substrate D-luciferin into the product oxyluciferin, and uses magnesium ATP and molecular oxygen as co-factors, resulting in the generation of AMP plus inorganic phosphate, plus carbon dioxide. The other main reporter luciferase is known as Renilla luciferase from *Renilla reniformis*. It catalyses the conversion of the substrate coelenterazine into the product coelenteramide using molecular oxygen as a co-factor, which is converted into carbon dioxide.

All commercially available luciferase reporter genes have been codon-optimized for use in mammalian cells, and have had as many known transcription factor binding sites removed by mutagenesis as can be achieved without destroying encoded enzyme activity, so that they will be as specific as possible for reporting control of the inserted promoter. Another modification that is available, is one that reduces the stability of the enzyme in the cell in which it is being expressed. This is important to consider, because many reporter enzymes, and particularly wild-type luciferases, suffer from being too stable, so not providing an accurate dynamic assessment of gene expression reductions that might occur in response to changes in physiological state of the cell. For example upon addition of a particular growth factor etc., the expression of the reporter gene might all but stop, but if the reporter enzyme hangs around it may not be immediately obvious that gene expression has reduced, never mind actually ceased. The use of unstable luciferases helps, because here, protein levels more accurately reflect combined transcription and translation rates *in vivo*. Clearly, the native gene, whose expression is controlled by the promoter, may be very stable, meaning that such acute control of transcription/translation rates are meaningless. However, it is not the role of promoter-probe reporters to mimic the stability of native genes and their protein products, but to report the expression dynamics. It must be left to end-product gene expression reporters, which include a large proportion of the native genes, to do this, as set out in Section 6.6.

Assay of luciferase activity

All luciferase enzymes generate one photon of light per catalytic turnover, which can be monitored using a luminometer. In its simplest guise, luciferase assays involve cell extracts containing the reporter enzyme being mixed with buffer containing the appropriate substrate and co-factors, and the mixture being incubated for about half an hour. This is necessary because, it takes quite a while for the steady state portion of the reaction to be reached, where the production of photons occurs at a constant rate dependent upon the amount of enzyme in the reaction. Next, the sample is moved into a luminometer, and the number of photons generated per second is calculated (usually the machine will be set to integrate around 30 s of data). This gives an appropriate V_{obs} for the enzyme reaction, which can be used to compare the amount of reporter enzyme in one cell extract with another.

The production of specific enzyme activity data for eukaryotic cells from raw V_{obs} data can be achieved in a number of ways. First, the density of each population of cells can be estimated by recording the turbidity of each cell suspension immediately prior to cell lysis. Second, an aliquot of cell extract can be kept back and used to determine protein concentration. Third, a control reporter gene, spliced to a constitutive promoter can be introduced into the same cells using a so-called control reporter vector, which will provide a baseline for protein levels. Nowadays, the control and test reporters can in fact both be luciferase genes, but those that encode enzymes that produce light at markedly different wavelengths, that can be quantified separately using a luminometer fitted with the appropriate interchangeable colored filter. Dual-reporter luciferase enzymes are made using the luciferase genes of the click beetle, *Pyrophorus plagiophalam*. This insect produces at least four subtly different luciferase enzymes, each generating light with a unique wavelength. Two of them have been modified for use as synthetic reporter genes in mammalian cells; one produces red light, the other produces green light. It is possible to choose which gene to use as the promoter-probe reporter, with the other being used as the control reporter, since both genes are available in both vector forms.

Luminometers come in all shapes and sizes. The simplest involves a charge coupled device at the end of a rubber housing into which a plastic bijou bottle carrying the reaction mixture is pushed. The operator then starts the machine that simply counts the photons until the user, or a timer, tells it to stop. These are cheap, easy to maintain and to use, and are fine for experiments involving just a few samples. However, they can become very tedious if you have to get though dozens of samples. For larger scale experiments, you should consider investing in a luminometer that can deal with multi-well plates. If you do choose such a piece of apparatus, be prepared to spend a little bit extra and buy plates that are white-walled, or white-bottomed (depending on how your luminometer works) to reduce background luminescence leaking across the plate from one well to surrounding wells.

For the very impatient, it is now possible to assay luciferase activity without actually lysing cells. Indeed, if you use tissue-culture compatible multi-well plates, the cells can be grown, treated, and then luciferase within them assayed *in situ*, without even removing the growth medium, just by adding cell permeable substrate. For Renilla luciferase, it is not possible to use coelenterazine in this way, because it is very unstable within cells. Instead, a pre-substrate is provided, with the region within the molecule modified by luciferase protected by a large group joined by an ester bond. Once within cells, the ester bond in the pre-substrate is cleaved by intracellular nonspecific esterase enzymes, releasing the true substrate, which is converted to product by reporter luciferase with the generation of light. In such *in vivo* measurements of luciferase activity, it is equally important to allow the reaction to enter the steady state (usually takes 30 min to 1 h) prior to reading enzyme activity with a luminometer. Such *in vivo* measurements, therefore, must not be taken too seriously, because the incubation of the cells in the presence of assay reagents for up to several hours prior to assay may well produce nonspecific, and more worryingly and less easily dealt with, specific changes in test promoter activity, which may well affect the apparent luciferase reporter gene expression level. You should also remember that if you are treating cells with a growth factor etc. for a set amount of time, the substance must be in contact with the cells right up to the time the cells are placed into the luminometer. Otherwise, removal of the growth factor during the pre-incubation step may well entirely or partly reverse its effects on gene expression. Therefore, take into consideration the pre-incubation time when deciding how long to incubate the cells with the growth factor prior to starting pre-incubation. Another major problem with such *in vivo* assays of luciferase activity is that they are prone to significant system variability. Different batches of growth medium, serum and other additives etc. may well affect the apparent amount of light recorded by the luminometer. Hence, when comparing data from two experiments, make sure that as few changes are made as possible or at least accept that the data generated will only represent pointers concerning changes in gene expression, and will only be meaningful if the apparent changes observed are large.

Chloramphenicol acetyl transferase

This enzyme, encoded by the *CAT* gene, confers resistance to chloramphenicol in many different bacteria. It works by modifying the drug. Specifically, an acetyl-CoA co-factor is transferred onto the drug, preventing it from interacting with its target.

CAT is a solely bacterial enzyme, and as such its gene is not found in eukaryotic cells. Whilst many manufacturers provide codon-optimized, and transcription factor binding site free *CAT* genes for use as promoter-probes, the fact that this enzyme is not native to eukaryotic cells means that it should only be used as an indicator of gene expression in these cells. In bacteria, however, and particularly enteric bacteria, *CAT* is a very popular reporter gene.

Assay of CAT

As well as using acetyl-CoA to modify chloramphenicol, CAT can use other longer chain CoA derivatives. In the CAT assay (Seed and Sheen, 1988) the chloramphenicol substrate is provided radioactive (usually ^{14}C or ^{3}H) and the co-factor used is *n*-butyryl-CoA. When CAT transfers this butyryl moiety onto chloramphenicol, it becomes soluble in xylene, wherease unmodified chloramphenicol is not. Hence the addition of xylene to stop the reaction will also allow separation of modified chloramphenicol from unmodified chloramphenicol which stays in the aqueous phase.

Specifically, for a 125 µl assay, cell extract (50–100 µl) is incubated with 5 µl of 5 mg ml^{-1} *n*-butyryl-CoA and [^{14}C]chloramphenicol (at 25 mM final and 44.4 kBq ml^{-1} (1.2 µCi ml^{-1}) final specific radioactivity made up in cell extraction buffer (250 mM Tris-HCl, pH 8.0)) at 37°C for 30 min to 30 h (start at 3 h and alter if the amount of product formation is too low or too high.

Stop the reaction by adding 300 µl of mixed xylenes (Sigma Aldrich 247642). Centrifuge for 3 min and remove 200 µl of the top (xylene) phase and count the radioactivity in this phase using scintillation counting according to your own preferred protocol. To get a curve of enzyme activity over time, it is best to set up several reactions at the same time, and stop each one after a different period of time.

6.5   Using promoter-probe reporter vectors

It is highly unlikely that you will construct your own promoter-probe reporter vector. However, it is possible to ligate a promoterless gene, with its ribosome binding sequence into a cloning vector, leaving space for the insertion of a promoter upstream. This is particularly useful if you happen to work on an organism in which commercially available reporter vectors will not replicate. The reporter gene should be PCR amplified and ligated into your vector of choice, and large amounts of vector recovered before ligation of the promoter region, again amplified by PCR. If the cloning vector you are using has a good multiple cloning site then try and put restriction enzyme sites into the PCR primers for both the reporter and promoter, and use these restriction sites to clone them in. If you use different restriction sites at each end of each fragment, then you will be able to put them in the correct orientation. Try and get the reporter gene into one of the 3′ proximal restriction sites, so that there will be plenty of available sites upstream for ligating the promoter. The promoter itself can be amplified in the presence of a certain amount of up and downstream sequence (and this will be essential if you do not know exactly where the promoter is). Try not to include the ribosome binding sequence or translational initiation codon within the region cloned, however, since this may well affect the translation of the reporter protein. *Figure 6.3* gives an overview of the steps to take in constructing your own reporter.

There are a large number of commercially available reporter vectors (*Table 6.3* lists some of these) and all come with a multiple cloning site upstream of the promoter probe reporter gene. To use these, it is best to PCR amplify the promoter region using primers that introduce restriction sites into the product, allowing for it to be ligated into the vector.

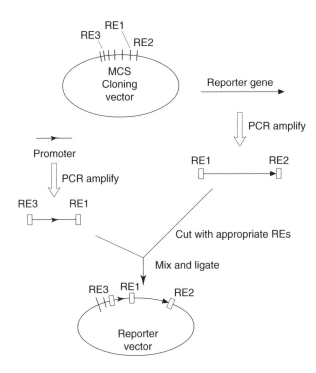

Figure 6.3

Making promoter-probe reporter constructs. The reporter gene must be amplified using PCR, with primers designed to add on a different restriction site at each end of the gene. The sites must not be present within the gene. The test gene's upstream sequence is similarly amplified, though in this case, the restriction enzyme sites chosen allow joining of the promoter with the reporter gene, both in the correct orientation and adjacent to each other in the multiple cloning site of the cloning vector of choice. If using ready-made promoter–probe reporter vectors, then all that will be required will be amplification of the promoter region with appropriate restriction sites to allow ligation into the vector.

Table 6.3 Selected promoter-probe reporter vectors

Manufacturer	Vector	Reporter	Comments
Promega	pGL4	Luciferase (renilla or firefly)	Numerous configurations. Different vector origins, selectable markers, etc. Eukaryotic cells only
	pGL2	Firefly luciferase	Available as promoter probe, or to look at the activity of enhancers (i.e. already carry a strong promoter)
	pGL3	Firefly luciferase	As above, but luciferase gene has been codon optimized
	Chroma-luc	Click beetle luciferase	Three different colored luciferases available, both in promoter-probe and transfection control (i.e. with a strong promoter); can be used simultaneously
	phMGFP	Monster GFP	Modified protein is brighter and less toxic than wild-type GFP. Can be used as promoter-probe or end-product gene expression vector
	pCAT3	CAT	As pGL3, but with the CAT reporter – can be used in *E. coli*
Invitrogen	pBLUE	Beta-galactosidase	TA cloning site within MCS, allows direct cloning of PCR product containing promoter region. Vector allows eukaryotic and *E. coli* use
	pGLOW	Cycle-3 GFP	As above. Mutant GFP is brighter than wild type, but has the same spectral properties

6.6 End-product gene expression reporter assays

End-product gene expression reporters can be made from any gene used as a promoter-probe vector (see above). Many other reporter genes exist, which are not always available as constituents of ready-made promoter-probe reporter vectors, but which have proven efficacy as reporter genes for end-product gene expression studies. A number of these are described below. However, if you have an particular enzyme of your own, which you are used to and are confident about the stability of the assay you use to quantify it, and about its stability and compatibility with the specific cells you want to measure gene expression within, then you should seriously consider developing your own end-product gene expression reporter system. How to go about this will be discussed in Section 6.7, below. There is only one real reason for sticking with other reporter genes, and that is to make it easier to compare your experiments directly with those from other research groups, and to make it easier for others to repeat your experiments. However, reporter genes are for experiments what horses are for courses, and if you are working with a poorly characterized organism where existing reporters may not be well tolerated, why not create your own cell-specific reporter? If others follow your example, it might be the best-cited paper you ever publish!

Beta-galactosidase in eukaryotic cells

LacZ from *E. coli* can theoretically be used as both a promoter-probe and end-product gene expression reporter in eukaryotic cells. However, the use of LacZ as a promoter-probe reporter in eukaryotic cells is very rare, and I could find only one commercially available *lacZ*-containing promoter-probe vector for eukaryotic use. *lacZ* genes under the control of strong eukaryotic promoters are more often found on commercial eukaryotic vectors where the reporter is used as a transfection control. This is possible, since vector copy number (dictated by transfection efficiency) is proportional with LacZ activity.

In eukaryotic cells, therefore, the major use of LacZ as a reporter in gene expression studies, is as a do-it-yourself end-product gene expression reporter. The enzyme is large, meaning that use of the *lacZ* gene in this way will result in the generation of a hybrid gene with a very large exon (see Section 6.7, below). Whilst this may not affect transcription, splicing and nuclear export rates of the hybrid gene, there is always the possibility that it could. Furthermore, the use of any non-native reporter gene in eukaryotic cells could be fraught with other problems. The gene may have transcription factor and transcriptional regulator binding sites within its sequence, which, if proteins actually bind to these sequences, could affect the rate of transcription of the gene (either positively or negatively). Many commercially available promoter-probe vectors have reporter genes where known transcription factor binding sites etc. have been removed by site-directed mutagenesis, which can be used in end-product gene expression studies as well. However, I am not aware of a *lacZ* gene that has been modified in this way. Thus when performing end-product gene expression studies with LacZ as the reporter, it is always advisable to check the hybrid RNA level using a probe or primers that target the non-*lacZ* part of the hybrid and compare this to RNA levels for that gene in cells that have not had the gene modified to rule out the possibility that the sequence of the reporter portion of the hybrid gene is affecting expression of the hybrid gene, giving false gene expression data.

Since *lacZ* is an *E. coli* gene, it has a guanine–cytosine content and codon usage profile (i.e. the profile of codons preferred over others to encode a particular amino acid, where there is a choice) fairly similar to those of most higher eukaryotic cells. Therefore, the fusion of *lacZ* with a eukaryotic gene should not affect the rate of translation of the mRNA product, because the ratio of available tRNAs for each codon will be similar to that in *E. coli*.

Measurement of LacZ activity in eukaryotic tissue culture cells

There are many different methods for assaying LacZ activity in eukaryotic cell lysates. It is quite possible to use ONPG, and perform assays very similar to that described for bacterial cells (Section 6.4) using cell lysates prepared from the tissue culture cells. Here, if there is a need to standardize for protein concentration per unit of cell extract (which is less of a problem with tissue culture cells, since two dishes of confluent cells will have a very similar number of cells, and so a very similar amount of total protein) then a Bradford assay or equivalent will be necessary.

Other LacZ substrates that are easier to assay are available. For example, the Beta-Glow reagent from Promega. This is a galactose sugar linked to a luciferin moiety. Once cleaved by LacZ in the presence of luciferase enzyme, the released luciferin is converted to oxy-luciferin and light is emitted. This end product is stable for many hours, so the LacZ assay is allowed to run for a set length of time, usually around 30–60 min and then V_{obs} for the enzymatic reaction is determined by a measurement of light output over a given time by using a luminometer, as for assay of luciferase in Section 6.4. The reagent has been formulated so that it can be used to quantify LacZ in tissue culture cells without having to prepare lysates, or even to remove the culture medium. The reagent is simply added to cells grown in wells on a multi-well plate, and the entire plate is incubated, and then transferred to a luminometer capable of taking multi-well plates. There are many factors that can affect the performance of the Beta-Glow reagent, including specific culture medium and additives. Thus, it is not advisable to compare results from separate experiments, and certainly not to compare results from different cell types or different media. Internal comparison between cells within different wells on the same plate can be compared, however, and relative differences in *LacZ* production calculated for each separate experiment, each using a different plate. This limits the usefulness of this reagent in reporter assays, because often, one wants to test the effect of different additives (e.g. growth factors) on gene expression. Thus if you want to do this sort of experiment, you will have to confirm that the additive does not simply affect the assay, rather than the level of LacZ by repeating the experiment using a *lacZ* transfection control vector with a constitutive promoter that should be expressed in a manner that is not significantly affected by the growth factor. Though of course there is always the potential for some growth factors to up- or down-regulate transcription and/or translation even of supposedly constitutive genes, so care must be taken.

Another possibility when using LacZ as a reporter in eukaryotic cells is that LacZ activity can be monitored in a single cell, or group of individual cells, rather than in a mixed population. The way to do this is to fix the cells onto a microscope slide with glutaraldehyde, and then wash over a solution of the chromogenic LacZ substrate XGAL (5-bromo-4-chloro-3-indoyl-β-D-galactopyranoside). Like ONPG, this is a dye molecule coupled to galactose by a beta-glycosidic bond, and when this bond is cleaved by LacZ is releases the dye, which in this case is blue, and precipitates. Thus as the reagent enters eukaryotic cells and is cleaved by LacZ, the cells become gradually more blue. This means that if the reaction is allowed to proceed for a set amount of time, and then the XGAL reagent is washed away, the cells can be visualized by using microscopy, the image captured digitally, and the intensity of the blue color quantified in each cell using a digital densitometer device, such as those used to measure band intensity on gels.

In all protocols where LacZ production is being quantified in eukaryotic cells, it is important to remember that enzymes with significant beta-galactosidase activity are commonly found in such cells. Therefore, it is important to use control cells that have not been transfected with the LacZ encoding reporter vector, to take into consideration intrinsic beta-galactosidase activity in all beta-galactosidase assays.

Beta-lactamase

My own personal favorite end-product gene expression reporter enzyme is the beta-lactamase. There are very many beta-lactamases known, representing a vast array of enzymes with different beta-lactam substrate profiles, though the best known is the TEM beta-lactamase, which represents the ampicillin resistant determinant on many cloning vectors, so this *bla* gene is accessible to most laboratories for reporter-fusion production in do-it-yourself end-product gene expression studies. They are not yet commercially available for use as promoter-probe reporters, and indeed many promoter-probe vectors encode the TEM beta-lactamase as a dominant selectable marker for manipulation in bacteria, which might well ruin the experiment. Beta-lactamases hydrolyze the beta-lactam bond that characterizes the beta-lactam group of antimicrobial drugs. One derivative of these drugs, which is simply a laboratory tool for the detection of beta-lactamase enzyme activity, is the cephalosporin compound nitrocefin. This is essentially a cephalosporin ring coupled with a large dye group, and hydrolysis of the beta-lactam bond in nitrocefin causes a rearrangement of electrons, and makes the dye change color. The pale yellow substrate is converted into a bright red product, meaning that beta-lactam bond cleaving activity can be monitored using nitrocefin as an increase in red color per unit time using a principle very similar to that used with beta-galactosidase. Redness is monitored using a spectrophotometer since red solutions absorb to the greatest extent light with a wavelength of 482 nm. The assay is usually monitored in real time, rather than taking readings every few minutes, since the color change associated with nitrocefin hydrolysis is generally more rapid per unit of enzyme than that seen with ONPG hydrolysis by beta-galactosidase. This means that beta-lactamase assays can be conducted over shorter periods of time. The result is the same, however, with both these assays; enzyme activity is measured by the change of absorbance at 482 nm per unit of time. With nitrocefin, this can accurately be converted into an enzyme activity of nanomoles of substrate hydrolyzed per minute in the reaction by dividing the increase in absorbance at 482 nm per minute by 17.4×10^{-6}, which is the extinction coefficient for the AU value per nmol of nitrocefin hydrolysis product.

Whereas the classical beta-galactosidase assay involves measuring the density of cells extracted to make up the assay, assuming that each cell carries an equivalent amount of protein, I would advocate a determination of the concentration (or at least relative concentration) of protein in each cell extract for accurate beta-lactamase reporter assays. See Section 6.2 above to find out how this can be done.

Apart from the excellent dynamic range and accurately quantifiable activity, beta-lactamase has the advantage of being a periplasmic protein in Gram-negative bacteria. This means that it is easier to get the protein into an assayable state, without using harsh treatments aimed at disrupting the cell entirely. This can be a disadvantage, however, because in Gram positive bacteria the protein may be secreted, and in eukaryotes, it may change the membrane association properties of a hybrid protein that contains it. I would advise, therefore, that if you use beta-lactamase reporters in organisms other than Gram-negative bacteria, you should only splice on the portion of the gene encoding the 'mature' protein (i.e. missing the leader sequence). In this way, the reporter will follow the native protein to which it has been added. As with LacZ, TEM beta-lactamases are encoded by bacterial genes with a 50% guanine–cytosine ratio. There is always the potential that the presence of a *bla* gene as part of a hybrid will make the rate of transcription of the hybrid different to that of the native test gene. To confirm that this is not the case, transcripts for the native gene (i.e. in a cell that does not have the hybrid gene) and for the *bla* portion of the hybrid gene should be quantified and compared. The only other real problem with using beta-lactamase reporters is that the substrate nitrocefin is very expensive.

Green fluorescent protein

Green fluorescent protein (GFP) is commonly used to monitor gene expression and protein trafficking within intact cells. GFP fusion proteins are easily visualized by standard fluorescence microscopy to track real time subcellular localization of a protein of interest. There are many different mutant variants of the *gfp* gene from *Montastrea cavernosa* (and even some that are not green!), but the point of all of them is the same: the expressed protein is tagged with GFP and can be visualized *in situ*. The proteins are fluorescent without the need for a substrate, making them ideal for *in vivo* use where substrate permeation might not be ideal. They contain chromophores that are excited by laser light, and emit a different wavelength of light that can be detected. In the case of measuring gene expression, this is usually fluorescence in a well containing tissue culture cells, though if you want to look at differences between different cells in a population, or to look at the locations of proteins in cells, you will need to use fluorescence microscopy. I will not dwell on the assay of GFP in cells, since the use of microscopy is beyond the scope of this book.

An example of a modified GFP commercially available for reporter gene use is the Monster GFP from Promega. The gene has been codon optimized for eukaryotic use, and as a product of this, it can be used in *E. coli* (since codon usage is very similar to that in mammalian cells). Other modifications to the gene include the removal of known consensus transcription factor binding sites, and the removal of commonly used restriction enzyme sites, which increases the opportunity to insert a gene of interest upstream.

Epitope tags for end-product gene expression analysis

There are a large number of commercially available epitope tags. These have historically been used to report the level of production of recombinant proteins in *E. coli* or other heterologous hosts. They are found on expression vectors, usually as C-terminal tags. Examples include the myc tag, which has the sequence EQKLISEEDL from human c-myc. An example of an N-terminal tag would be the polyHis tag (usually six to eight histidine residues). Commercially available primary antibodies are available against both of these tags, and also a wide variety of other expression tags. These can be used in western blot studies to determine protein expression level. The beauty of epitope tags is that they are short, and have a minimal effect on protein folding and translational control, therefore have much less opportunity to introduce artifacts into gene expression studies.

6.7 Making reporter gene fusions for end-product gene expression studies

I do not intend to give a detailed exposition of the molecular biology approaches used to make end-product gene expression reporter vectors. However, I will give some pointers. First, you must have a vector that replicates and can be selected for in the cell you want to work with. Next, you must amplify the gene whose expression you want to monitor (the 'test gene') as a PCR amplicon that is large enough to encompass up- and downstream regulatory elements, and the gene must be ligated into the cloning vector. The easiest way of doing this is to introduce restriction sites at the ends of the amplicon using mutagenic primers. Remember, if you work with eukaryotes, the gene must be amplified from chromosomal DNA and not cDNA, since part of the experiment is to look at the effect of splicing on gene expression.

Now, you will need to generate the reporter fusion. It is best to locate a suitable restriction site in the 3′ end of the cloned test gene (definitely in the last exon of the gene), and amplify the reporter/epitope tag gene using PCR primers that incorporate the chosen restriction site at each end. The restriction site must be chosen with caution. When inserted, the reporter/epitope tag gene must be in the same reading frame as the test gene. It will be possible to alter the reading frame by changing the PCR primer binding sites used to amplify the reporter/epitope tag gene, but make sure that the extra bases do not generate a stop codon when present within the fusion gene. It is not so critical that the gene continues in frame at the end of the reporter/epitope tag section. Indeed, it might be best to put an in frame-stop codon after the inserted section (i.e. just before the restriction enzyme recognition site). This is your call, but do not leave a downstream frameshift, or the possible resulting product might be degraded. The restriction enzyme site must clearly not be anywhere else on the recombinant vector containing the test gene. The use of a single restriction site will mean that the reporter gene can be inserted into the test gene in either direction. It will only report gene expression if inserted in one of these directions, so make sure you perform a PCR screen for the orientation of the insert. (See *Figure 6.4* for a description of this process).

References

Bradford, M (1976) A rapid and sensitive method for the quantitation of microgram quantities of protein utilizing the principle of protein-dye binding. *Anal Biochem* 72: 248–254.

Miller, JH (1972) *Experiments in Molecular Genetics*. Cold Spring Harbor Press, Cold Spring Harbor, New York.

Seed, B and Sheen, J-Y (1988) A simple phase-extraction assay for chloramphenicol acetyltransferase activity. *Gene* 67: 271–277.

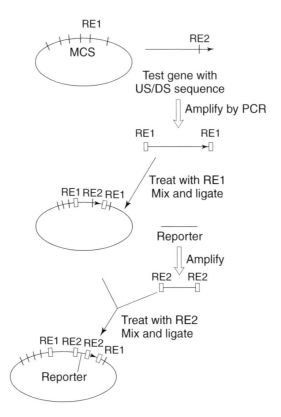

Figure 6.4

Making end-product gene expression reporter constructs. These are not usually available commercially, and must be made in a two-step process. First, the test gene, whose expression is to be measured is amplified by PCR in such a way that restriction sites are incorporated at each end, and the product can be ligated into a cloning vector. It does not matter which way around the insert goes in. Next, a restriction site only found at the 3' end of the gene (i.e. not anywhere else in the recombinant test gene-containing vector) is used to ligate in the reporter gene. This will only work if the reporter gene's reading frame is contiguous with that of the test gene, and if the reporter gene inserts in the correct orientation.

Protocol 6.1 The Bradford protein assay

EQUIPMENT

Graduated flasks
Dark bottles, for storage
Plastic cuvettes
UV–visible spectrophotometer

MATERIALS AND METHODS

1. Make up dye stock of Coomassie blue G (100 mg) dissolved in 50 ml of methanol. Add to 100 ml of 85% v/v H_3PO_4 in water, and then dilute to 200 ml with water. The solution is stable indefinitely in a dark bottle at 4°C.

2. Make up assay reagent as required by diluting one volume of the dye stock with four volumes of distilled water. The solution should appear brown, and have a pH of 1.1. It is stable for weeks in a dark bottle at 4°C.

3. In a 1-ml plastic cuvette, mix 10 µl of cell extract with 100 µl of assay reagent. Add 890 µl of water, place parafilm over the top and invert to mix. Make up a no-protein control, using 10 µl of water instead of cell extract.

4. Leave the samples at room temperature for 10 min.

5. Measure the absorbance of each sample at 595 nm, using the no-protein control to zero the spectrophotometer. If the values are <0.1 AU, then repeat with more cell extract (and appropriately less water). If the samples give an AU >1.0, then dilute an aliquot of the cell extract appropriately, and repeat the assay.

6. Protein standards should be prepared in the same buffer as the cell extracts were made in. A standard curve of concentration versus AU_{595} can be made using bovine serum albumin (BSA) with concentrations of 0, 250, 500, 1000, 1500, 2000 µg ml^{-1}.

Protocol 6.2 Simplified assay of β-galactosidase activity

EQUIPMENT

Microfuge tubes
Thermostatically controlled heater
UV–visible spectrophotometer
1-ml cuvette

MATERIALS AND METHODS

1. Grow cultures under whatever conditions you wish to test.
2. During growth, pre-measure 80 µl aliquots of permeabilization solution (100 mM Na_2HPO_4, 20 mM KCl, 2 mM $MgSO_4$, 0.8 mg ml^{-1} CTAB (hexadecyltrimethylammonium bromide), 0.4 mg ml^{-1} sodium deoxycholate, 5.4 µl ml^{-1} β-mercaptoethanol) into 1.5 ml microfuge tubes and close them.
3. Before taking each aliquot of cells, measure and write down the AU_{600} for the culture.
4. Remove a 20 µl aliquot of the culture and add it to the 80 µl of permeabilization solution. The sample will be stable for several hours.
5. When you are ready to assay, add 600 µl of substrate solution (60 mM Na_2HPO_4, 40 mM NaH_2PO_4, 1 mg ml^{-1} ONPG, 2.7 µl ml^{-1} β-mercaptoethanol) to each tube and note the time of addition.
6. Incubate at 30°C.
7. After sufficient color has developed, add 700 µl of 1 M Na_2CO_3, mix well, and note the stop time.
8. After stopping the last sample (some may take longer than others, but generally they take 30–90 min), transfer the tubes to a bench-top microfuge and spin for 5–10 min at full speed.
9. Transfer 750 µl of each sample (i.e. avoiding any cell debris) to a 1 ml cuvette.
10. Record the AU_{420}. This should be less than 1 and greater than 0.05.
11. Calculate Miller units as $1000 \times (AU_{420})/((AU_{600}$ of culture sampled) $\times$ (volume of culture [0.02 ml]) $\times$ (reaction time)).

Analysis of the proteome

7

7.1 Direct methods for calculating the relative amounts of a known protein in different cell extracts

As discussed in the previous chapter, the only true measure of gene expression is a measure of protein production. In some situations, post-transcriptional control events mean that changes in protein level do not run parallel with changes in transcript level (this is more commonly the case in eukaryotes than in prokaryotes). It would be surprising to find situations where the transcript level for a gene increased whilst the protein level decreased, and the reverse is even less likely, but to find no change, or more usually a disproportionate change in protein levels compared with transcript levels is common. Thus, interpreting the results of experiments that have determined transcript levels can be difficult, and can sometimes lead to incorrect conclusions concerning changes in gene expression. To give an accurate picture of gene expression, therefore, and if you do not want to use reporter gene approaches as described in Chapter 6, you will have to measure protein levels directly. If all you want to do is confirm transcriptomics experiments, and have therefore identified a small number of test proteins whose abundances you want to measure in a pair of cell extracts, one relative to the other, then a number of different possible methods exist that will allow you to do this. However, all these methods require some degree of prior knowledge concerning each test protein. You either need to have an antibody that recognizes the protein (Section 6.3 and below) or you need to know its ligand binding properties (the focus of this section) or have an assay for its enzyme activity if any (see Section 6.2). If none of this prior knowledge applies, then you might have to use proteomics to separate all the proteins in a cell extract, locate the spot representing the test protein, and then quantify the intensity of this protein spot when different cell extracts, each made from cells growing in a different physiological state, are subjected to the same separation protocols.

Antibody-based methods for determining the relative amount of protein

If you have raised antibodies against the test protein, you can use western blotting, as set out in Section 6.3, to determine relative amounts of that protein in two or more cell extracts. The antibody can also be used to perform ELISA analysis of test protein levels. Here, proteins from cell extracts are first fixed to the surfaces of plastic wells in micro-titer plates. Next, primary and then enzyme-tagged secondary antibodies are washed over, and the tagged enzyme produces a chromogenic product from a colorless substrate. The amount of colored product produced over a set amount of time is proportional to the amount of antibody bound to the surface of the well and so, also, the amount of test protein in the cell extract (*Figure 7.1*). Comparison of the amounts of a given protein in different cell extracts can be performed by simply using the ELISA assay values raw. It should be noted, however, that it is important when comparing samples that the ELISA experiment is performed in an identical manner for all cell extracts. In an ideal situation, samples of all the cell extracts being compared would be analyzed in parallel using the same micro-titer plate. This will reduce the amount that experimental variability affects the apparent differences in protein levels observed. It is also important to perform multiple experimental replicates for each cell extract to minimize the inherent experimental variability associated with the ELISA technique. It is sensible to repeat the ELISA analysis using different dilutions of cell extract, because high total protein concentrations can affect the interaction of primary antibody with its target protein.

In order to raise an antibody against a protein, you need to know at least part of its amino acid sequence so that a peptide epitope can be synthesized and used to stimulate an immune response in an animal in the presence of a large molecular weight adjuvant such as keyhole limpet hemocyanin. It is not quite so straightforward though, since to allow the generation of an antibody that recognizes a protein using a peptide antigen, you must be certain that the peptide is folded and presented as an epitope that would be identical in the native protein. If you cannot know this for sure, then the best way around the problem is to make an anti-protein antibody, which will in fact consist of a whole group of IgG molecules, each targeted to a different epitope on the surface of the protein. The problem with doing this is that the protein must be purified first, which is more easily said that done.

Affinity chromatography to purify specific proteins

Another way of measuring the relative amounts of a known test protein in two cell extracts would be to purify the protein from both extracts and determine the total amount of protein purified from each. The purification of proteins using various nonspecific chromatographic techniques is possible, but usually the protein must be tracked during the purification procedure using some unique property that it possesses. This is normally an enzyme activity, or reactivity with an antibody. Thus, this type of approach is not particularly beneficial for measuring protein amount, since both enzyme activity and reactivity with an antibody can be used for relative quantification of proteins in cell extracts, meaning that the purification itself is a waste of effort. Furthermore, a multi-step purification process means multiple opportunities for loss of protein, and there will be different amounts of loss during different experiments, which is particularly unwelcome if determining the amount of protein purified is the goal. There is one type of chromatography that might be useful in this scenario, however: affinity chromatography.

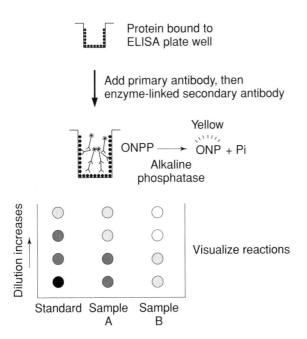

Figure 7.1

The principle of ELISA to determine the amount of a given protein in a cell extract. Total cell proteins are bound to the surface of the ELISA plate well and primary IgG is used to target the protein for enzyme labeling via a second antibody that recognizes the common region of the primary IgG. The immobilized enzyme is usually alkaline phosphatase, and the colorless substrate *ortho*-nitrophenyl phosphate (ONPP) can be used to determine the amount of alkaline phosphatase in the well, because it is converted into the yellow product *ortho*-nitrophenol (ONP). Following each antibody addition to the wells, wash steps are used to remove unbound antibody. Following incubation with ONPP for a set amount of time, the degree of yellow color developed can be determined spectrophotometrically. This can be compared between protein samples, and can be quantified more accurately using a standard concentration of pure test protein. In this example, there is more test protein in sample A than in sample B. In order to prevent saturation of the assay from potentially affecting the results, it is best to use serial dilutions of sample.

Many proteins have known binding ligands, and it is possible to immobilize these ligands onto, for example, Sepharose beads in a chromatography column. Thus, when a cell extract is passed through the column matrix, only the test protein will bind and the contaminants can be washed away. Next, the pure test protein can be eluted, either with a solution of the binding ligand more concentrated than the ligand concentration on the column, or with salt, which disrupts charge–charge interactions between the protein and its ligand. A classical example of this would be purification of the cap-binding translation initiation factor eIF4E using m^{7}GTP–Sepharose (the mRNA cap is m^{7}GTP; see Section 1.3). One problem with using affinity chromatography is that you may not only purify the test protein, but also proteins to which it tightly binds. For example, the purification of eIF4E from translationally quiescent cells may co-purify one or more 4E-BPs (*Figure 7.2*). One other potential problem with affinity chromatography is that it is possible for the affinity resin to become saturated with ligand-binding protein. The result of this can be an underestimation of the difference between protein amount in two different cell extracts.

Once purified, the amount of pure protein in each preparation can be determined by the Bradford method (see Section 6.2). Alternatively, each sample can be subjected to SDS-PAGE in parallel and the protein bands stained with Coomassie blue dye (see Section 6.3) to reveal their intensities (and to confirm purity if co-purification of binding proteins is suspected), which are proportional to their amounts. The stained gel can be scanned or photographed and the intensities of the protein bands from different cell extracts can be determined from the digital gel image files using band-analysis software identical to that described in Section 3.7.

It should be remembered that both these approaches allow only fairly rough quantification of protein amounts since Coomassie blue dye (which is at the center of both techniques) only interacts with certain amino acids, meaning that proteins with an over- or under-abundance of these amino

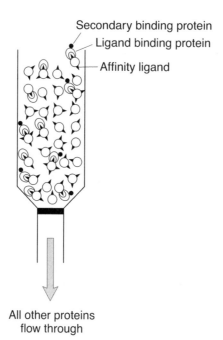

Secondary binding protein
Ligand binding protein
Affinity ligand

All other proteins
flow through

Figure 7.2

Affinity chromatography to purify proteins. In this figure, Sepharose beads are coated with an affinity ligand. Total protein is passed through the column and affinity ligand binding proteins will become fixed to the column, with remaining proteins passing through. The figure illustrates that any protein that binds tightly to the affinity ligand binding protein will also be purified using this method.

acids stain differently from the norm. Hence, for measuring the relative abundance of the same protein in two preparations these approaches work, but when comparing the relative amounts of two different proteins in the same cell extract, or indeed, when accurately calculating the concentration of a single protein, these methods may well give incorrect results. Silver staining of proteins separated by polyacrylamide gel electrophoresis allows more accurate relative quantification of proteins and will be discussed in Section 7.2.

Measuring the amount and purity of a protein by using reversed phase HPLC

The best way of determining protein amount accurately is to use the absorbance of the solution at 212 nm. It is not suitable to use 280 nm, which only gives a general feel of protein amount given that it only measures the absorbance of aromatic amino acids, and not all proteins have the same proportion of such amino acids. Light having a wavelength of 212 nm is specifically absorbed by peptide bonds, and so the amino acid composition of a protein does not affect its absorbance at 212 nm. Of course the size of a protein affects its absorbance, because the more amino acids, the more peptide bonds there are in a protein, but this can be taken into consideration when calculating relative amounts of a known protein in two preparations. Therefore, rather than using a protein dye-based method of determining protein concentration, if studying purified proteins, I would definitely recommend the use of direct spectrophotometric approaches. However, if there is any possibility that the purified protein preparation might be contaminated, for example with proteins that bind to the test protein, you must separate out the protein constituents before measuring test protein concentration. This is because you cannot assume that the amount of binding protein per unit of test protein will be identical in all physiological states.

One of the best ways of separating individual proteins from simple mixtures is the use of reversed phase high-performance liquid chromatography (HPLC). Samples are usually made up in water containing 0.1% v/v trifluoroacetic acid, or some other acidic buffer, meaning that proteins are in a charged (polar) liquid phase. This solution is then passed over a noncharged (apolar) solid phase consisting of beads coated in hydrophobic molecules (hence the term 'reversed phase'). For protein purification, the typical column would be a C_4 matrix (this refers to the length of the hydrocarbon chain attached to the beads, and so the level of hydrophobicity). The more hydrophobic the surface, the stronger the interaction between it and proteins passing over it, so for small proteins, a C_8 column might be used. For small peptides and individual amino acids, a C_{18} column would be used. Proteins are separated based on the strength of the hydrophobic interaction between them and the column. Normally, the larger the protein, the stronger its interaction with the matrix, but this is only a rough guide. Proteins are eluted from the column using a gradient of organic solvent such as acetonitrile, which breaks the hydrophobic interaction between proteins and the matrix. A typical gradient would be 5–75% acetonitrile in water plus 0.1% trifluoroacetic acid over 35 min at a flow rate of 200 μl min⁻¹. The 'retention time' of a protein on a reversed phase HPLC column refers to how much acetonitrile is required to elute it, given that the concentration of acetonitrile increases with time as the gradient is run (*Figure 7.3*). The elution of proteins from the column can be monitored in real time if the sample is passed through a spectrophotometric cell. If set at 212 nm, the readings can be used to plot a graph, effectively representing how many peptide bonds are present in each protein peak. If the size (number of amino acids) of the test protein is known, this can be used to normalize the signal to give a measure of the amount of protein present in each sample. If the machine has been calibrated with samples of known concentrations of standard proteins (this must be done with a number of proteins having different molecular weights) then effectively an extinction

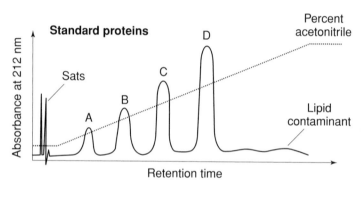

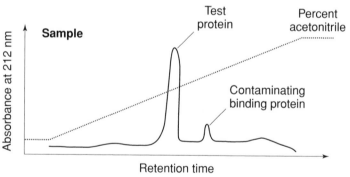

Figure 7.3

Reversed-phase HPLC separation of proteins. Proteins are bound to the HPLC resin, and are eluted with an increasing concentration of acetonitrile. Their absorbance is measured at 212 nm (specific for peptide bonds). In the top panel, a known concentration of four standard proteins has been run. The area under each peak is different due to the different numbers of peptide bonds present (which is dependent both upon concentration and molecular mass). These data can be used to calibrate the system in order to determine the concentration of any pure protein of known molecular weight. The HPLC conditions used are denaturing, which means co-purified proteins will be separated, and can be individually quantified (as well as their binding ratio).

coefficient can be determined for the area under the absorbance peak curve representing a protein, and the absolute concentration of peptide bonds (and so by reference to the length of the protein the absolute concentration of test protein). This 'apparent extinction coefficient' must be determined by you using your specific HPLC equipment, because lots of factors can affect it (*Figure 7.3*).

7.2 Separating a test protein from the rest of the proteome using two-dimensional gel electrophoresis

If there is no specific method for quantifying your chosen test protein, and no simple affinity method for purifying it, then you must find a way of separating the protein from others in a cell extract. There are likely to be thousands of different proteins in each cell extract (though perhaps only hundreds in a bacterial cell extract) and the best way of separating proteins one from the other is on the basis of their physical properties. The most commonly used way of attempting to separate all the constituents of a proteome is two-dimensional gel electrophoresis. In this approach, proteins are separated on the basis of their isoelectric point (using isoelectric focusing gel electrophoresis) in the first dimension and their molecular size (using SDS-PAGE) in the second dimension. SDS-PAGE has been discussed at length in Section 6.3. Isoelectric focusing will be explained below. The combination of these two separation approaches has the potential to break up the proteome into many discrete spots. However, since a large proportion of proteins within a proteome are 'average', both in terms of size and isoelectric point, there will always be a congested area within any two-dimensional gel, where proteins overlap, and are difficult to individually quantify. Approaches for minimizing this problem are described in Section 7.3.

Isoelectric focusing: what it is and how to do it

All proteins are made up of amino acids with different physical properties. Just by chance (or through natural selection), some proteins have an over-abundance of amino acids with acid side chains, and some have an over-abundance of amino acids with basic side chains. At neutral pH, acid side chains are deprotonated (and so are negatively charged) and basic side chains are protonated (and so are positively charged). Thus the balance between the number of acidic and basic amino acids dictates the net charge of a protein at neutral pH. Some proteins will have equal numbers of acidic and basic amino acids, meaning that at neutral pH (pH 7.0), they will not be charged at all. Thus for these proteins, pH 7.0 is said to be their isoelectric point (pI), the pH at which they have no net charge. The more the balance swings to the presence of basic amino acids, the higher the pI will be, and the more there is a preponderance of acidic amino acids, the lower the pI will be. If proteins have a net charge, they will move in an electric current towards an electrode with the opposite charge. Hence, if proteins are loaded onto a polyacrylamide gel in the absence of SDS (i.e. they retain their native net charge) they will move towards whichever electrode is oppositely charged to their net charge at the pH within the gel (*Figure 7.4*).

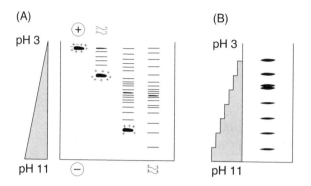

Figure 7.4

The principle of isoelectric focusing gel electrophoresis. In (A), a continuous pH gradient has been established across the gel using pre-focused ampholytes. Proteins are loaded in a buffer at one pH extremity. In this example, most proteins are positively charged, though a few are negatively charged because their pI values are less than 3.0 and these float off into the anode buffer. The proteins move through the gel, their charge eventually neutralizing and their positions becoming fixed. Those proteins with pI values greater than 11.0 will remain positively charged at the end of the gel and will continue into the cathode buffer. In (B) an immobilized pH gradient has been generated in steps. The same sample of proteins illustrated in (A) is also illustrated in (B), showing that proteins cluster at the interphase between pairs of pH regions, lowering the absolute resolving power. In practice, commercial immobilized pH gradient strips would have far more individual steps.

If, however, the gel contains a gradient of different pH regions, then proteins will move in the electric current based on the net charge at the pH found in one region, and will therefore enter another region with a different pH, resulting in a different net charge on the proteins. If one of the regions has a pH which is equal to the pI of a protein, then the protein will become uncharged within that region of the gel, and will cease to move with the electric current. It will become fixed within the gel (*Figure 7.4*).

Isoelectric focusing gel electrophoresis starts with the establishment of a pH gradient across the gel. A typical wide gradient would be from pH 3.0 to pH 11.0. The highest pH is at the bottom of the gel, where the negative electrode resides. Proteins are loaded at the top of the gel, with the most acidic pH. Hence, proteins with pI values > 3.0 will be positively charged at pH 3.0, and will move towards the negative electrode until the point when the surrounding pH equals their pI, when they will become uncharged, and will stop moving. Proteins with a pI of < 3.0 will be negatively charged at pH 3.0, and will thus run towards the positive electrode, out of the wells and will be lost. Proteins with a pI of > 11.0 will remain positively charged at pH 11.0, and so will continue to run off the end of the gel and will be lost (*Figure 7.4*). The majority of proteins will be 'average' however, with neutral pI values, and so there will be a cluster of proteins in the middle of the gel. IEF is always run in denaturing conditions, meaning that the protein samples are made up in urea, and urea is also present in the gel.

Denaturation of proteins prevents secondary structure formation, which can affect the reproducibility of IEF separation. It also exposes all charged amino acids to the gel, revealing the true pI of a protein. Another important step in blocking secondary structure is to reduce thiol groups within cysteine and methionine residues. This is done with a reducing agent such as dithiothreitol.

There are two main ways of establishing a pH gradient across a polyacrylamide gel. The first is to add to the gel before pouring it a mixture of chemical carrier ampholytes (Robertson *et al.*, 1987). These are molecules with a certain balance of acid and base constituents, meaning that they carry with them a particular pH. The gel is 'pre-focused' before proteins are loaded, and the ampholytes move based on their charge at the original pH of the gel matrix. As they move, they establish the pH gradient. After pre-focusing, the protein mixture is loaded onto the gel and the pH gradient produced by the ampholytes allows for isoelectric focusing of the proteins. However, the big problem with this approach is that the ampholytes are not fixed in space, and can move slightly during electrophoresis (generally towards the cathode). This means that the pH gradient can become distorted during each gel run; more importantly distortion is experiment-specific, making comparisons between patterns of protein bands obtained following different experiments sometimes difficult.

The other main method for establishing a pH gradient is to make a series of different PAGE gel solutions, each containing a different buffering agent, at a different pH, that becomes chemically cross-linked into the gel. Small amounts of each gel solution are poured into the gel sequentially, each being allowed to set before the next layer is applied. The end result is a series of pH zones rather than a gradual gradient across the gel. Thus whilst the strong advantage of this approach is that the gradient is spatially fixed, the disadvantage is that there is slightly less resolving ability, since proteins tend to group together at the interface between two zones (*Figure 7.4*). Because of the difficulty in making these 'immobilized pH gradient' gels, and the length of time it takes to do so, it is perhaps best to purchase them ready made by robots.

However the pH gradient has been established, different buffers at the cathode and anode, each with a pH equal to that at the extremes of the pH gradient in the gel must be used. The use of a single, intermediate pH buffer at each end of the gel would cause neutralization of the pH gradient, and would ruin the whole separation process.

Two-dimensional gel electrophoresis

Whilst isoelectric focusing gel electrophoresis is often used to separate proteins within a mixture, it has poorer absolute resolving power than SDS-PAGE. Hence, on its own it is not suitable for separating your test protein of interest from the proteomic pack. However, coupled with SDS-PAGE as a two-dimensional approach to protein separation the resolving power increases significantly, potentially allowing separation of your test protein from the rest (O'Farrell, 1975). The beauty of using these two physical separation methods is that there is no link between the physical properties of size and pI within proteins, so they are truly independent separation approaches meaning that proteins which are 'average' for one, so resolve away poorly from the proteome as a whole have every chance of being 'unusual' for the other.

Traditionally, the IEF dimension is run first, and then each 'lane' of the IEF gel is placed on top of an SDS-PAGE gel which is made without wells, and the proteins are separated according to mass. There are a number of complexities. First, each lane of proteins on the IEF gel must be thin enough to be placed onto the top of an SDS-PAGE gel, and must be easily separated from other lanes, which will be separated on other SDS-PAGE gels. For those who make their own IEF gels using ampholytes to establish a pH gradient, the IEF dimension is frequently run on a so-called 'tube gel'. Simply, the gel solution is mixed and a capillary tube (around 2 mm in diameter) is stood vertically in the solution, which travels up the tube by capillary action. Once the gel has set, the bottom of the tube is dipped into a reservoir of cathode buffer at the bottom of a cylindrical container. A 'collar' is placed onto the tube, sealing the sides of the cylinder at the same time, and anode buffer is poured on top, flooding the top of the tube. It is imperative that the collar is properly sealed because the leakage of anode buffer might cause

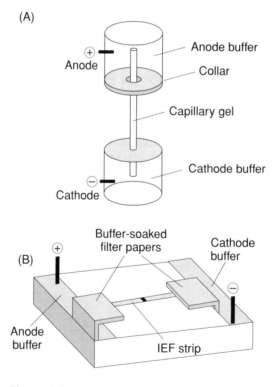

Figure 7.5

Features of IEF gel apparatus. Panel (A) illustrates a typical capillary tube gel apparatus, with two reservoirs of buffer being separated by a plastic collar. In (B) a typical immobilized pH gradient IEF strip 'flat bed' electrophoresis apparatus is illustrated. The buffers are provided soaked into filter papers that are placed in reservoirs. The sample is spotted in the middle of the strip.

the buffer level to drop below the end of the gel, stopping the flow of electric current, and might also result in significant neutralization of the cathode buffer, and a resultant distortion of the pH gradient within the gel. The protein sample is simply injected onto the top of the gel within the tube using a Hamilton syringe and an electric current applied (*Figure 7.5*).

Once run, the tube gel is extruded from the capillary tube using a syringe to force it out with air pressure. The syringe is attached to the end of the capillary tube with a piece of silicone tubing to create an airtight seal. Once extruded, the tube gel is equilibrated in buffer containing SDS and bromophenol blue. This step is to neutralize the pH, coat all proteins with negative charge so that they move towards the anode in SDS-PAGE, and also to allow visualization of protein migration in the form of a dye-front of bromophenol blue on the SDS-PAGE gel, which runs at the same speed as a protein with a molecular weight of around 14 kDa. After equilibration, the tube gel is carefully placed onto the surface of an SDS-PAGE gel which has been poured so that there is only a slight gap at the top (*Figure 7.6*). It probably appears clear to you now why there is considerable experimental variability when running two-dimensional gel electrophoresis experiments using this method for the first dimension. The tube gel can become distorted during the extrusion process, and when being forced onto the top of the SDS-PAGE gel, and this can lead to subtle differences in apparent migration of the proteins within the gel. This, coupled with the inherent problems associated with ampholyte maintained pH gradients (above), make this sort of first dimension method far from ideal for proteomics (Section 7.3). It is useful, however, for the separation of one known protein from the rest of the proteome.

Several manufacturers now make strip gels with immobilized pH gradients within them. These gels are slightly thicker than the average tube gel, and tend to have plastic backings. All in all, therefore, they are far more

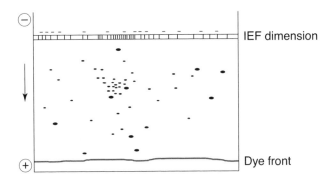

IEF dimension

Dye front

Figure 7.6

Two-dimensional gel electrophoresis. The IEF gel (capillary or strip gel) is placed onto the surface of an SDS-PAGE gel, following soaking of the IEF gel in buffer (to neutralize pH) containing SDS and a bromophenol blue dye. The positions of proteins are illustrated to show there is a clustering within the intermediate pH region of the gel. Once separated according to SDS-PAGE, again there is a clustering of proteins at the intermediate molecular weight position. The bromophenol blue dye front (illustrated) is more usually run off the bottom of the gel.

robust and are not susceptible to significant distortion when loading onto the SDS-PAGE gel. Strip gels tend to be run horizontally on a flat surface with cathode and anode buffers being provided soaked onto filter papers at each end. Another difference compared with the use of tube gels is that proteins are applied in the middle of the pH gradient. Once electrophoresis has been completed, the strip gel is transferred to equilibration buffer and slotted in between the plates of an SDS-PAGE gel, bringing it into direct contact with the surface. If the surface of the SDS-PAGE gel has not set perfectly flat and level, this can cause significant problems with the second dimension. Use a leveling agent such as water saturated butanol and make sure the gel casting apparatus is perfectly level in both directions with a spirit level, using tissue paper to pack the bottom of the apparatus if necessary (see Section 6.3 for details about making and running SDS-PAGE gels).

The fact that immobilized pH gradient strip gels are manufactured by robots means it is less likely for there to be problems with variability between the compositions of gels made on different days by different scientists. It also allows more gradual gradients to be produced than would be possible by hand. I would strongly recommend going for this strip IEF gel option (*Figure 7.5*). One example would be the Immobiline Dry Strip gels from GE Life Sciences. The gels are 3 mm by 0.5 mm, and come in a variety of lengths depending upon the size of the SDS-PAGE dimension (70–240 mm). The gels are further strengthened by being fixed onto a polyester film backing. They are available in wide range pH gradient form, for example pH 3 to pH 11, right down to very narrow range gels, for example pH 5.0 to pH 6.0 depending upon the resolution you require (see Section 7.3).

There is no need for a stacking gel in the SDS-PAGE dimension of two-dimensional gel electrophoresis. The gel is clipped into running apparatus, the reservoirs filled with SDS-PAGE running buffer (see Section 6.3) and an electric current applied to separate proteins (*Figure 7.6*). More information about preparing samples for two-dimensional gel electrophoresis and about running these gels is presented in Section 7.3.

Visualizing proteins following two-dimensional gel electrophoresis

As with standard SDS-PAGE gels, it is possible to stain proteins within two-dimensional gels with Coomassie blue dye. This stain is not particularly strong, though, with its limit of detection being around 10 ng mm^{-2} in a gel. If your test protein is high abundance, you may well see a distinct spot, but if the protein is low abundance, or if it is poor in the particular amino acids to which Coomassie blue dye binds, you will probably need a more sensitive stain. For this eventuality, there are three main options. First, is the traditional silver staining approach, which is very popular for proteomics experiments. There are many different types of silver staining techniques but all fall into one of two categories: negative and positive staining (Rabilloud, 1990).

Silver ions bind to proteins through a number of different interactions, and associate with several different amino acids (the exact interactions made will depend upon the pH of the solution), making them bind in a manner that does not differentially stain proteins with an over- or under-abundance of certain amino acids (as does Coomassie blue dye). The development of all silver stain protocols involves the reduction of silver ions. When complexed with proteins, silver ions are less reducible than when complexed with polyacrylamide gels. Therefore, if a solution of silver is washed into a gel, it has the potential to mark the positions of the proteins, because, when a reducing agent is applied, the background gel will 'develop' reduced silver (and so will go brown), whilst the sliver bound to proteins will remain uncolored for longer. This is the basis of the negative stain, and is very tricky to perform because the difference in rate of reduction between protein and polyacrylamide-bound silver ions is not great, meaning the timing has to be absolutely correct. Therefore, positive silver staining methods are almost exclusively used. Here, a silver ion solution is washed into the gel, and binds to proteins tightly, and the gel weakly. A wash step is included to rinse silver out of the gel, and then a strong reduction step is used to reduce the silver that remains bound to proteins, and so reveal them as brown bands or spots. Other complexities of the various silver staining procedures are there to either enhance the binding of silver to proteins, or reduce the silver ion concentration in the gel. Therefore, they enhance the signal strength and the contrast between protein spots and the background gel.

Silver staining is tricky. The reagents used must be of the very highest quality and purity, and the procedure is very sensitive to experimental variation. Differences in temperature and light intensity (remember, silver reduction is the whole basis of photography!) during the reaction can lead to very different staining results on different occasions. Dirt is also a real worry, introduced onto the gel through gel plates, dirty fingers or more likely from the containers used in the staining procedure. Make sure you use gloves, clean your gel equipment very well, and try to perform the staining protocols in exactly the same way each time. Some manufacturers make gel staining machines, which add reagents in the appropriate order for an appropriate amount of time, and keep temperature and light levels constant. Only consider investing in such a machine if you will be producing large numbers of two-dimensional gels. Otherwise, the best advice I can give is to stain the gels containing your two comparator preparations of cellular protein in parallel, making one double batch of each reagent and splitting it in two, incubating both gels at the same time and in the same place. Thus for each experimental replicate, it is possible to confidently compare spot intensities between the two comparators.

There is now a whole host of chemical dyes that can be used to stain proteins in SDS-PAGE gels. They have the distinct advantage over silver staining that they are simple one-step staining procedures, and are far less likely to suffer from the problems of experimental variability. An example of such a dye would be the SYPRO dyes, which are fluorescent, and interact with proteins that have been coated in SDS. Different SYPRO dyes have different emission spectra. The sensitivity of SYPRO is thought to be equivalent to that of silver staining (Berggren *et al.*, 2001).

The third approach to visualizing proteins in gels is to tag all the proteins in a preparation with a fluorescent dye before running the IEF dimension. This means there is no possibility of differential staining of proteins within the gel due to, for example, differences in diffusion rates in different parts of the gel. The main advantage of this approach is that two preparations of proteins can be fluorescently labeled using different dyes in parallel, and then be mixed and separated at the same time on the same gel. This will allow differences in protein spot intensity to be revealed as an over-abundance of one fluorescent label in a given spot on the gel. This is the basis of two-dimensional difference gel electrophoresis (Unlu *et al.*, 1997), which will be discussed in a little more detail in Section 7.3.

How do you know which protein spot contains your test protein?

Following the completion of a two-dimensional gel electrophoresis run, and the staining of proteins within the gel, you will be met by a whole host of spots. Section 7.3 will describe in more detail the problems of separating protein spots, and what can be done to help, but for this section, let us assume that you have been able to separate your test protein from all the others. But how do you know which spot it represents in order to be able to quantify the relative abundance of the protein in two different preparations of proteins, each having been resolved in an identical manner by two-dimensional gel electrophoresis?

Well, this is a bit of a 'speculate to accumulate' problem. You need to know a fair bit about your test protein if you are to easily find it on a two-dimensional gel. It would be best if you knew its predicted amino acid sequence; that is, derived from the cDNA sequence (and not the genomic sequence of the gene, if eukaryotic, since differential splicing might significantly affect the sequence of the protein product). From this amino acid sequence, you can calculate the predicted pI and molecular weight of the protein product. There are a number of computer programs that can do this calculation, such as 'Protean' from the *DNA Star* suite of programs. This information will allow you to predict where on the gel the protein spot should be. However, this will only be possible if the gel has been calibrated using standard proteins having known molecular weights and pI values. These standard proteins are mixed with the test protein sample and are provided in relatively large amounts so that they stand out on the gel. They are designed to run at widely spaced positions, allowing calibration of both the pI and molecular weight dimensions, since neither can be assumed to be a simple gradient (except for the IEF dimension when using an immobilized pH gradient gel). Two-dimensional gel electrophoresis calibration standards are available from a number of manufacturers, for example from BioRad (161-0310), where the standards cover a pI range of 8.5 to 4.5 and a molecular weight range of 76 kDa to 17 kDa.

If you find a protein spot at the expected size and pI, it may not be your protein. Indeed, there might be several spots in the vicinity. One of the simplest ways of differentiating between a small number of proteins, and so finding which protein spot represents your test protein is to determine a proteolytic cleavage pattern for each protein, and match the patterns to the predicted protease cleavage signature of the test protein, according to its amino acid sequence. The most commonly used proteases for this purpose are trypsin, which cleaves after lysine and arginine residues (unless they are immediately followed by a proline), Glu-C, which cleaves after glutamic acid residues, and Asp-N, which cleaves just before aspartic acid residues.

Protein spots are cut out of the gel and proteins are eluted from gel pieces into water containing a 1 mM dithiothreitol. Proteins are then acetone precipitated (add 4 volumes of 100% acetone and place at –20°C overnight) to remove impurities from the gel and protein dye, which stay in the supernatant, and the pellet is made up in the appropriate protease reaction buffer prior to digestion (50 mM NaH_2PO_4, pH 7.8 for Asp-N; 25 mM $(NH_4)_2CO_3$, pH 7.8 containing 5% acetonitrile for Glu-C; 50 mM NH_4HCO_3, pH. 8.0 containing 5% acetonitrile for trypsin) using a 10:1 protease to protein w/w ratio. Once digested, samples are acidified by the addition of 5% acetonitrile in water containing 0.1% v/v trifluoroacetic acid, and are separated using reversed phase HPLC Typically, a C_8 column is used, running a 35 min gradient of 5–75% acetonitrile in water containing 0.1% v/v trifluoroacetic acid with a flow rate of around 200 µl min^{-1}. The trace will reveal how many peptides are produced using each protease, and their respective retention times will give you the approximate relative sizes of the individual peptides. Whichever protein gives a pattern of peptides predicted for the test protein from the cDNA sequence encoding it is highly likely to be the true test protein. In fact, often simply the number of peptides is sufficient information to identify the protein. If by very bad luck, two proteins give the same number and patterns of peaks predicted for the test protein with one protease, try a second protease, since the chances of the same pattern being produced upon digestion of two different proteins with two different proteases is very small indeed.

As a final test for whether a spot contains your test protein, the protein can be subjected to total acid hydrolysis (95 min, 150°C in the presence of 6 M HCl). The free amino acids are then derivatized by reacting them with phenylisothiocyanate and the amino acid composition can be determined using a HPLC system that has been calibrated with pure derivatized amino acids to reveal their individual retention times. The HPLC column is a C_{18} matrix, and the initial liquid phase buffer (buffer A) is 50 mM sodium acetate pH 5.45. Derivatized amino acids are eluted using a gradient 7–60% buffer B (70% acetonitrile/32 mM sodium phosphate pH 6.1) over 30 min, 300 µl min^{-1} (all derivatized amino acids have a retention time <25 min under these conditions). The relative sizes of the peaks representing each amino acid (measured at 254 nm, which is the wavelength absorbed by the phenylisothiocyanate chromophore) reveals the relative abundance of each amino acid in the protein. Whilst the overall abundance of some amino acids in proteins is fairly stable, others have dramatically different abundances, and so this approach will give a fingerprint for each candidate test protein, and the known sequence of the test protein will enable you to locate the real test protein amongst all the candidates. Acid hydrolysis completely destroys cystine and tryptophan, and converts asparagine and glutamine to aspartic acid and glutamic acid, respectively, so this will affect the results and should be taken into consideration when producing a predicted composition for the test protein. Machines are available for amino acid analysis. They are essentially HPLC systems with sophisticated detection apparatus and a computer that calculates amino acid composition. For example the Applied Biosystems

130A, which comes with an automatic phenylisothiocyanate derivatization chamber.

7.3 Determining changes in the proteome

So far we have discussed methods by which a test protein can be separated from the rest of the proteome and quantified, either absolutely, or in one protein extract relative to another. More and more, however, scientists wish to take a big picture, and actually look at all the proteins in the proteome as a whole. Then, when analyzing proteomic complements of two cell extracts, each derived from cells growing in a different physiological state, it is hoped that differently produced proteins, encoded by genes expressed at different levels in the two physiological states, can be located. The basic method for separating the proteomic complement is two-dimensional gel electrophoresis as described in Section 7.2, above.

Getting the required resolution

It is obvious that the degree of separation and the resolution required for separating an entire proteome is far greater than when simply looking at a test protein spot. A eukaryotic cell may produce 10 000 to 50 000 different proteins at any one time. Thus the biggest problem is to separate out all these proteins into discrete spots. If you produced the world's largest IEF and SDS-PAGE gels, then clearly this would be possible, but most of us only have limited laboratory space, so we need to find alternative ways of improving the level of separation we can achieve.

The IEF dimension is perhaps the easiest to manipulate. It is possible to produce a 24 cm strip IEF gel with an immobilized pH gradient of 3.0 to 11.0, but it is equally possible to produce a similar strip gel with a pH gradient of 6.8 to 7.0. What would happen if proteins were run on the latter gel is that many would remain charged wherever they were in the gel and would simply move off into the buffer at one end or the other. Thus the resolving power of the gel would dramatically increase, but the number of proteins that could be resolved would be very small. In reality, the smallest pH gradient I have found in a commercially available strip gel is one pH unit, and it is possible to divide a protein sample up into aliquots and run each aliquot on a different single pH range gradient gel. Thus if covering pH 3.0 to pH 11.0 in total, eight separate gels would be run, increasing the effective IEF gel length from 24 cm (the standard large strip gel length) to more than 1.5 meters. In practice, it is probably not necessary to do this, since there will be very few proteins with pI values at the extremities. One might choose a gradient of pH 3.0 to pH 5.0, then three single unit pH gradients covering pH 5.0 to pH 8.0, with a final single gel having a gradient of pH 8.0 to pH 11.0.

Similarly, it is possible to produce SDS-PAGE gels having different percentage acrylamide and different acrylamide:bis-acrylamide ratios. An 8% acrylamide gel with a 50:1 acrylamide:bis-acrylamide ratio, if subjected to electrophoresis for the appropriate amount of time, will effectively separate large molecular weight proteins, whilst small and even medium sized proteins will run off the anode end of the gel. Likewise, a 15% gel with an acrylamide:bis-acrylamide ratio of 37.5:1 will separate small proteins, and if run long enough, medium-sized proteins, whilst large proteins are trapped at the top of the gel. This can cause problems, though, since a large protein smear at the top of the gel might mask spots further down. So, one possible alternative is to pass the protein sample through a size exclusion matrix in order to separate small from large molecular weight proteins. A crude way of doing this might be to pass the sample through a filter having a very strictly defined pore size (for example a Centricon spin filter with a 30 kDa size cut-off). Small proteins would pass through the filter, proteins > 30 kDa would stay above the filter. Both pools could be subjected to two-dimensional gel electrophoresis separately. The crudity of this approach is because of two main facts. First, some proteins behave strangely when being passed through these types of filters, depending upon secondary structure, though this problem can be reduced by using urea and dithiothreitol to denature the proteins and reduce disulfide bond formation, thereby removing the vast majority of secondary structure. Second (and this problem actually increases in denaturing conditions) many proteins stick to the filter membrane, and to a greater or lesser extent, are subtracted from the sample. Thus in terms of proteomics the possible differential loss of unknown numbers of proteins is far from ideal.

A better approach to size fractionation of proteins prior to two-dimensional gel electrophoresis is the use of size exclusion chromatography (*Figure 7.7*). Here, a mixture of proteins in solution (and it is best to have them devoid of secondary structure) is loaded onto a column containing beads with pores in them of different sizes (for example Sephadex G-25). Small proteins can fit into the pores, and so take a long time to pass out of the other end of the column. Large proteins do not fit into the pores, and

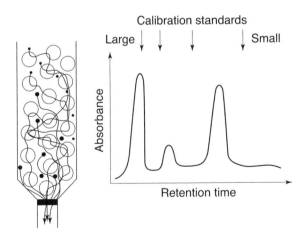

Figure 7.7

Size exclusion chromatography. The Sephadex beads have pores; the smaller a protein, the more likely it is to become diverted into these pores as it passes down the column, and so the longer it will take to emerge from the bottom. This retention time can be calibrated for a given chromatography apparatus using known molecular weight standards.

pass around the outside of the beads, taking a much shorter amount of time to reach the end of the column. Thus, by simply collecting fractions of the column eluent over various periods of time, one can effectively split the sample into groups of proteins selected on the basis of their similar size. Each fraction can be separately run on a two-dimensional gel system, providing a snapshot of the protein abundances in that group of proteins. Furthermore, dividing each fraction into several aliquots, and running each through a different combination of IEF and SDS-PAGE conditions will mean that the resolving power of the experiment as a whole is enough to see virtually all proteins as discrete spots. The downside, of course, is that you will need to prepare larger amounts of total cellular protein to start with, since aliquotting a diluted small sample is likely to result in excellent resolution of proteins whose abundances are below the level of detection by the stain you choose to visualize them. Getting hold of large amounts of protein might not be particularly easy, depending upon your chosen tissue or cell type of interest.

Preparing protein samples prior to proteomics

I do not have space to describe all possible extraction methods, or indeed extraction buffers that could be used to prepare cell extracts of a large variety of different tissues. The best way of producing a sample of total protein (i.e. also including very hydrophobic proteins including membrane proteins) is to treat the cells with a solution of 7 M urea, 2 M thiourea, 40 mM Tris base and 4% w/v of a zwitterionic detergent such as CHAPS. A strong reducing agent will be required, such 2 mM tributylphosphene (which significantly helps to solubilize proteins). Iodoacetamide (15 mM) can be added to alkylate proteins, thus preventing disulfide bonds reforming, though be careful with this chemical as it is very unstable (Herbert *et al.*, 2001). Around 1 ml of extraction buffer per 100 mg of tissue or 50 µl of cell pellet is about right, but for plant tissue at least double the amount of buffer. Always err on the side of too much rather than too little buffer since proteins can be concentrated later on, and incomplete lysis and solubilization of proteins will result from too little buffer, and will totally ruin the experiment. Tissues and cells must be homogenized, possibly using a bead grinder, though it is probably best to sonicate them. Keep the sample cold (if the sample is warmed above 30°C, the urea can precipitate), and I would definitely recommend the use of protease inhibitors such as pepstatin, antipain and leupeptin (1 µg ml^{-1} each) and 0.1 mM phenylmethylsulfonyl fluoride. Centrifuge the sample at 14 000 × g for 25 min to pellet cell debris and use the supernatant directly for IEF analysis (though see below for hints on clean-up).

If you want to differentially extract cytoplasmic and membrane proteins, you should initially extract cells using a buffer containing only 8 M urea and 2% w/v CHAPS (i.e. with no reducing agent). The supernatant following centrifugation will contain soluble proteins and some weakly hydrophobic membrane proteins, and the pellet will contain membrane proteins as well as cell debris. This pellet should then be treated with 7 M urea, 2 M thiourea, 40 mM Tris base and 4% w/v CHAPS (1 ml of this for every 1 ml of cell extraction reagent used initially) and the sample re-centrifuged to release a membrane protein containing supernatant. If proteins in the two supernatants are not going to be precipitated (see below) add reducing and alkylating agents at this stage (2 mM tributylphosphene and 15 mM iodoacetamide).

There may be some very stubborn, often large membrane proteins remaining in the second pellet. One way of getting them into solution (and cutting them into a more manageable size) is to treat the pellet with cyanogen bromide, which cleaves the proteins after methionine residues. This will release smaller, and more soluble fragments which must be acetone precipitated (see below) to remove contaminating chemicals. When this fraction of the cells is separated using two-dimensional gel electrophoresis, the result will be a number of spots containing large fragments of membrane proteins. Thus, if the expression of one protein changes in the condition being analyzed, the intensity of a number of spots will change, but all can be identified as set out in Section 7.4 for whole, soluble proteins.

There are a large number of problems that you might encounter during sample preparation that will impact on the quality of the two-dimensional gels you will produce, some of these are described below.

Contamination with nucleic acids

It is important to thoroughly homogenize the sample, or sonicate it so that chromosomal DNA is sheared into small pieces. Nucleic acid contamination is then removed by adding 80 U of DNase and 1.5 U of RNase (both from bovine pancreas) to every 1 ml of cell extract, and incubating the sample for 1 h at 30°C. Alternatively, if your nucleases have been derived from other sources, refer to the instructions that accompany them.

Contamination with salts

Salts can severely affect the IEF dimension of two-dimensional gel electrophoresis and must be removed. This can be through size exclusion chromatography with a 6 kDa cut-off, meaning that all molecules >6 kDa pass through the column quickly and all small molecules rattle around in the pores of the beads and are separated (*Figure 7.7*). Alternatively, the sample can be treated with acetone/trichloroacetic acid (4 parts of 100% acetone containing 10% w/v trichloroacetic acid and 20 mM dithiothreitol per one part sample by volume) and the mixture incubated for 1 h on ice. Proteins will precipitate and salts will remain in the supernatant, which is removed following centrifugation of the sample (14 000 × g for 10 min at 4°C).

Other contaminants

Other contaminants including detergents, lipids and phenolic compounds (which are particularly a problem with plant tissues) may also be separated from proteins using acetone/TCA precipitation as above. If present in a protein preparation, they will cause smearing, usually in the IEF dimension of a two-dimensional gel. For this reason, they should be removed when they are suspected of being present in significant amounts (and this will probably be a matter of experience with your own chosen cell or tissue type).

A brief discussion of two-dimensional gel equipment

Once you have a pellet of protein (or cyanogens bromide treated protein fragments), it can be re-solubilized in an appropriate volume of 7 M urea, 2 M thiourea, 40 mM Tris base and 4% w/v CHAPS containing 2 mM tributylphosphene (or 5 mM dithiothreitol) together with iodoacetamide (15 mM) and protease inhibitors (above) if required. You should aim to produce a protein concentration of around 400 µg ml^{-1} (if silver or SYPRO staining is being used to visualize proteins), with approximately 400 µl (150 µg) being loaded onto a 24 cm IEF immobilized pH gradient strip gel (with proportionately less if smaller strips are being used). An identical amount of total protein must be used for two-dimensional gel electrophoresis when comparing proteomic complements in two different preparations of cells. Total protein concentration can be adequately determined using a protein dye-based method such as the Bradford method described in Section 6.2.

Many different suppliers make two-dimensional gel electrophoresis apparatus. Each will come with detailed instructions concerning its use, and the voltages and times required to resolve proteins. Stick to these instructions and do not try to run gels at a higher voltage than is recommended, in order to get the result more quickly. This course of action is likely to cause the gels to overheat, which may distort the pattern of protein spots. Thus what time you save in running the gels, you lose when coming to analyze the results.

I do not have time to go through all the possibilities for gel apparatus, though my discussions above will hopefully have pushed you away from the use of tube IEF gels and towards the use of horizontal strip immobilized pH gradient gels. I must also encourage you to purchase the largest SDS-PAGE system you can afford. Even if you currently do not require 24 cm gel separation, there may well come a time in the future when you will, so go for the largest gels now and you can expand into them. Shop around, because there is considerable competition in this sector.

Whatever gel apparatus you choose, keep it clean, and keep it in good order. Make sure electrodes are tightened and sanded regularly to keep a good electrical contact. Replace the wires every few years in case their insulation has become damaged. Replace gel plates if they become chipped or cracked, or even if they become rough to the touch, which can happen with repeated cleaning due to the abrasive action of dried polyacrylamide gel matrix during cleaning. Make fresh electrophoresis buffers regularly, and check their pH before using them. All these steps will improve your chances of getting reproducible separation of proteins over many years of happy proteomics.

Visualizing and calculating proteomic differences

Spots on a two-dimensional gel are usually visualized with silver or fluorescent stains (see Section 7.2) because they are more sensitive to small amounts of protein than dyes like Coomassie blue. Once the proteins within a gel have been stained, the pattern of spots can be captured as a digital image. If using silver staining, then the gel can be scanned. If using a fluorescent stain, then usually an image station is employed to both excite the fluorophore, and record the image of emitted fluorescence on a charge coupled device. A similar approach is used to capture an image of target:probe hybridization to spots on a micro-array (Section 3.12). However, the situation with protein spots is far more complex. Arrays are ordered regular patterns of probe DNA spots, two-dimensional gel electrophoresis involves a separation process, and as such, the spots may become differentially separated on different gel runs. Therefore, direct comparison of spot intensities on a digital image using image analysis software may not be possible. All the images must be distorted so that all the proteins spots overlap. To do this, a computer algorithm must detect the positions of a group of control spots (these are used to calibrate the IEF and SDS-PAGE dimensions, as set out in Section 7.2), and all gels to be compared are stretched digitally so that the control spots overlap. The remaining spots are then called by the computer and their intensities compared (*Figure 7.8*).

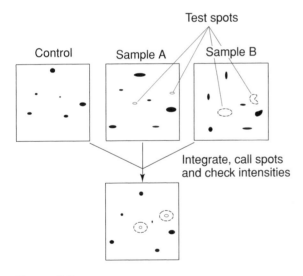

Figure 7.8

Overcoming distortion in two-dimensional gels. Two protein samples are run in parallel, having first being spiked with a number of standard proteins. Sample A has been run slightly too long in the IEF dimension (not possible with an immobilized pH gradient); sample B has been run too long in the SDS-PAGE dimension, and there is a distortion on the gel in the far right. A third gel must be run to calibrate the system, in order that the gel analysis software learns where the standard proteins run. These protein spots are then located in all sample gels, and their positions normalized by the software, in order that the positions of the test protein spots will be similarly normalized. In the figure, only two test protein spots are illustrated, and both are considerably more intense in sample B (dotted line) than in sample A (solid line).

One important point to note is that when the computer flags up a spot whose intensity is different in one comparator gel than the rest, have a look at the images and make sure that the real reason why the computer is calling the difference is not some mark on the gel, or some slight localized distortion on the gel. Spot analysis software is available from a number of suppliers, and most people purchase it alongside the hardware required to run the gel.

Two-dimensional difference gel electrophoresis

The basis of two-dimensional difference gel electrophoresis (2D-DiGE) is to overcome issues of poor reproducibility of separation leading to difficulties in matching protein spots from two different comparator preparations of protein by mixing the two comparators in a 1:1 ratio and subjecting the mixture to electrophoresis. The key is to be able to visualize for each spot, how much protein comes from comparator preparation A and how much comes from preparation B. The way to do this is to differentially label all the proteins in each preparation, meaning that a fluorescent signal can be read from the gel to indicate the ratio of a given protein in the two comparator preparations. The approach to visualization of two mixed comparators is identical to that seen with dual channel micro-array hybridization analysis, where each of a pair of comparator cDNAs is labeled with a different dye (Section 3.12). However, for micro-arrays, the use of this mixed label approach is falling out of favor because the labeling step introduces potential experimental error. Here, however, there is no problem associated with array spot distortion, and differential separation in different experiments, and for proteomics it is likely that the protein labeling step will introduce significantly less experimental error into the overall experiment than the use of two separate electrophoresis runs, one for each comparator.

The most commonly used 2D-DiGE dyes (as with dual channel micro-array hybridization experiments) are Cy3 (green) and Cy5 (red). These are described in Section 3.12 in detail, but to recap, they are chemically and physically almost identical (so will not differentially affect the migration of proteins to which they have been linked) though with quite different fluorescent properties. It should be remembered, however, that whilst there is no difference in the effect each dye will have on protein migration, there is very likely to be an effect on migration when either dye is linked to a protein, compared with unlabelled protein, so the use of these dyes will stop you from easily identifying protein spots using two-dimensional gel image maps (see Section 7.4) since the vast majority of these maps have been produced using unlabelled proteins.

The labeling of proteins with Cy dyes is through NHS–ester linkage. The dyes are purchased with an attached NHS reactive group (e.g. CyDye™ DIGE Fluor dyes from GE Biosciences). The NHS facilitates covalent attachment of the Cy dyes to the epsilon amino group of lysine of proteins via an amide linkage. Given the relatively low pI of this amino group, the reaction must proceed at alkaline pH, where the epsilon amino group will be deprotonated. Accordingly, proteins should be acetone precipitated (above) and made up to around 50 mg ml^{-1} in suitable alkaline buffer (10 mM Tris pH

8.5, 5 mM magnesium acetate, 8 M urea, 4% w/v CHAPS). The ratio of dye to protein has to be closely monitored. The recommendation is that 50 µg of protein be labeled with 400 pmol of Cy dye for 30 min on ice in the dark. After this, the reaction is stopped by the addition of 0.2 mM (final) lysine This will ensure that approximately 3% of the available proteins are labeled and then only on a single lysine per protein (i.e. one dye per protein, or minimal labeling).

The amino acid lysine in proteins carries an intrinsic single positive charge at neutral or acidic pH. Cy dyes also carry a single positive charge which, when coupled to the lysine, replaces the single positive charge of the lysine with its own, ensuring that the pI of the protein does not significantly alter compared with the same unlabeled protein. A reciprocal gel is run where the dye labels on the comparator protein samples are reversed to account for any differences in reactivity between dyes for the proteins. The Cy dye adds around 500 Da onto the apparent molecular weight of a protein. Given that only around 3% of proteins are labeled in the ideal reaction, if the gel is treated with a standard protein stain, a double spot will be seen for each separated, with the lower molecular weight spot being greatly more abundant than the higher molecular weight, fluorescent spot. It is this nonfluorescent spot that should be picked for identification purposes (see Section 7.4) because it will provide a more useful amount of material, and the Cy modification might well affect the identification; not least because modification of lysine residues will block trypsin cleavage, which normally occurs after lysine and arginine (Section 7.4).

7.4 Identifying proteins

The aim of proteomics is to determine which proteins are differentially abundant in two different preparations of protein, each from a cell type growing in a different physiological state. In section 7.3 I have explained how it is possible to find differentially abundant spots. However, how do you find out which protein is represented by which spot? In Section 7.1 I described ways of confirming that a protein spot contains a test protein, but here, you have to know the amino acid sequence of the test proteins in order to predict which spot it makes up, and to confirm that the spot does in fact contain the test protein. It is not practical to go through all the cDNAs representing the gene expression pattern of a cell type in a given physiological state, and systematically find which spot represents each of the amino acid sequences predicted from the cDNA sequences using the approaches set out in Section 7.1. So methods need to be used that determine a fingerprint for the protein in each spot, and then match it up to the predicted fingerprints of all possible proteins produced by a genome. By definition, therefore, you will need to have the sequence of the genome, or at least a large proportion of it before spot identification by fingerprint analysis will be possible. However, it is important to note that once a spot has been identified, this need never be done again, for providing the same two-dimensional gel electrophoresis protocol is followed on each occasion, the pattern of spots will remain the same, and any computer software designed to capture the image will be able to integrate previous spot identification information into the picture. Furthermore, these patterns are fairly robust and the protocols are fairly portable, meaning that it is possible to compare patterns of spots generated in different laboratories, using different equipment, so long as the same gel matrices and pH gradients are used. Because of this, picture archives exist showing patterns of protein spots from different tissues and cells, with a number of the spots being marked with the proteins they represent. One example can be found using the url: http://www.expasy.org/ch2d/. If you are lucky enough to be able to match your pattern of protein spots to one that is already available (and you trust the people that have deposited the information!) you may never actually need to identify the proteins within the spots yourself, and can simply skip to the conclusion: protein x is produced at greater or lesser extent in condition y than in condition z. However, we can't always be that lucky, so below, I have set out the two main ways of identifying proteins. Both require different degrees of prior information.

Identifying proteins based on the masses of the tryptic peptides they break up into

As already described in Section 7.1 the pattern of proteolytic fragments generated from a protein can be used as a fingerprint, and since a predicted fingerprint can be drawn up for every protein potentially encoded by a genome (provided the genome sequence is known) it is theoretically possible to identify an unknown protein by matching its proteolytic fingerprint with one predicted fingerprint. However, the HPLC approach to deriving a proteolytic fingerprint described in Section 7.1 is not appropriate here, because it is not possible to predict the exact HPLC fingerprint that each putative protein from a cell might possess. This is because the HPLC trace is far more than a measure of the number and absolute size of all the fragments produced upon proteolytic digestion. Accordingly, some more discriminatory method is required.

The method chosen is mass spectrometry. Mass spectrometers measure molecular weight. To do this, molecules are ionized, and then separated according to the mass to charge ratio (m/z). A single charged form will travel one distance unit, a doubly charged form will travel double the distance, etc. in a given period of time. Each atom has a defined mass in daltons (1 dalton is the mass one twelfth that of carbon-12), and so the mass of a molecule is its combined atomic mass. A mixture of different molecules (e.g. proteolytic fragments of a protein) can be separated, giving a fingerprint of peptide masses. This is sufficient to identify a protein.

Ionization of sample molecules prior to mass spectrometry is performed in two main ways. Matrix-assisted laser desorption/ionization (MALDI) and electrospray ionization (ESI). MALDI works by pulsing a laser onto the sample which has been spotted onto a metal surface. The energy from the laser causes electrons and protons to be transferred from the metal into the sample, producing a protonated form. The vast majority of ions produced in this way will only have one positive charge. For MALDI, the sample is mixed with a low molecular weight matrix molecule such as α-cyano-4-hydroxycinnamic acid, which is present simply to protect the sample from the laser energy. In ESI, a liquid sample is subjected to a strong electric field (up to 6000 V) through a metal wire immersed in the sample. The result is multiple ionized forms of the sample: the number of charged forms depends on its size. In both cases, the ionized sample is vaporized, and is passed into the detection chamber of the mass spectrometer using a gas carrier at a set velocity. Once they enter the detection apparatus, ionized molecules move in a vacuum along a barrel towards an oppositely charged detector. Ions move at a speed dependent upon their mass/charge ratio, and so they hit the detector after a period of time that is proportional to this mass/charge ratio. There are a number of other approaches to detection (e.g. see *Figure 7.9*), and to getting a spectrum of mass/charge ratios from a sample, and a lot of complicated mathematics, but I do not need to go into more detail here.

The result of mass spectrometry analysis of a proteolytically cleaved protein sample is a fragment pattern showing a number of peaks, each representing a different peptide having a different and defined average mass. By average mass, of course we mean that the presence of low abundance naturally occurring isotopes means that the actual mass of each amino acid could vary depending upon which isotope is actually present in each atomic position. This fact needs to be taken into consideration when calculating a mono-isotopic mass (i.e. if the most common isotope were used at each atomic position). This fragment pattern can be matched *in silico* to all the predicted mono-isotopic mass fragment patterns for all the

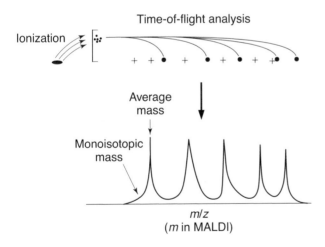

Figure 7.9

Mass spectrometric analysis of peptide masses. In the figure, peptides are ionized as single negatively charged ions (this is probably MALDI). They are passed into a vacuum tube, and all have an identical starting velocity. The side of the tube contains a strong electromagnet, which attracts the ions. The greater an ion's mass, the shorter its time of flight before hitting the side of the tube, where its presence is detected, and the information is used to generate a series of peaks, each representing the mass of a peptide. These masses can be used to identify the protein from which the peptides have been derived.

possible proteins in a genome. There are even proteomic signature databases coupled with pattern matching search interfaces available online (e.g. http://www.expasy.org/tools/aldente/) that are updated as and when new genome sequences are deposited, so that you do not have to hold large databases or pattern matching software in your computer. Simply submit a list of peptide sizes, and the protease used, and wait for the result. Like all pattern matching approaches, however, the result will be a little more complex than a single result. All potential proteolytic fragment patterns will be matched to the actual pattern submitted, and given a matching score. The output will be a list of the top few matching scores.

This all sounds very simple doesn't it? Well, like so many technologies discussed in this book, it suffers from potential pitfalls. There are two main complications to be very wary of. First, and most common, is contamination of the protein sample, meaning that there is a mixed proteolytic fragment pattern in the mass spectrum, and multiple, not particularly strong hits to more than one protein in the database. It is possible to have two proteins in the same two-dimensional gel electrophoresis spot (i.e. they have identical pI values and molecular weights but different amino acid sequences), but the most common source of contamination is the environment. If cutting spots out of two-dimensional gels by hand, be very careful not to contaminate the sample with pieces of yourself. The number of times I have seen a mass spectrum of human keratin is legion. Tiny flakes of skin seem to get everywhere, and this is why serious proteomic facilities use robotic spot pickers and clean rooms.

If you do not have access to such facilities, try to avoid contamination as much as possible. Carry out all procedures wearing latex or nitrile gloves, preferably the unpowdered versions. Store gel samples in Eppendorf tubes that have been previously washed in methanol and distilled water and make sure that the containers used for staining and destaining gels are cleaned in a similar fashion and have not previously been used for western blotting (otherwise it is likely that you will get contamination with casein or BSA). Carry out as many procedures as possible in a laminar flow cabinet.

When subjecting a spot to procedures aimed at identification of the protein contained within, it is sensible to also run a similar-sized piece of gel from an unstained portion of the whole gel to act as a control, since it will be possible to subtract mass spectrometer peaks representing general contaminants.

The second problem is that proteins are often modified *in vivo*. Glycosylation, and the addition of lipid moieties, or even the chemical modification of certain amino acids can affect the actual masses of the proteolytic fragments of the proteins, or in some cases will affect proteolytic digestion. Therefore, not all the peptide masses will match the database fragmentation pattern for a protein. However, provided there are enough fragments remaining that do match, and provided you are not unlucky enough that the modified peptides have masses that happen to match with the predicted fragmentation pattern of another protein, the fact that a series of possible matches are provided by the pattern matching software, in order of the match strength, means that you can probably identify the protein, or at least narrow down the possibilities. To absolutely identify the protein, you may need to use peptide sequencing as set out below.

Peptide sequencing by Edman degradation and amino acid analysis

Perhaps the best way to identify a protein is to obtain amino acid sequence sufficient to search the protein database. The classical way of sequencing polypeptides from their N-termini is Edman degredation. In this reaction phenylisothiocyanate reacts with the amino acid residue at the amino terminus under basic conditions (provided by *n*-methylpiperidine/ methanol/water) to form a phenylthiocarbamyl derivative. Trifluoroacetic acid then cleaves off the first amino acid as its anilinothiazolinone derivative (ATZ-amino acid) and leaves the new amino terminus for the next degradation cycle. The ATZ-amino acid is then removed by extraction with *n*-butyl chloride and converted to a phenylthiohydantoin derivative (PTH-amino acid) with 25% v/v trifluoroacetic acid in water. The PTH-amino acid is transferred to a reverse-phase C_{18} HPLC column for detection at 270 nm. A standard mixture of PTH-derivatized amino acids is used to calibrate the HPLC and so following each round of Edman degradation, the amino acid released can be identified. This chemistry, and the amino acid identification step, is coupled in an automatic peptide sequencer, such as the Perkin Elmer Applied Biosystems 494.

One major problem with Edman degradation as a protein sequencing tool comes from the fact that sequencing is from the N-terminus, and in many cases eukaryotic proteins are blocked at their N-termini by the presence of lipid groups, which prevent action of phenylisothiocyanate on the N-terminus of the protein, so sequencing cannot commence. This problem can be overcome by the use of HPLC separated peptides as substrates for sequencing. These are generated by proteolysis of the test protein, and can be simply an aliquot of tryptically digested peptides taken from the sample prepared for mass spectrometry. The N-terminal peptide is likely to be hydrophobic because of the lipid block, so it is advisable to pick peptides with short retention times. It is almost guaranteed that sequencing two separate 10 amino acid peptides from a single protein will provide enough specific sequence information to allow unequivocal identification of the protein.

Whilst automated Edman degradation is the 'gold standard' for sequencing peptides, there are some very sophisticated mass spectrometry techniques being developed that allow peptide sequencing. The principle is to subject the peptides to a single mass spectrometry run to determine their masses. The peptides are then subjected to fragmentation procedures, which break up the peptides into clusters of shorter peptides by shearing peptide bonds at random positions. The clusters are then passed through a second mass spectrometry run (hence the name tandem mass spectrometry, MS–MS, for this approach) in order to determine the masses of the fragmentation products. From the second mass data, it is possible to extrapolate back to the order of amino acids in the peptide. The maths is incredibly complicated, and you are highly unlikely to take this approach yourself. Most proteomics facilities collaborate with mass spectrometry experts who are able to perform and interpret the results generated by this technique, and it is only important that you know it exists.

Further reading

Lorkawski, S and Cullen, P (Eds) (2003) *Analysing Gene Expression, A Handbook of Methods, Possibilities and Pitfalls*. Wiley VCH, Weinheim, Germany, pp. 623–655.

References

Berggren, KN, Chernokalskya, E, Lopez, MF, Beecham, JM and Patton, WF (2001) Comparison of three different fluorescent vizualization strategies for detecting *Escherichia coli* ATP synthase subunits after sodium dodecyl sulfate–polyacrylamide gel electrophoresis. *Proteomics* 1: 54–65.

Herbert, B *et al.* (2001) Reduction and alkylation of proteins in preparation of two-dimensional map analysis: Why, when and how? *Electrophoresis* 22: 2046–2057.

O'Farrell, PH (1975) High resolution two-dimensional electrophoresis of proteins. *J Biol Chem* 250: 4007–4021.

Rabilloud, T (1990) Mechanisms of protein silver staining in polyacrylamide gels: a 10 year synthesis. *Electrophoresis* 11: 785–794.

Robertson, EF, Dannelly, HK, Mallow, PJ and Reeves, HC (1987) Rapid isoelectric focusing in a vertical polyacrylamide minigel system. *Anal Biochem* 167: 290–294.

Unlu, M, Morgan, ME and Minden, JS (1997) Difference gel electrophoresis: a single gel method for detecting changes in protein extracts. *Electrophoresis* 18: 2071–2077.

Protocol 7.1 ELISA analysis

EQUIPMENT

96-well plates
UV–visible spectrophotometer

MATERIALS AND METHODS

1. Add 50 μl of protein antigen diluted in ELISA coat buffer (2.93 g l^{-1} $NaHCO_3$, 1.59 g l^{-1} Na_2CO_3, pH 9.6) to 96-well polyvinyl chloride (PVC) plates. Use 50 μg ml^{-1} and then several 1:10 serial dilutions.
2. Incubate for 2 h to overnight at 37°C.
3. Aspirate the antigen solution from the sample wells (plates can be stored at –20°C for a month after this stage).
4. Add 100 μl of PBS–Tween buffer (137 mM NaCl, 2.7 mM KCl, 4.3 mM Na_2HPO_4, 1.8 mM KH_2PO_4, pH 7.2 containing 0.05% v/v TWEEN-20) to each well, mix the solution, and aspirate the PBS–Tween buffer from the well.
5. Add 200 μl of ELISA coat buffer containing 3% w/v bovine serum albumin to each well and incubate the plate at room temperature for 2 h.
6. Wash sample wells with 100 μl of PBS–Tween buffer.
7. Repeat the BSA–ELISA coat buffer wash.
8. Add 50 μl of undiluted monoclonal supernatant or diluted serum to each well. The appropriate dilution should be 1:500 to 1:10 000 for polyclonal antisera and about 100-fold less for monoclonal supernatant (dilute in PBS–Tween buffer).
9. Incubate plate at room temperature for 4 h to overnight at 4°C.
10. Wash sample wells three times with 100 μl of PBS–Tween buffer (as in step 4).
11. Add 50 μl of a 1:2500 to 1:10 000 (v/v) dilution (in PBS–Tween buffer) of alkaline phosphatase-conjugated secondary antibody (anti-mouse IgG for monoclonal, or anti-whatever IgG depending upon whatever animal you used to make the polyclonal antisera).
12. Incubate the plate at room temperature for 2 h.
13. Wash the sample wells three times with 100 μl of PBS–Tween buffer (as in step 4).
14. Wash the sample wells three times with 100 μl 0.15 M NaCl.
15. Add 100 μl of ONPP substrate (1 mM ONPP in 100 mM Tris-HCl, pH 8.7 containing 1 mM $MgCl_2$) to each well and watch for the development of a color.
16. Measure the absorbance at 405 nm using a microtiter plate reader after a set amount of time. Take more than one reading at different time points to confirm that the reaction is in the linear phase (see Section 6.2).

Protocol 7.2 Cyanogen bromide cleavage of insoluble proteins

EQUIPMENT

Glass tubes
Thermostatically controlled heating block
Fume hood

MATERIALS AND METHODS

1. To the protein pellet from 100 mg of cells, add 1 ml of 0.1 M Tris-HCl, pH 8 containing 3% w/v SDS. Add dithiothreitol to 2 mM and 2-hydroxy-ethyldisulfide (or cystamine) to 20 mM. Heat for 2 min at 100°C. (This step blocks SH groups.)
2. To one 30-μl aliquot, add 70 μl of formic acid containing 5 mg cyanogen bromide (CNBr). To two other aliquots add 70 μl formic acid with 0.7 mg CNBr or 70 μl with 0.1 mg CNBr.
3. Incubate the samples for 4 h at room temperature then heat for 5 min at 50°C in a fume hood.
4. Neutralize 100-μl samples with 200 μl N-ethylmorpholine on ice.
5. Mix solution thoroughly with 5–10 volumes of reagent grade acetone, store for 1 h at −80°C or overnight at −20°C.

Statistical analysis of gene expression data

8

8.1 Statistical analysis: what is the point?

Sorry, but it has to be done. If you want to do experiments where gene expression is being measured in a cell or tissue in two different physiological states in order to determine the effect of those states relative to one another, you will need to do some statistical analysis of the results to confirm that the differences seen are significant. I aim here to give a basic outline of statistical approaches that are routinely used. If you need more complex analysis, then might I direct you to a statistics textbook?

Let us take a typical example. You measure the expression of a gene in a cell type under what you call 'control' conditions, and again under 'experimental' conditions. The experiment might be the addition of a drug or the presence of a stress, or it might be where a mutation has been made in the genome of the cell type. Whatever, the analysis of the data is the same: is the expression of your gene of interest different in one of the conditions than it is in the other? In an ideal world, you do the experiment three times and get exactly the same gene expression level data for both conditions on each occasion and the difference between the two is 10-fold or more. But it never works like that, does it? There is always a level of variation in gene expression in cells from each condition.

If you measured the expression of the gene of interest in cells growing in the two conditions, known as condition A and condition B hundreds of times, you would expect the values obtained, when plotted on a graph of 'value' versus 'number of times that value is observed' to give two 'distributions', one for each set of data. Each distribution would have a peak on the graph representing the modal (most common) value obtained for each data set. Each value is noted symbolically as A_i or B_i, depending upon which data set it is from, and the total number of values in the samples is called N_a or N_b. Thus we all know that the arithmetic mean of each set of values (μ_a or μ_b) would be the sum (Σ) of all the values in that set divided by the number of values:

$$\mu_a = \frac{\Sigma A_i}{N_a}$$

and

$$\mu_b = \frac{\Sigma B_i}{N_b}$$

or generically if the data set does not have a 'name':

$$\mu_x = \frac{\sum X_i}{N_x}.$$

So when you are doing statistical analysis on two sets of data, you are really saying: 'Is there a difference between their means, and how significant is that difference?' In fact, you are effectively saying how much of an overlap is there between the distributions of the two sets of data. If you obtain measurements of 1, 2, 2, 2, 3 gene expression units for the control condition and 11, 22, 33, 33, 44 gene expression units for the experimental condition, then you can see without high-powered statistics that the distributions of these data are unlikely to overlap very much and the difference between the mean values is likely to be significant. Statistical analysis is not really about validating 10-fold differences in gene expression, though, but more likely, less than five-fold changes. Of course, if you collected enough data so that a proper distribution curve can be drawn for the two conditions, it will be very obvious if the curves overlap, or if they are offset. However, we usually only collect a handful of data points, and not the hundreds it would need to give valid distribution curves. So in statistical analysis we are simply extrapolating the data available, and coming up with an estimate concerning how likely there is to be a real difference between the distributions. Clearly, the more data points we have, the more sure we can be about our conclusion. Thus for small differences in gene expression, there is no substitute for large numbers of replicate experiments but for large differences in gene expression, small data sets are adequate.

8.2 Standard deviations

For each value (X_i) in a set of data, the 'deviate' is the amount it differs from the mean of the entire set of data (μ_x):

$$\text{deviate} = X_i - \mu_x.$$

The overall amount by which individual values in a set differ among themselves is in direct proportion to the aggregate deviate for all the values. However, since some values will be less than the mean, and some will be greater than the mean, it is likely that simply summing the deviate values will give an aggregate of very close to zero. Therefore, a 'squared deviate' is derived for each value in a set of data, since squaring the deviates makes them all positive.

$$\text{Squared deviate} = (X_i - \mu_x)^2.$$

When squared deviate values are calculated for each value in a data set and are all added together, this gives the sum of squared deviates (SS) for the entire set of data:

$$SS = \sum (X_i - \mu_x)^2.$$

The variance (s^2) (also known as the mean square) amongst a group of values is calculated as the average squared deviate of all of the values:

$$s^2 = \frac{SS}{N_x} = \frac{\sum(X_i - \mu_x)^2}{N_x}.$$

More often, however, variance is calculated as $^{SS}/_{(N_x-1)}$ because this allows a small data set to be used as being representative of an entire population. In measurements of gene expression, therefore, this is the best way of calculating s^2 for a set of data.

The standard deviation (s) (also known as the root mean square) of a set of data is the square root of the variance:

$$s = \sqrt{\frac{SS}{N_x - 1}}$$

$$= \sqrt{\frac{\sum(X_i - \mu_x)^2}{N - 1}}.$$

8.3   The normal distribution

If a distribution of values is what is known as a 'normal distribution' the modal value is equal to the arithmetic mean. In simple terms, this means that most commonly, the values are all pretty much the same; very occasionally, they are wildly different, but there would be no bias towards the prevalence of values that are higher or lower than the arithmetic mean. Strictly speaking, 68% of the data points should fall within 1 standard deviation (s – calculated as set out above) of the mean, 95% of the data should fall within 2 standard deviations ($2s$) of the mean and 99.7% of the data should fall within 3 standard deviations ($3s$) of the mean. For data spread over normal distributions, 'parametric' statistical analysis can be performed. This means that the absolute values of the data are used in the statistical analysis.

If the distribution of data is biased such that the modal value is not equal to the mean or some other property of the normal distribution is not applicable (even if you simply suspect the data is not normal) then nonparametric statistics should be used. Here, the absolute data are not used, but the data are ranked, and the differences between the mean rankings of the two experimental conditions are assessed.

8.4   Simple parametric statistics: the *t*-test

The most common statistical test is the *t*-test. It can be used to test whether there is a difference between the distributions of two sets of data provided that the following three requirements have been met.

1. *The two samples should be randomly and independently drawn from the source population.*

When analyzing gene expression data where, for each experiment, a separate preparation of cells or tissue is made and divided in two, with one portion being treated with a drug, or exposed to some other experimental condition, the first requirement for the use of a *t*-test is met. Furthermore, if you want to compare cells of two different types, for example normal and tumor cells, even if one cell type is a mutant derivative of the other, then the t-test can still be used.

2. *The scale of measurement has the properties of an equal-interval scale.*

Since gene expression data is always presented as a numerical scale, usually of band intensity etc. then this second requirement is always going to be met.

3. *The source populations can be reasonably supposed to have normal distributions.*

This is the difficult one. Whether the data come from normal distributions is impossible to know for sure when only limited data are available. Usually you will have 10 or fewer data points for each condition. However, even with this you can get an idea of non-normality. Divide each set of data up into clusters (e.g. <10, 10–20, 20–30, 30–40 etc., as appropriate for your data range). Plot a very simple bar graph of the number of values falling into each cluster. If the graph looks obviously non-normal, then use a nonparametric statistical test. If the graph gives you no reason to believe that the data are not normal, then you should be alright with a parametric test. Most biological data sets are normal. The exceptions are where you have not controlled a variable in the experiment, and so there are actually multiple sets of data within what you think is just one.

Performing an unpaired *t*-test

In the vast majority of cases, an unpaired t-test is used, since the two sets of data being compared are independent of each other, representing different preparations of cells or tissues. This is the test described here. However, on rare occasions, you will only have one single piece of tissue, which you will divide into several portions, half of which will be treated differently to the other half. In this case, a paired t-test should be used, since the original samples are identical; truly paired. For the description of a paired t-test, you will need to look elsewhere.

In this example of an unpaired t-test, you are comparing two sets of data, which we shall call A and B. The t-test works by calculating the ratio of the difference between the mean values of the two sets of data ($\mu_a - \mu_b$) and an estimate of the variability in all the data collected, called the 'standard deviation of the sampling distribution of sample-mean differences' (σ_{M-M}).

To estimate σ_{M-M}, the first thing you need to do is to calculate the sum of squared deviates (SS) for each set of data (SS$_a$ and SS$_b$) with N_a or N_b data points (N_a and N_b do not have to be identical). SS values are calculated as set out above. From these values, you can calculate the pooled variance (s^2_p) of all the data in the test (i.e. from both experimental conditions). This is determined by calculating s^2 for each data set (as described above) and adding the two results together:

$$s^2_p = s^2_a + s^2_b.$$

From this, the standard deviation of the sampling distribution of sample-mean differences can be estimated as:

$$\text{est.}\sigma_{M-M} = \left(\frac{s^2_p}{N_a}\right) + \left(\frac{s^2_p}{N_b}\right).$$

From this, the unpaired *t*-test formula is:

$$t = \frac{\mu_a - \mu_b}{\text{est.}\sigma_{M-M}}.$$

Interpretation of the *t*-test

If the *t* value calculated is a negative value, then the expression of your gene is higher in condition B than in condition A, and vice versa if *t* is positive.

To find out whether the difference is significant, you need to decide what level of significance (the critical *p* value) you are using. The standard critical *p* value is 0.05. This literally means that the difference between the means of the two groups of data observed in your experiment will have occurred simply by chance 5% of the time. As the *t* value changes, the *p* value changes, and so the percentage of times the difference between the means seen might be simply due to chance will change accordingly.

To convert a *t* value into a *p* value, you need to know how many degrees of freedom (*df*) you have to work with. This is basically the amount of data you have to work with:

$$df = (N_a - 1) + (N_b - 1).$$

Values of *t* representing different critical *p* values have been determined for all possible *df* values and are presented in statistical tables (some are presented in *Table 8.1*). Simply decide on your required critical *p* value and read off the *t* value needed to give that *p* value for the particular *df* value you have to work with. If the *t* value you calculate for your data sets is greater (forget about the sign) than the *t* value quoted in the table, then the *p* value for the difference in means you see is more significant than the chosen cut-off. By looking at the *t* value in the table most similar to the *t* value you calculate, you can estimate the actual *p* value for your data.

8.5 Simple nonparametric statistics: the Mann–Whitney test

If something seems not quite 'normal' about your data, then you are best performing nonparametric statistics. The nonparametric equivalent of the unpaired *t*-test is the Mann–Whitney test. This is a test where all the data in both sets being compared are combined in order to rank them. Rank number 1 is assigned to the data point with the lowest value and so on. When two or more data-point values are identical, they are each given a rank which is the average of all the ranks they share. Once this is done, the data points, now converted to ranks, are returned to their two individual groups, and the comparison of the difference between the sets reverts to a comparison of the ranks. If all the original data are mixed up, with no real difference between them, the ranks they are ascribed will be mixed up as well. Thus, when all the ranks in each group are summed, the sum from both sets of data would be about the same. What you are looking for is a significant difference between the two sums of ranked data. The level of significance ascribed to any difference between the summed ranks is always related to the maximum obtainable difference of summed rank, which of course is dependent upon the total number of data points. Thus again, the more data you obtain, the more significant the result you obtain can be.

To explain the Mann–Whitney method, let us take an example experiment. The control condition (condition A) gives gene expression data of:

$$10, 13, 14, 18, 19, 23, 26.$$

For the experimental data (condition B) the gene expression values obtained are:

$$15, 19, 20, 28, 30, 31, 32.$$

The data look different, but is the difference statistically significant? First, combine the data in rank order:

Group: A,   A,   A,   B,   A,   A,   B,   B,   A,   A,   B,   B,   B,   B

Data: 10, 13, 14, 15, 18, 19,   19,   20, 23, 26, 28, 30, 31, 32

Rank: 1,   2,   3,   4,   5,   6.5, 6.5, 8,   9,   10, 11, 12, 13, 14.

Next, you need to sum the ranks (R) of the two groups. For group A,

$$R_A \text{ is } 1 + 2 + 3 + 5 + 6.5 + 9 + 10 = 36.5.$$

For group B,

$$R_B \text{ is } 4 + 6.5 + 8 + 11 + 12 + 13 + 14 = 68.5.$$

The values are clearly different, but to determine the significance of the difference you have to work out the maximal possible R value (R_{max}) for each group and relate the actual R value to R_{max}. If the data were perfect, and every value in group B was higher than the values in group A, then

$$R_A \text{ would equal } 1 + 2 + 3 + 4 + 5 + 6 + 7 = 28$$

and

$$R_B \text{ would equal } 8 + 9 + 10 + 11 + 12 + 13 + 14 = 77.$$

So R_{max} in this experiment would be 77. You don't have to add up the ranks yourself every time, however, because, for any number of data points (N) in each of two groups of data, you can calculate R_{max} for each group with the formula:

$$R_{A,max} = (N_A N_B) + \frac{N_A(N_A + 1)}{2}$$

and

$$R_{B,max} = (N_B N_A) + \frac{N_B(N_B + 1)}{2}.$$

In my example, because N_A and N_B are equal, the R_{max} values would be both the same:

$$R_{max} = (7 \times 7) + \frac{7 \times 8}{2}$$

$$= 49 + \frac{56}{2}$$

$$= 77.$$

See above to prove to yourself that this is correct!

The two N values do not have to be equal, however, which means that the two R_{max} values can be different. The higher of the two R_{max} values will go to the group with the larger N value.

To do the statistical analysis, you need to work out how the observed R value for each set of data relates to the maximal theoretical R value. This statistical measure is known as U:

$$U_x = R_{x,max} - R_x.$$

In our example

$$U_A = 77 - 36.5 = 40.5$$

and

$$U_B = 77 - 68.5 = 8.5.$$

Interpretation of the Mann–Whitney test

If there were no difference between the data in the two groups in an experiment, then U_A would equal U_B. The fact that there is a difference between the U values, however, does not prove there is a statistically significant difference between the data. The difference has to be greater than could occur simply by chance, allowing for a critical threshold of chance (a p value) exactly as for the t-test. The required U values to obtain critical p values depend on the amount of data you have. The required values are determined using a complex calculation, which you don't need to understand. See *Table 8.2* for U value thresholds at different critical values for different sizes of data sets.

For our example, where both N_A and N_B are 7, the U values required for a critical p value of 0.05 are 41 and 8. Since the actual U values obtained in the example experiment are 40.5 and 8.5 (above), then the difference between them has a p value >0.05, so would come below the threshold of significance needed for us to accept that the two sets of data are different. The difference between the data is, however, significant with a p value of <0.1 (where the critical values are 38 and 11). This means that the likelihood that the difference between the data sets seen is entirely due to random chance is $>5\%$ but $<10\%$. Would you accept this level of risk that you are incorrect? After all, it will be up to you to decide.

Table 8.1 *t*-values required giving different threshold *p*-values when there are different degrees of freedom (*df*) in the data sets

df	Level of significance (p-value) for a non-directional t-test				
	0.1	0.05	0.02	0.01	0.001
1	6.31	12.71	31.82	63.66	636.58
2	2.92	4.3	6.96	9.92	31.6
3	2.35	3.18	4.54	5.84	12.92
4	2.13	2.78	3.75	4.6	8.61
5	2.02	2.57	3.36	4.03	6.87
6	1.94	2.45	3.14	3.71	5.96
7	1.89	2.36	3	3.5	5.41
8	1.86	2.31	2.9	3.36	5.04
9	1.83	2.26	2.82	3.25	4.78
10	1.81	2.23	2.76	3.17	4.59
11	1.8	2.2	2.72	3.11	4.44
12	1.78	2.18	2.68	3.05	4.32
13	1.77	2.16	2.65	3.01	4.22
14	1.76	2.14	2.62	2.98	4.14
15	1.75	2.13	2.6	2.95	4.07
16	1.75	2.12	2.58	2.92	4.01
17	1.74	2.11	2.57	2.9	3.97
18	1.73	2.1	2.55	2.88	3.92
19	1.73	2.09	2.54	2.86	3.88
20	1.72	2.09	2.53	2.85	3.85

Table 8.2 *U* values required to give various critical *p*-values depending upon the number of data points

Sizes (N) of the two data sets	Upper/lower limits of U for p = 0.1	Upper/lower limits of U for p = 0.05	Upper/lower limits of U for p = 0.02
5, 5	4 / 21	2 / 23	1 / 24
6, 5	5 / 25	3 / 27	2 / 28
7, 5	6 / 29	5 / 30	3 / 32
8, 5	8 / 32	6 / 34	4 / 36
9, 5	9 / 36	7 / 38	5 / 40
10, 5	11 / 39	8 / 42	6 / 44
6, 6	7 / 29	5 / 31	3 / 33
7, 6	8 / 34	6 / 36	4 / 38
8, 6	10 / 38	8 / 40	6 / 42
9, 6	12 / 42	10 / 44	7 / 47
10, 6	14 / 46	11 / 49	8 / 52
7, 7	11 / 38	8 / 41	6 / 43
8, 7	13 / 43	10 / 46	7 / 49
9, 7	15 / 48	12 / 51	9 / 54
10, 7	17 / 53	14 / 56	11 / 59
8, 8	15 / 49	12 / 52	9 / 55
9, 8	18 / 54	15 / 57	11 / 61
10, 8	20 / 60	17 / 63	13 / 67
9, 9	21 / 60	17 / 64	14 / 67
10, 9	24 / 66	20 / 70	16 / 74
10, 10	27 / 73	23 / 77	19 / 81

Index